Legal Aspects of Satellite Remote Sensing

Studies in Space Law

VOLUME 5

Legal Aspects of Satellite Remote Sensing

By

Atsuyo Ito

MARTINUS
NIJHOFF
PUBLISHERS

LEIDEN • BOSTON
2011

This book is printed on acid-free paper.

Library of Congress Cataloging-in-Publication Data

Ito, Atsuyo.
Legal aspects of satellite remote sensing / By Atsuyo Ito.
p. cm. -- (Studies in space law ; 5)
Includes bibliographical references and index.
ISBN 978-90-04-19032-0 (hardback : alk. paper) 1. Artificial satellites in remote sensing--Law and legislation. 2. Liability (Law) 3. Law--Data processing. 4. Remote-sensing images--Law and legislation. I. Title.
KZD5632.I86 2011
344'.095--dc22

2010053961

ISBN 1871-7659
ISBN 9789004190320

Contents

Acknowledgements

This book is based upon a thesis successfully submitted for a PhD degree at University of Paris XI*. It has been made possible with the help and support of many people and I would like to use this opportunity to express my sincere gratitude to all who have helped me in various ways to make this book happen.

Publication was made possible with funding of Secure World Foundation; I am most grateful to Prof. Frans von der Dunk and Prof. Ray Williamson for arranging this essential support.

My heartfelt appreciation goes to Prof. Sir Martin Sweeting whose influence is considerable in starting this research, and whose constant encouragement and help in every possible way were invaluable. I am indebted to Dr. Philippe Achilleas for supervision during my PhD studies and Prof. Armel Kerrest for being on the viva voce committee. I would like to thank Mr. Marcio Barbosa and Dr. Mario Hernandez for their support for this book.

I would like to express my gratitude to Prof. Isabella Diederiks-Verschoor for her warm encouragement and kindness in accessing to her library. I owe a very important debt to Mr. Stéphane Dupont for his kind help, and his suggestions and shared insights during the interview were invaluable. I would like to thank Dr. Marco Ferrazzani and Ms. Marie-Angèle Lemoine for their kind assistance at ESA. I am grateful for Prof. Joanne Gabrynowicz for providing me generous help. Sincere thanks goes to Ms. Annie Bernard for doing everything she could to help me during the preparation of my doctoral thesis.

* A summary of this book was originally published in 34 Journal of Space Law (2008). Published with permission.

I would like to acknowledge the opportunities and resources provided by UNESCO, UN OOSA and ESA which were an enormous help to me.

Needless to say, any shortcomings in this book are entirely my own.

Special thanks go to Ms. Carol Carnet who has been a dedicated and encouraging editor for this book and doing everything she could to convey exactly what I wanted to say.

Finally, I would like to thank my mother without whose love and support I could not have completed this work.

List of Abbreviations

ABM	Anti-Ballistic Missile System
ALOS	Advanced Land Observation Satellite
ASAT	Anti-satellite weapon
ASEAN	Association of South East Asian Nations
AVHRR	Advanced Very High Resolution Radiometer
BNSC	British National Space Council
CNES	Centre National d'Etudes Spatiales
CNSA	China National Space Administration
COFUR	Cost of Fulfilling User Request
CONAE	Comisión Nacional de Actividades Espaciales
COPUOS	Committee on the Peaceful Uses of Outer Space
CSA	Canadian Space Agency
DEM	Digital Elevation Model
DLR	German Space Agency
DOD	Department of Defense
DMC	Disaster Monitoring Constellation
DMCii	Disaster Monitoring Constellation International Imaging
EADS	European Aeronautic Defence and Space Company
ECHO	European Community Humanitarian Office
EO	Earth Observation
EOSAT	Earth Observation Satellite Corporation
ERS	Earth Resource Satellite
ESA	European Space Agency
EUMETSAT	European Meteorological Satellite Organization
FAO	Food and Agriculture Organization
FEMA	Federal Emergency Management Agency
FTP	File Transfer Protocol

GEO	Group on Earth Observation
GHG	Greenhouse Gases
GIS	Geographical Information System
GNSS	Global Navigation Satellite System
GMES	Global Monitoring for the Environment and Security
GSD	Ground Sample Distance
HRC	High Resolution Image Clearing Committee
ICAO	International Civil Aviation Organization
ICBM	Inter-Continental Ballistic Missile
ICJ	International Court of Justice
IGO	International Intergovernmental Organization
INTERCOSMOS	Council on International Cooperation in the Study and Utilisation of Outer Space
INTELSAT	International Telecommunications Satellite Organization
IMO	International Maritime Organization
IRS	Indian Remote Sensing Satellite
ISO	International Standardization Organization
ISRO	Indian Space Research Organization
ITU	International Telecommunication Union
JAXA	Japanese Aerospace Exploration Agency
MEA	Multilateral Environmental Agreement
MODIS	Moderate Resolution Imaging Spectroradiometer
MOS	Marine Observation Satellite
MOU	Memorandum of Understanding
NASA	National Aeronautics and Space Administration
NATO	North Atlantic Treaty Organization
NGO	Non-governmental Organization
NIMA	National Imagery and Mapping Agency
NOAA	National Oceanic and Atmospheric Administration
NRSC	National Remote Sensing Centre
PDD	Presidential Decision Directive
PPP	Public-Private Partnership
PTWC	Pacific Tsunami Warning Centre
RCA	Radio Corporation of America
SALT	Strategic Arms Limitation Treaty
SAR	Synthetic Aperture Radar
SDGN	Secretariat General de la Defence Nationale
SERTIT	Service Régional de Traitement d'Image et de Télédétection
SGD	Singapore Dollar
SNCF	Société Nationale des Chemins de fer Français
SPOT	Satellite Pour l'Observation de la Terre
SSTL	Surrey Satellite Technology Limited
TESEO	Treaty Enforcement Services Using Earth Observation
TIROS	Television and Infrared Observation Satellite
UN	United Nations
UNDRO	Office of the United Nations Disaster Relief Coordinator
UNHCR	United Nations High Commissioner for Refugees

UNDP	United National Development Programme
UNEP	United Nations Environmental Programme
UNESCO	United Nations Education Science and Cultural Organization
UNGA	United Nations General Assembly
UNICEF	United Nations Children's Fund
UNISPACE	United Nations Conference on the Exploration and Peaceful Use of Outer Space
UNITAR	United Nations Institute for Training and Research
UNOCHA	Office for the Coordination of Humanitarian Affairs
UNOOSA	United Nations Office of Outer Space Affairs
USGS	United States Geological Survey
UV	Ultra Violet
VAP	Value-Added Product
WIPO	World Intellectual Property Organization
WFP	World Food Programme
WHO	World Health Organization
WMO	World Meteorological Organization

Part I

Clarification of the Current Regime

Chapter One

Introduction

Human beings are increasingly exposed to the degradation of the environment and are vulnerable to the risks of natural disasters. Yet mankind's capabilities to deal with such risks have broadened and enhanced, thanks to the advance of technology. Today space technologies supply an effective means of gathering information for use in addressing environmental problems and disasters. Remote sensing from the high vantage point of space provides a powerful tool to gain understanding of the ever-changing environment on the Earth. Satellites can make repeated observations of specific areas of interest; thus they are well-suited for monitoring the progress and effects of natural and man-made disasters, as well as assessing harmful interference to the regional and global environment. Not only is satellite remote sensing effective for the assessment and mitigation of disaster impact, it is also appropriate for the verification of a range of international treaties, including claims related to environmental protection. Satellite imagery can be used to assist policy makers, governments, researchers, and other relevant users in proposing and implementing more effective solutions and, subsequently, to assess the results of these applications.

1.1. Definition

Remote sensing is "the science of extracting information from an object through the analysis of data acquired by a sensor that is not in direct contact with that area."[1] The technology of remote sensing rests on "making use of the properties of electromagnetic waves emitted, reflected or diffracted by the sensed objects."[2] All physical objects interact with electromagnetic waves in different wavelengths and this enables us to detect and differentiate the object by analyzing the reflected or emitted energy. In short, remote sensing is a means to obtain the information on distant objects without having a direct physical interaction with them.

The technology of remote sensing originated with the invention of the camera more than 150 years ago.[3] Aerial views of the Earth's surface were recorded by cameras mounted on tethered balloons as early as 1858.[4] Instead of a camera that uses film, satellite remote sensing today uses electronic photosensitive sensors to take imagery in a digital format.[5] Unlike a film that needs to be retrieved physically for processing, the information collected through this electronic process is assigned a digital value and transmitted via a radio communications link to the ground where it is reconstructed into imagery.[6]

This book addresses only modern electronic satellite remote sensing,[7] as opposed to aerial remote sensing. A key advantage that satellite remote sensing has over aerial photography is that it enables repeated observations to be made more easily and unobtrusively whilst covering large areas across national boundaries. Indeed, satellites orbiting around the Earth at altitudes of 300–1000 kilometres or higher can map 20,000 km^2 per second, roughly 2,000 times more than what can be covered by high-flying aircraft.[8]

---

[1] NPA Group, *Final Report BNSC Sectors Studies Programme Applications of Earth Observation to Legal Sector*, BNSC, London, 2001, 16.

[2] Principles Relating to Remote Sensing of the Earth from Outer Space, UNGA Resolution 41/65 of 3 December 1986; UN Doc. A/AC.105/572. [Hereafter: UN Remote Sensing Principles]

[3] Crowsey Incorporated, *Legal Assistant Guide to Legal Applications for geospatial information*, June 10, 2002, 24. Used to be available on the Internet at http://www.crowsey.com/pdf/spatialInformation.pdf. Present location unknown.

[4] See L. Schmidt, New Tools for Diplomacy: Remote Sensing in International Law, Available on the Internet at http://earthobservatory.nasa.gov/Features/Diplomacy/.

[5] See R.J. Aamoth, Use of Remote Sensing Imagery by the News Media, *Heaven and Earth: Civilian Uses of Near-Earth Space*, Martinus Nijhoff Publishers, Hague, 1997, 139.

[6] See Y. Sneifer, The Implications of National Security Safeguards on the Commercialization of Remote Sensing Imagery, 19 *Seattle University Law Review*, 1996, 543. Electro-optical technology introduced in the early eighties broadened the sensing capability, which was limited in film-based sensors.

[7] Satellite remote sensing is defined under the UN General Assembly Resolution specifically as "sensing of the Earth's surface from space by making use of the properties of electromagnetic waves emitted, reflected or diffracted by the sensed objects."

[8] See D.H. Staelin & J. Kerekes, Remote Sensing Capabilities, *Heaven and Earth: Civilian Uses of New Earth Space*, Martinus Nijhoff Publishers, Hague, 1997, 162.

1.2. The Mechanism of Remote Sensing

Data is collected through either passive or active instruments, which differ fundamentally in their processes for measurement. Passive instruments detect the radiation reflected off or emitted by the sensed objects from a source other than the instruments, such as sunlight.[9] An example of a passive instrument is an optical sensor. It relies on recording of reflected solar energy; thus, it is only able to take imagery during the day. On the other hand, active instruments transmit energy toward the target and then record the reflected or backscattered energy from the object.[10] Active sensors such as radar,[11] including synthetic aperture radar (SAR),[12] have the advantage over passive remote sensing because they enable monitoring even at night, can penetrate cloud cover and rain, and are largely unaffected by the weather. Furthermore, SARs are particularly well suited for studying geological structures, partly because of their ability to penetrate foliage and view the underlying surface formations.[13] Active sensors are valuable tools for such applications as monitoring forest clearings[14] in very cloudy regions.

Although satellite remote sensing includes those satellites circling around other celestial bodies such as the Moon, this book focuses on the remote sensing of the Earth. Satellites need to be placed in certain orbital trajectories around the Earth to acquire remote sensing data. Remote sensing is conducted principally from two different types of orbits: the geostationary orbit and low-Earth orbit.[15] Satellites in polar orbits can cover the entire surface of the globe over a period of time as the Earth turns below them.[16] The altitude of the orbit determines the size of objects that can be identified: the lower the altitude, the smaller the objects that can be discerned.[17]

---

[9] See NPA Group, *Final Report BNSC Sectors Studies Programme Applications of Earth Observation to Legal Sector*, BNSC, London, 2001, 25.

[10] See D.H. Staelin & J. Kerekes, Remote Sensing Capabilities, *Heaven and Earth: Civilian Uses of New Earth Space*, Martinus Nijhoff Publishers, Hague, 1997, 165.

[11] Radar uses a transmitter operating at either radio or microwave frequencies.

[12] SAR uses the movement of the satellite and sophisticated processing to synthesize a larger antenna and to create a very high resolution image from the backscattered radiation.

[13] See D.H. Staelin & J. Kerekes, Remote Sensing Capabilities, *Heaven and Earth: Civilian Uses of New Earth Space*, Martinus Nijhoff Publishers, Hague, 1997, 169.

[14] See K.J. Markowitz, Legal Challenge and Market Rewards to the Use and Acceptance of Remote Sensing and Digital Information as Evidence, 12 *Duke Environmental Law and Policy Forum* (Sept. 2002), 230.

[15] See D.H. Staelin & J. Kerekes, Remote Sensing Capabilities, *Heaven and Earth: Civilian Uses of New Earth Space*, Martinus Nijhoff Publishers, Hague, 1997, 163.

[16] See D.H. Staelin & J. Kerekes, Remote Sensing Capabilities, *Heaven and Earth: Civilian Uses of New Earth Space*, Martinus Nijhoff Publishers, Hague, 1997, 163.

[17] See A. Florini, The Opening Skies: Third Party Imaging Satellites and US Security, 13-2 *International Security*, 1998, 94–95..

The smallest size of an object that can be identifiable in an image is referred to as spatial resolution, which is a major element in determining the capabilities of the satellite. An image with a spatial resolution of 1 metre means that the smallest object that can be discernible in the image is approximately 1 m, depending upon such factors as contrast ratio or signal-to-noise ratios; hence the geometrically defined term, Ground Sample Distance (GSD), is more frequently used. Of course, images with higher spatial resolutions give the users enhanced capabilities to assess the specific areas of interest in detail; however, there are other factors that are also important in determining the capability for remote sensing, such as image swath width, spectral resolution, and temporal resolution. Spectral resolution enables objects of similar sizes but different material to be distinguished, whereas temporal resolution – also referred to as 'revisit time' – defines how often a satellite can observe the same location and measure the change in observed phenomena. A temporal resolution/revisit of 16 days means that a satellite passes over the same area for sensing every 16 days, which is suitable for monitoring slowly changing or seasonal phenomena, but insufficient for disaster monitoring, which requires observations of the same location at least daily to monitor the progress of the disaster and its aftermath.

The width of the Earth's surface that a satellite's sensors can observe is called the swath width. Spatial resolution and swath width are typically linked. For a low spatial resolution sensor with the ability to cover a wide swath, high temporal resolution can often be achieved with a single satellite. These satellites are best suited for large-scale environmental measurements.[18] High resolution sensors limit the swath width to tens of kilometres, so a constellation of satellites is needed to achieve the combined capabilities of high spatial resolution and high temporal resolution simultaneously. With the advent of small lower cost satellites, such constellations are increasingly used for different applications.[19]

The operational flow of remote sensing activities includes all of the processes from data collection to data handling activities. The Scientific and Technical Sub-Committee of the United Nations Committee on the Peaceful Uses of Outer Space (COPUOS) identified the following phases involved in remote sensing:[20]

1. Data acquisition;
2. Data reception;

---

[18] See NPA Group, *Final Report BNSC Sectors Studies Programme Applications of Earth Observation to Legal Sector*, BNSC, London, 2001, 17.

[19] Examples of constellations include the Disaster Monitoring Constellation, the German Rapid Eye for agricultural application, the German SAR-lupe constellation of radar reconnaissance satellites, and the Italian constellation Cosmo-Skymed. See, in general, T. Hitchens, International Satellite Innovation and Cooperation, *Military Satellites 2006*, April 18 2006, 5–25.

[20] U.N.Doc. A/AC/.105/170, 12 April, 1976, p. 8.

3. Data pre-processing;
4. Data storage;
5. Data analysis; and
6. Information Utilization.

Wherever data is collected, as the satellite acquires information about the observed phenomena it is transmitted to a ground station, where data is received and processed by computers for human viewing and interpretation.[21] Each receiving station has a set of processes to generate a certified product.[22] Once the data has gone through pre-processing, it is analyzed so as to deliver useful information: the data may simply be enhanced, or further value may be added to create maps.[23] Advanced data analysis and value-adding are the processes that convert data into relevant knowledge ready for immediate use; thus, data goes through various stages of handling operations before it is delivered to the end users.

The Principles Relating to Remote Sensing of the Earth from Outer Space (hereafter: UN Remote Sensing Principles),[24] which is the only international document addressing the sensing activities from space, categorizes three different types of data products: primary data,[25] processed data,[26] and analyzed information.[27] In practice, however, various other terms are used: for instance, 'raw data' or 'unenhanced data' for primary data; 'enhanced data' for processed data; and 'derived products' and 'value added products' for analyzed information. Furthermore, different processing levels are adopted by different entities.[28]

---

[21] See L.S. Spector, The Not- so Open Skies, *Commercial Observation Satellites and International Security*, Macmillan Press, Hampshire, 1990, 168.

[22] See NPA Group, *Final Report BNSC Sectors Studies Programme Applications of Earth Observation to Legal Sector*, BNSC, London, 2001, 31.

[23] See K.J. Markowitz, Legal Challenge and Market Rewards to the Use and Acceptance of Remote Sensing and Digital Information as Evidence, 12 *Duke Environmental Law and Policy Forum* (Sept. 2002), 232.

[24] Principles Relating to Remote Sensing of the Earth from Outer Space adopted on 3 December 1986.

[25] Primary data is defined as "those raw data that are acquired by remote sensors borne by a space object and that are transmitted or delivered to the ground from space by telemetry in the form of electromagnetic signals, by photographic film, magnetic tape or any other means" under Principle I (b) of the UN Remote Sensing Principles.

[26] Processed data is defined as "the products resulting from the processing of the primary data, needed to make such data usable" under Principle I(c) of the UN Remote Sensing Principles.

[27] Analyzed data is defined as "the information resulting from the interpretation of processed data, inputs of data and knowledge from other source" under Principle I(d) of the UN Remote Sensing Principles.

[28] For example, Landsat categorize different level/type of products into L0R, L1R, L1G, L1Gt, L1T whereas Spot Image categorize them into 1A, 1B, 2A, 2B.

No uniformity has yet been achieved for referring to different types of products from different systems, and this generates complexities for both data comparison and data integration by different systems into a final product.

1.3. Applications

Once data or derived products are delivered to the users, they are utilized for a wide range of applications. Applications include, but are not limited to: map making, land use, crop monitoring, natural resource management, registry of lands, urban planning, studies of environment, and verification of claims. Amongst other uses, satellite remote sensing is suited for monitoring and protection of the environment. This book focuses on the environmental and disaster management applications.

The Earth's environment includes the atmosphere, land, and oceans. Environmental applications include the assessment of land mass (including forests), and air and marine pollution. Global warming is an imminent environmental problem; satellite remote sensing is useful in monitoring the oceans, ice, atmosphere, and forests to provide key information on the environmental components, as well as evidence of global warming. Present data and past data can be compared to assess the extent and rate of deforestation, desertification, and loss of polar ice. Deforestation in the Amazon, for instance, has been the subject of many studies.[29] Satellite images are also used for verification of environmental harm in specific areas of interest. Many lawsuits centre on the presence of toxic material in the air, water, or soil and satellite data is an effective tool to verify such claims.[30] In particular, detection of oil slicks is a major application for verification purposes.

Satellites have provided data on earthquakes, floods, landslides, and fires. Satellites can monitor the progress of hurricanes and fires as well as measure the extent of damage from earthquakes and floods in the aftermath. The data becomes a vital tool to identify the damaged areas for an effective disaster response. Indeed, following the 2004 Indian Ocean Tsunami, a number of satellite images were provided and used to assess the location and the extent of the devastation, and to direct relief services and aid agencies most effectively by comparing these images with those taken before the tsunami. Critical decisions were made based on the satellite images to identify the overall extent of the disaster, as well as to conduct street level damage assessment to give the necessary direction for aid.

---

[29] Studies on the Amazon include a series of Landsat images from the 1970s to present day showing the extent of deforestation.

[30] See Crowsey Incorporated, *Legal Assistant Guide to Legal Applications for geospatial information*, June 10, 2002, 41.

1.4. Overview of the Historical Development

An overview of the history of applications and practices in remote sensing can provide a better understanding of its development, which can roughly be divided into three eras: 1) the era of reconnaissance and weather monitoring by space powers; 2) the era of civil remote sensing programmes by spacefaring nations; and 3) the era of commercial and operational remote sensing.

1.4.1. *Era of Reconnaissance and Weather Monitoring by Space Powers: 1957 to 1972*

Remote sensing is rooted in military use. Remote sensing technology was originally developed by the early space powers – the United States (US) and the Union of Soviet Socialist Republics (USSR) – for the purposes of defence planning and intelligence gathering.[31] During the Cold War, US and USSR reconnaissance satellites observed each other's territories extensively at altitudes between 150km and 350km.[32] The US Corona,[33] a reconnaissance programme that continued until 1972, retrieved its film capsule in 1960 for the first time.[34] Although the resulting picture was fuzzy, it captured an air field with a resolution of 8m.[35] This signifies the fact that a relatively high resolution remote sensing satellite was already utilised in military applications back in 1960.

Weather monitoring was the other earliest application of remote sensing. The first US weather satellite, Television and Infrared Observation Satellite (TIROS-1), was launched in 1960 and provided important information on cloud cover and weather conditions.[36] The first important use of weather data was for weather forecasting and monitoring severe storms such as hurricanes.[37]

---

[31] See Y. Sneifer, The Implications of National Security Safeguards on the Commercialization of Remote Sensing Imagery, 19 *Seattle University Law Review*, 1996, 546.

[32] See B. Jasani, Security – A New Role for Civil Remote Sensing Satellites, *Air and Space Law in the 21st Century*, Heymann Publisher, Köln, 2001, 339.

[33] See for detail, D.A. Day & J.M. Logsdon (Eds.), *Eye in the Sky: Story of the Corona Spy Satellites*, Smithsonian Inst Press, 1999.

[34] D.G. Clarke, Access Control of Remote Sensing Satellites, *Commercial Satellite Imagery and the UN Peacekeeping*, Ashgate, Aldershot, 2003, 171. The Corona programme later achieved a resolution of two meters.

[35] D.G. Clarke, Access Control of Remote Sensing Satellites, *Commercial Satellite Imagery and the UN Peacekeeping*, Ashgate, Aldershot, 2003, 171.

[36] See C.Q. Christol, Remote Sensing and International Law, 5 *Annals of Air and Space Law*, 1980, 382.

[37] See D.H. Staelin & J. Kerekes J, Remote Sensing Capabilities, *Heaven and Earth: Civilian Uses of Near Earth Space*, Martinus Nijhoff Publishers, Hague, 1997, 170.

Satellite remote sensing contributed to improved weather predictions, which could be made up to several days in advance.[38] Geostationary weather satellites emerged in the late 1970s and since then, satellites have been widely utilized in daily weather forecasting.

1.4.2. *Era of Civil Remote Sensing Programmes by Spacefaring Nations: 1972 to mid-1990*

Land observation for civil purposes started with the 1972 launch of the US Earth Resource Technology Satellite (later renamed as Landsat-1), which was developed by the National Aeronautics and Space Administration (NASA) and later operated by the National Oceanic and Atmospheric Administration (NOAA). The Landsat programme[39] demonstrated the utility of remote sensing in different applications, including land use, natural resource management, agriculture, studying land cover change, and monitoring such Earth environment aspects as deforestation.[40] The United States permitted data reception by ground stations of other nations through agreements;[41] indeed, in the 1970s raw data already had been collected by ground stations in a number of countries, including Canada, Italy, Brazil, Sweden, Japan, India, Australia, and Argentina; other nations followed later on.

Spacefaring nations as well as intergovernmental organizations started to engage in civil remote sensing programmes one after another from the mid-1980s onward. In 1986, France launched the civil remote sensing Satellite Pour l'Observation de la Terre (SPOT), which was developed by the Centre National d'Etudes Spatiales (CNES), the French Space Agency. France's launch of SPOT put an end to the US/Soviet "duopoly" of remote sensing by space powers,[42] and SPOT since remains a primarily competitor for Landsat.[43] Japan initiated

---

[38] See D.H. Staelin & J. Kerekes J, Remote Sensing Capabilities, *Heaven and Earth: Civilian Uses of Near Earth Space*, Martinus Nijhoff Publishers, Hague, 1997, 170.

[39] For the Landsat program in general, see C.M.Hayward, Remote Sensing: Terrestrial Laws for Celestial Activities, 8 *Boston University International Law Journal,* 1989, 166.

[40] A.L. Brigham, Natural Resources: View from Above, 55-2 *Journal of Air Law and Commerce*, 1989, 518.

[41] NASA entered into agreements with foreign ground stations that allowed them to collect data directly from the satellite as it passed overhead for an access fee of $200,000 per year. See M.A. Roberts, US. Remote Sensing Data from Earth Observation – Law Policy and Practice, *Proceedings of the Thirty-Ninth Colloquium on the Law of Outer Space*, IISL, AIAA,1996, 111.

[42] H. De Santis, Satellites, Alliance, Relations and the Developing World, *Commercial Observation Satellites and International Security,* Hampshire, the Macmillan Press, 1990, 78.

[43] See R. Bender, *Launching and Operating Satellites Legal Issues*, Martinus Nijhoff Publisher, Hague, 1997, 222.

its remote sensing programme in 1987 with the launch of the Marine Observation Satellite (MOS)-1, dedicated to observation of oceans. India joined the club of remote sensing nations with the launch of its Indian Remote Sensing Satellite (IRS) -IA in 1988, developed by the Indian Space Research Organization (ISRO).[44] Early civil earth observation systems were developed and utilized primarily for national advancement, but later began to be used for purposes beyond national needs. In the 1990s, the European Space Agency (ESA) – the international intergovernmental organization that coordinates the European space program with the contribution from each of its member states – launched its first remote sensing satellite, the Earth Resource Satellite (ERS-1) in 1991.

During the early phase of this era, remote sensing was reserved by spacefaring nations primarily as a governmental activity to fulfil their own needs, as well as to observe the state of the Earth's environment. Remote sensing programmes were conducted typically under the auspices of the governmental space agency that developed and operated such systems. In the 70s and first half of the 80s, other nations accessed data through intergovernmental ground station arrangements. A step towards commercialization has been taken by national space agencies, most notably through data sales from governmental remote sensing satellites, such as the commercial marketing of Earth Observation (EO) data from the SPOT-1 satellite begun in 1987 by Spot Image – then a quasi-private entity controlled by CNES, and now a private company under the European Aeronautic Defence and Space Company group.[45] In addition to the commercial sale of SPOT images, some photos taken from the Russian military reconnaissance satellites were made available for purchase in the early 1990s.[46] This practice of the government developing and launching remote sensing satellites and then delegating data distribution to a private entity soon became widespread and heralded a degree of competition in the data sales market.

1.4.3. *Era of Commercial and Operational Remote Sensing: Late 1990 to Present*

With the spread of data sales, the commercialization of remote sensing advanced rapidly. Remote sensing data was no longer limited to governmental uses but opened up to a wider user community and range of applications. The advent of the era of commercial remote sensing was confirmed with the emergence of

---

[44] IRS-1A was followed by a series of IRS satellites dedicated for different purposes including agriculture, natural resources management, ocean monitoring, and mapping.

[45] See C.M. Hayward, Remote Sensing: Terrestrial Laws for Celestial Activities, 8 *Boston University International Law Journal* (1989), 168.

[46] G. Steinberg, Dual Use Aspects of Commercial High-Resolution Imaging Satellites, 37 *Mideast Security and Policy Studies* (Feb. 1998). Available on the Internet at http://www.biu.ac.il/Besa/books/37pub.html.

private actors in remote sensing, and – along with the end of the Cold War – was boosted by the release of imagery from a reconnaissance satellite by a Russian company in the United States in 1993. Private entities such as Space Imaging, Digital Globe, and OrbImage started to launch very high resolution satellites in the late 1990s. In 1999, Space Imaging was the first to release imagery with the spatial resolution of 1 m, revealing remarkable detail of the Washington DC area.[47] Because of the enhanced spatial resolutions of 1 m or higher, EO data started to have high commercial value.

A growing number of states and private entities started to recognize the utility of remote sensing and began to seek access to remote sensing data. Since 2000, more and more non-spacefaring nations had begun to have remote sensing satellites developed by foreign private manufacturers.[48]

Funding for EO systems from the private sector has increased. Public-private partnerships have become more common in recent years. RADARSAT-2[49] is a public-private partnership between the Canadian Space Agency and MacDonald Dettwiller and Associates, Ltd. Rapid Eye, a constellation of 5 optical EO satellites designed for agricultural imaging and mapping, has public-private financing. TerraSAR-X, a high resolution satellite launched in 2007, is a public-private partnership amongst the German Space Agency, DLR, and EADS-Astrium. Commercial nature of data sale means since private investment has increased, the trend for return of investment through data sale accelerated.(rather than marginal cost) This era also saw the emergence of coordinated services based on operational remote sensing; that is, a shift from data provision via a single satellite to data provision from integrated systems through international collaborative efforts, particularly for the purpose of disaster management. Using a constellation of satellites or combination of several satellites that operate independently, different satellite operators cooperate to provide services in fulfilling the common objective of the coordinated EO programme for environmental protection and disaster response.

1.5. The Current Environment Surrounding Remote Sensing Data

The current environment surrounding the supply and use of EO data is quite different from that of the earlier era where civil remote sensing programmes

[47] D.G. Clarke, Access Control of Remote Sensing Satellites, *Commercial Satellite Imagery and the UN Peacekeeping*, Ashgate, Aldershot, 2003, 171.

[48] Examples include the SSTL for development of Alsat-1 for Algeria, Astrium for development of Formsat-1 for Taiwan, and SatReci for development of X-Sat for Singapore.

[49] RADARSAT-2 is used for enhancing marine surveillance, ice monitoring, disaster management, environmental monitoring, resource management, and mapping in Canada and around the world.

were the preserve of spacefaring nations, and where suppliers and users comprised a relatively small number of governments. Today, satellite operators that generate EO data have greatly increased in numbers – now nearly 30 nations have remote sensing satellites. Data from the governmental space systems as well as from private entities are widely disseminated both on a commercial and non-commercial basis.

The data supply environment has become complex, involving various entities in the course of operations that include data collection, processing, value adding, and distribution – often referred to as the 'data supply chain'.[50] Entities that are involved in each phase of the data supply chain include data generators, image processing wholesalers, value-added providers, and distributors.

Data generators, such as space agencies, own and operate satellites and conduct the data collection. The image processing wholesaler conducts the processing of data to more advanced levels, including geometric and radiometric calibration. Value-added entities[51] (also referred to as value-added resellers) conduct information extraction services that enhance the original data to produce derived products of greater value and utility to the end-user, such as maps and digital elevation models. Finally, data distributors disseminate the resulting products to the end users. Networks of distributors have been established, particularly amongst private entities and data owners who can exploit their assets through licensing distributors to sell data in the different regions.[52] Data and derived products are distributed with various pricing schemes, depending upon the user. The availability, pricing, and other conditions associated with the data sale are known as 'data policy,' and affect the availability and accessibility of data in the user community. There are increasing numbers of players in each category; consequently, there is a wide variety and divergence of data policies.

Various data sources and different processing procedures are now combined to produce final products. The resulting products may be a synthesis of satellite data and aircraft data as well as other sources of information such as the Geographical Information System (GIS), GPS, and population data. Various sources can combine to provide useful knowledge to support critical decision

---

[50] For data chain, see Crowsey Incorporated, *Legal Assistant Guide to Legal Applications for Geospatial Information*, June 10, 2002, 24.

[51] Examples of value adding entities include SERTIT, GAF, GEO Image, and Digitec International. In addition, there are a limited number of private or privatized value adding companies with low vertical integration, such as Geomet, Geosys of France, Nordpointer of Sweden, Infocarto of Spain, Novosat of Finland, and Infoterra of the UK. See D. Rosenholm & R. Harris, Structural analysis of the competitive environment of Earth observation industry, *Earth Observation Data Policy and Europe,* A.A. Balkema Publishers, Lisse, 2002, 67.

[52] For example, Spot Image has established a world-wide network of distribution of data by licensing regional distributors to sell their data in different regions including the United States, Canada, Japan, Brazil, Australia, and Saudi Arabia.

making; for instance, following Hurricane Katrina in 2005, a combination of satellite images and GPS technology was used to provide the necessary information to aid workers during the rescue operations to help them locate the affected population, even though the visible landmarks were erased by severe flooding. In such cases, practices of different entities with no uniform standards for conducting data handling operations may raise issues both as to the integration of data by those entities into final products, and as to the resulting product quality. These complexities make it difficult for the concerned parties in the data supply chain to verify the reliability of each operation or to allocate the responsibilities and benefits properly amongst themselves. This creates an uncertain environment for users, who do not receive clear information about the operations involved before delivery of the final products and the division of responsibilities. For instance, data suppliers may be outsourcing a part of data handling operations, and end-users would not know where the error involved in the product originated.

Users have continued to diversify and increase in numbers. Today users can be anyone as long as they can afford the data. Users may include, but are not limited to, environmental organizations, aid agencies, the United Nations, NGOs, insurance companies, law firms, and a variety of individuals including expert witnesses and researchers. The accessibility of the satellite data has been enhanced dramatically with the spread of the Internet. Today EO data is easily retrievable by communities around the globe with Internet access. Satellite data is widespread and easily accessible to the public: indeed, anyone can view a variety of satellite data and derived products that are placed on the Internet. Internet-based data viewers such as Google Earth, or Google Maps (that emerged in 2005), surprised society at large by putting together different satellite images and aerial photos varying in spatial resolution to offer an interactive map of the whole globe that allowed any person to see images of any location with zoom, on demand.

With the dramatic changes in the circumstances surrounding EO data supply and use over the last decade – particularly in the last two to three years – and in preparation for the new era to come, one can easily see the urgent need for a relevant legal regime to correspond to the fast moving field.

Today satellite remote sensing is addressed only under the international space law that was formulated in 1967, and through a 1986 UN General Assembly Resolution specifically dealing with satellite remote sensing; so the question should be posed: is the current legal framework for remote sensing still adequate?

Considering the current EO trend, issues are increasingly being raised with regard to: the availability and accessibility of data in terms of different uses; the rights of all parties concerned over the data and derived products; the reliability and quality of data; and the liability arising from the supply and use of data. Without a doubt, data that is incorrect may generate damage to the end-users as well as to a third party that relies on such data. Users can use the data in a manner that is detrimental to the interest of a data supplier as well as that of a third party. Moreover, data is now available on the Internet and can easily be reproduced; thus users can infringe the rights of the data suppliers. The role of the legal framework is more critical than ever to regulate the supply and use of remote sensing data, as well as to serve as a stimulus for remote sensing applications and related services. In light of these realities, the issues subject for examination under this book lie in three areas.

The first issue to be addressed concerns data policy, which articulates terms and conditions of supply and use of data. Data policies are usually set separately by the different entities and there is no unified mechanism for data authentication and validation. The data policy issues concerned are related to the overall framework addressing the access and pricing policies for public benefit, commercial distribution, and intellectual property rights.

The second issue to be addressed concerns data accuracy and authenticity. The lack of objective criteria for data assessment creates obstacles to balancing the interests of different stakeholders, and for ensuring both the integrity of data and the reliability of the resulting information products.

The third issue to be addressed concerns liability. There is presently an ambiguity over the responsibility and liability arising from supply and use, or misuse, of the data and resulting products. As the data supply grows more complex, the need for clear data policies with consequent responsibilities becomes even more critical. A more precise and comprehensive legal regime is required to: achieve easier access to the data for environmental protection and disaster management; provide clarification of the data suppliers' responsibilities; and create user confidence in the data itself.

By taking a practical approach to topical issues that are not addressed in the current legal framework of satellite remote sensing, this book aims both to present a clear picture of the reality of the legal aspects of satellite remote sensing, which have been familiar to a relatively small EO community, and to make a contribution to better regulation and promotion of satellite remote sensing by first identifying gaps and then filling existing gaps. It undertakes a review of the current legal framework surrounding remote sensing; identifies any legal shortcomings in the current legal and policy regimes of space law, environmental law, and disaster management; examines what needs to be rectified; and addresses how improvements can be brought to the existing problems.

Part I sets forth the existing regime as a point of departure for the analysis of improvements and recommendations to be provided in Part II. Part I, which consists of four chapters, clarifies the legal environment surrounding satellite remote sensing, defines problems, discusses their impact, and demonstrates the need for improvements. Chapter 2 provides an overview of the existing space law regime, and identifies the key issues and ambiguities of the present framework. Chapters 3 and 4 address the precise understanding of the problem in terms of its impact on real remote sensing applications, environmental protection, and disaster management, respectively. These chapters provide an analysis of how the legal shortcomings create barriers to effective environmental protection and disaster management, demonstrating plainly the need for a comprehensive reassessment of – and call for improvements to – the present remote sensing legal regime.

Subsequent to the issues raised in Part I, Part II explores issues to seek solutions to better address the problems identified. Part II then provides feasible recommendations for each area that needs rectification. Chapter 5 explores the issues of data policies, including licensing and intellectual property rights of remote sensing data, authentication of data, and validation of data accuracy, by looking at the current practice of data suppliers and analyzing the overall trends. Chapter 6 addresses the issues of responsibilities and liabilities by examining liability in the context of both contract law and tort law, as well as the different bases for establishing liability. Chapter 7 provides a conclusion reaffirming the problems and presenting recommendations, a proposal of a new regime, and finally closes with a discussion on how proposed recommendations should be implemented, and how such implementation will result in key improvements to the state-of-the-art knowledge and practice in this field.

Chapter Two

Space Law

2.1. Introduction

Space law is a newly developed field of law with a relatively short history. The discussion of space law began in the 1950s following related technological developments achieved by the end of World War II. Space development originated with the invention of a long-range liquid fuel rocket by the German army. The German V-2 rocket, completed in 1942, was able to fly outside the Earth's atmosphere and became the foundation for the first artificial satellite.[1] The incentive for space technology was then driven forward by the subsequent Cold War, in the form of an arms race between the Soviet Union and the United States.

On October 4, 1957, the Soviet Union took the lead in the race – first by placing Sputnik-I, the first man-made satellite, successfully into orbit in outer space. Following the launch of Sputnik, the need for a legal regime governing outer space became apparent and the legal status of outer space suddenly became an urgent and real question to resolve – rather than merely a theoretical and speculative matter.[2]

The fundamental question was the legal status of outer space in terms of sovereignty: specifically, whether or not sovereignty recognized in airspace extended to outer space. US air law expert J.C. Cooper contended that

---

1 See G.H. Reynolds & R.P. Merges, *Outer Space Problems of Law and Policy,* Second Edition, Westview Press, Colorado, 1997, 3–4.

2 See W. Jenks, *Space Law*, Stevens Publisher, London, 1965, 3.

sovereignty did not extend to outer space;[3] this view prevailed, and the legal status of outer space based on that view continued to be recognized in the context of reconnaissance satellites.

The permissibility of observing foreign territories from space was first proposed by the United States. The official step toward legitimizing remote sensing for reconnaissance purposes was the proposal for an Open Skies Policy by President Eisenhower to allow the United States and the Soviet Union to monitor each other's territory to avoid a surprise attack.[4] Although rejected by Moscow at the time, afterwards the policy was accepted by the two superpowers on a *de facto* basis as they started to monitor each other's territories with reconnaissance satellites in the 1960s.[5]

The next question concerned where the boundary between airspace and outer space lay. There was no clear answer, since air law had never established a definition of airspace with a precise boundary.[6] After Sputnik, numerous theories were proposed in attempts to determine the demarcation of outer space. Cooper and common law experts believed that outer space simply stopped at the point below which an orbit could not be sustained.[7] The higher boundary of airspace meant more protection against unfriendly overflights,[8] as sovereignty was extended in airspace.[9] In the UN COPUOS, the Soviet Union proposed to fix the boundary line at an altitude of 100–120 km above sea level, and most delegates were in agreement.[10] At present, there is still no agreement as to the boundary of outer space.[11] From this history, one can clearly see that space law

---

[3] See J.C. Cooper, High Altitude Flight and National Sovereignty, 4 *International Law Quarterly Review* (1951), 411–418.

[4] See L.S. Spector, The Not- so Open Skies, *Commercial Observation Satellites and International Security*, Macmillan Press, Hampshire, 1990, 165. See also D.R. Terrill, *The Air Force Role In Developing International Outer Space Law*, Air University Press, Alabama, 1999, 9.

[5] See L.S. Spector, The Not- so Open Skies, *Commercial Observation Satellites and International Security*, Macmillan Press, Hampshire, 1990, 165.

[6] See I. Diederiks-Verschoor, *An Introduction to Space Law*, Kluwer Publisher, Hague, 1999, 17.

[7] See W.A. McDougall, *The Heaven and the Earth: A Political History of the Space Age*, Basic Books, New York, 1985, 186.

[8] See W.A. McDougall, *The Heaven and the Earth: A Political History of the Space Age*, Basic Books, New York, 1985, 186.

[9] Chicago Convention on International Civil Aviation 7 December. 1944, Article 44, 15 *UNTS* 295. [Hereafter: Chicago Convention].

[10] ITN Doc-A/AC-105/C-2/L-12.

[11] The critical reason for this is military interest. The higher boundary of national sovereignty means more protection against unfriendly overflight. Spacefaring nations had no incentives to clarify the ambiguities. Indeed, the principal focus of American policy was always protection of spy satellites.

developed in a political context, as the major space powers attempted to codify their particular interests.

Today there are two primary sources of space law under which satellite remote sensing activities are addressed: space law derived from international legal rules, and space law derived from national laws developed in several countries.[12] This Chapter provides an overview of the legal regime of outer space from the standpoint of satellite remote sensing by examining international multilateral treaties, national laws, and national policies. Those who are familiar with the background of international space law may skip the next section and proceed to Section 2.2.4.4, which analyzes the contents of rules on satellite remote sensing.

2.2. International Space Law

International space law is a product of the Space Age, an offspring of the scientific and technological revolution.[13] Unlike most other branches in international law, it has been developed in a relatively short period of time. Space law has been formulated by instantaneous custom rather than by customary law, which is formed over a long period of general state practice.[14] Indeed, the first treaty law of outer space was born in 1967, just ten years after the launch of Sputnik, and a set of international space laws including five conventions and three principles have developed in less than thirty years. In fact, sometimes the development of space law outpaced technology. For instance, law applicable to the Moon and other celestial bodies was established back in 1979 when most states hardly had access to such bodies. The Agreement Governing the Activities of States on the Moon and Other Celestial Bodies (hereafter: the Moon Agreement) had addressed issues regarding utilization of the Moon, such as the complete demilitarization of the Moon and the retrieval of samples from the Moon – ideas that were only theoretical back then.

International space law is unique in character: it is a special branch of international public law, and yet not autonomous.[15] The international law of outer

---

[12] Sweden, United States, United Kingdom, South Africa, Russia, Ukraine, Hong Kong, Australia, Brazil, Canada, Korea, and France.

[13] A.S. Piradov & Belitsky (trans.), *International Space Law*, University Press of Pacific, Hawaii, 2000, 67.

[14] See C.J. Cheng, New Sources of International Space Law, *The Use of Air and Outer Space Cooperation and Competition: Proceedings of the International Conference on Air and Outer Space at the Service of World Peace and Prosperity, Held in Beijing from 21–23 August 1995*, Kluwer Law International, The Hague-Boston-London, 1998, 207.

[15] See C.J. Cheng, New Sources of International Space Law, *The Use of Air and Outer Space Cooperation and Competition: Proceedings of the International Conference on Air and Outer Space at the Service of World Peace and Prosperity, Held in Beijing from 21–23 August 1995*, Kluwer Law International, The Hague-Boston-London, 1998, 209.

space is derived from several sources, particularly maritime law, air law, treaty law applicable to Antarctica, and legislation of the United Nations.[16] That space law has no autonomy is most notably illustrated by the fact that there is no institution mandated to regulate activities in outer space such as the International Civil Aviation Organization (ICAO), the specialized agency of the United Nations that governs air navigation.[17]

A brief analysis of the ICAO is useful to highlight the peculiarity of the regime governing space activities. ICAO has functions that are not a part of the structure in which space operations are conducted, such as law-making power[18] and competencies to exercise powers over member states. ICAO itself is a legal entity that has competency to enter into agreements with other legal entities, including other international organizations, and can also adopt binding regulations by majority decision.[19] It has played an important role in: addressing the issues on aviation insurance and establishing an insurance mechanism; performing mandatory audits on states in the fields of aviation safety and security; and establishing a funding mechanism to finance aviation safety projects.[20] In addition, the Council of ICAO is given the authority to monitor the conduct of contracting states[21] and can impose obligations upon disagreeing states.

In contrast, space law making and its implementation has taken a different path. Rather than establishing an agency within UN systems, the United Nations has served as a focal point for international cooperation on the use of outer space and the UN Committee on the Peaceful Uses of Outer Space (COPUOS) was established within the General Assembly as a forum for negotiation to develop a legal framework. COPUOS became the first UN body to adopt a method of consensus where member must reach unanimous agreement on the particular subject matter before any action or document can be

---

[16] See G.H. Reynolds & R.P. Merges, *Outer Space Problems of Law and Policy,* Second Edition, Westview Press, Colorado, 1997, 27.

[17] The ICAO, established by Article 44 of the Chicago Convention, is a specialized agency of the United Nations.

[18] See R. Abeyratne, ICAO's Involvement in Outer Space Affairs- a Need for closer scrutiny? 30 *Journal of Space Law* (2004), 189.

[19] See R. Abeyratne, The Legal Effect of ICAO Decisions and Empowerment of ICAO by Contracting States, 32 *Journal of Space Law* (2007), 525.

[20] See R. Abeyratne, The Legal Effect of ICAO Decisions and Empowerment of ICAO by Contracting States, 32 *Journal of Space Law* (2007), 525.

[21] Article 69 of the Chicago Convention provides that "if the Council is of the opinion that the airports or other air navigation facilities, including radio and meteorological services of a Contracting State are not reasonably adequate for the safe, regular, efficient and economical operation of international air services, present or contemplated, the Council shall consult with the State directly concerned, and other States affected, with a view to finding means by which the situation may be remedied, and may make recommendations for that purpose."

confirmed.[22] Consensus has been somewhat successful and has given birth to a series of legal instruments related to space.

The sources of international space law are primarily found in international legal rules and international custom, and the general principles of law fill the gap between the two in the absence of an exclusive source of law.[23] International space law framework has three components: the general regime of space law, the law governing certain space applications, and declarations. The first component is called *corpus juris spatialis*, a body of treaty law negotiated amongst states within the framework of the United Nations: the Treaty on Principles Governing the Activities of states in the Exploration and Use of Outer Space, including the Moon and Other Celestial Bodies;[24] the 1968 Agreement on the Rescue of Astronauts, the Return of Astronauts and the Return of Objects Launched Into Outer Space;[25] the 1972 Convention on International Liability for Damage Caused by Space Objects;[26] the 1974 Convention on Registration of Objects Launched Into Outer Space;[27] and the 1979 Agreement Governing the Activities of States on the Moon and Other Celestial Bodies.[28]

The second component is law concerning satellite applications: the Direct Broadcasting Principles of 1982;[29] the 1986 Principles of Remote Sensing;[30] and the Nuclear Power Source Principles of 1992.[31] The declarations component includes: the 1963 Declaration of Legal Principles governing the Activities

---

[22] See E. Galloway, Consensus Decision-making by the United Nations Committee on the Peaceful Uses of Outer Space, 7-1 *Journal of Space Law* (1979), 11.

[23] See C.J. Cheng, New Sources of International Space Law, *The Use of Air and Outer Space Cooperation and Competition: Proceedings of the International Conference on Air and Outer Space at the Service of World Peace and Prosperity, Held in Beijing from 21–23 August 1995*, Kluwer Law International, The Hague-Boston-London, 1998, 215.

[24] Treaty on Principles Governing the Activities of States in the Exploration and Use of Outer Space, including the Moon and Other Celestial Bodies, 27 Oct 1967, 610 *UNTS* 205. [Hereafter: Outer Space Treaty]

[25] Agreement on the Rescue of Astronauts, the Return of Astronauts and the Return of Objects Launched into Outer Space (1968), 672 *UNTS* 199. [Hereafter: Rescue and Return Agreement]

[26] Convention on International Liability for Damage Caused by Space Objects, 29 Nov 1971 24 *UST* 2389.

[27] Convention on Registration of Objects Launched Into Outer Space, 12 Nov 1974 28 *UST* 695. [Hereafter: Registration Convention]

[28] Agreement Governing the Activities of States on the Moon and Other Celestial Bodies,(1979) 18 *ILM* 1434. [Hereafter: Moon Agreement]

[29] UNGA Resolution 37/92 The Principles Governing the Use by States of Artificial Earth Satellites for International Direct Television Broadcasting adopted on 10 December 1982. [Hereafter: DBS Principles]

[30] Principles Relating to Remote Sensing of the Earth from Outer Space of adopted on 3 December 1986.

[31] UNGA Resolution 47/68 on the Principles Relevant to the Use of Nuclear Power Sources in Outer Space adopted on 14 December 1992.

of States in Exploration and Use of Outer Space;[32] and the 1996 Resolution on Declaration on International Cooperation in the Exploration and Use of Outer Space for the Benefit and in the Interests of All States, Taking into Particular Account the Needs of Developing Countries.[33]

A basic regime governing outer space is reflected in the five multilateral treaties that establish the basic principles governing outer space and the fundamental rules for engaging in space activities. It is crucial to provide an understanding of the legal framework and its implications for remote sensing activities; therefore, this section first examines the legal regime of outer space and then discusses laws specific to remote sensing.

2.2.1. *Toward Formulation of Rules Governing Outer Space*

As discussed, the background for the establishment of a space law regime was highly political in the vigorous competition between the United States and the Soviet Union for national pre-eminence in space development. As a consequence, the outer space legal regime is largely the result of negotiation and compromise over the views held by these early space powers.

The formulation of the Outer Space Treaty started right after the launch of Sputnik-1. Two years after the launch of Sputnik, the UN COPUOS was established as a permanent body of the United Nations and given the responsibility to negotiate for a legal framework for space activities.[34] The series of General Assembly Resolutions following the Resolution of 1958 that approved the establishment of the UN COPUOS became the foundation of space law,[35] in particular Resolutions 1721A (XVI) of 20 December 1961 and 1962 (XVIII) of 13 December 1963.[36]

[32] UNGA Resolution 1962 (XVIII) on Declaration of Legal Principles governing the Activities of States in Exploration and Use of Outer Space adopted on 13 December 1963.

[33] UNGA Resolution 51/122 on Declaration on International Cooperation in the Exploration and Use of Outer Space for the benefit and in the interests of All States, Taking into Particular Account the Needs of Developing Countries, adopted 13 December 1996.

[34] See C.M. Hayward, Remote Sensing: Terrestrial Laws for Celestial Activities, 8 *Boston University International Law Journal* (1989), 169. Establishment of the UN COPUOS was approved by the UNGA Resolution on 13 December 1958.

[35] UNGA Resolution 1348 (XIII) 13 December 1958, 1472 (XIV) 12 December 1959, 1721 (XVI) 20 December 1961, 1802 (XVII) 14 December 1962, 1962 (XVIII) 13 December 1963.

[36] See B. Cheng, *Studies in International Space Law,* Clarenton Press, Oxford, 1999, 125.

The freedom of use and exploration of outer space and, hence, the legality of the flight of space objects, was confirmed and the prohibition on the appropriation of outer space was formulated in the General Assembly Resolution 1721A (XVI) on December 20, 1961,[37] and reiterated in the General Assembly Resolution 1962 (XVIII) that also established other fundamental rules governing outer space entitled, the "Declaration of Legal Principles governing the Activities of States in Exploration and Use of Outer Space."[38] Indeed, much of the content of the Outer Space Treaty had been agreed upon in this Resolution, prior to its formal introduction as a binding, multilateral treaty in 1967.[39] Such was the birth of international space law by means of the Outer Space Treaty, which has been ratified by 98 states parties,[40] and is often referred to today as the 'Magna Charta' of space law.

2.2.2. *Outer Space Treaty*

It is important to have a clear understanding of what the Outer Space Treaty establishes and its position within international public law. The scope of the Outer Space Treaty consists of the nature of space activities, the subjects conducting space activities, and the applicability of other international law to the Outer Space Treaty.

Space activities can be roughly categorized into three types: scientific activities, military activities, and commercial activities. Article I of the Outer Space Treaty stipulates that "there shall be freedom of scientific investigation in outer space",[41] and therefore, space activities referred in the Outer Space Treaty undoubtedly include scientific activities. Various scientific programmes, such as planetary exploration and other Earth-focused scientific studies, have been extensively conducted in accordance with the Outer Space Treaty, and continue to be planned. The question arises as to whether the Outer Space Treaty also applies to military activities and commercial activities.

Military activities are not straightforward. The Outer Space Treaty has established two distinct regimes applicable to different parts of outer space – outer space in general and celestial bodies in particular, including the Moon. Article VI states that "the Moon and celestial bodies shall be used by all States Parties to the treaty exclusively for peaceful purposes."[42] Article VI specifically bans

---

[37] See A.S. Piradov & Belitsky (trans.), *International Space Law*, University Press of Pacific, Hawaii, 2000, 83–85.

[38] UNGA 1962 (XVIII) 13 December 1963.

[39] See B. Cheng, The 1967 Outer Space Treaty: Thirtieth Anniversary, 23-4/5 *Air and Space Law* (October 1998), 156.

[40] As of 1 January 2008.

[41] Article I of the Outer Space Treaty.

[42] Article IV of the Outer Space Treaty.

establishment of military bases, installations, and fortifications, the testing of any type of weapons, and the conduct of military manoeuvres on celestial bodies; however, it permits the use of military personnel for scientific research or other peaceful purposes.

The term 'peaceful purposes' needs closer attention. The definition is not provided under the Outer Space Treaty and it is worth noting that back in 1967 'peaceful purposes' had two interpretations: non-military on one hand, and non-aggressive on the other.[43] The Soviet Union insisted on banning all military activities, which is clearly interpreting the Outer Space Treaty as 'non-military'; but since the United States carried out satellite observation for military reconnaissance purposes, it interprets the Outer Space Treaty language as referring to 'non-aggressive'.[44] Through the examination of the provisions of the Outer Space Treaty, one can see that the meaning of 'peaceful purposes' at the time of drafting was perceived as 'non-military';[45] but what has gained general acceptance at the present time is the interpretation that 'non-aggressive' military uses are peaceful.[46]

With the exception of the Moon and celestial bodies, non-aggressive military uses are not prohibited in outer space. In fact, the ambit of permissible military applications under Article IV of the Outer Space Treaty is relatively broad. Under Article IV, military space activities that are banned specifically are: (1) placing objects carrying nuclear weapons and weapons of mass destruction, and (2) the installation or stationing of such weapons in outer space or on celestial bodies.[47]

Following the above line of argument, permissible military uses of space are not limited to defensive space applications such as reconnaissance satellites. Strictly speaking, anti satellite weapons (ASATs) are not prohibited under Article IV, as they neither fall under the 'nuclear weapon' nor 'weapon of mass

---

[43] See G.H. Reynolds & R.P. Merges, *Outer Space Problems of Law and Policy*, Second Edition, Westview Press, Colorado, 1997, 91. The issue of disarmament of outer space was addressed in the UNGA Resolution 1884 (XVII) of October 17, 1963, which stated that outer space is to be used exclusively for peaceful purposes.

[44] C.C. Okolie, *International Law of Satellite Remote Sensing and Outer Space*, Kendall/Hunt Publisher Co., Dubuque, 1989, 35.

[45] If "peaceful" means non-aggressive, then the provisions "establishment of military bases, installations and fortifications, the testing of any type of weapons and the conduct of military manoeuvres on celestial bodies shall be forbidden" but "use of military personnel for scientific research or for any other peaceful purposes shall not be prohibited" would be meaningless.

[46] C.Q. Christol, *The Modern International Law of Outer Space*, Pergamon Press, 1982, 22.

[47] Article IV of the Outer Space Treaty.

destruction' definitions.[48] On 11 January 2007, China tested an ASAT missile and destroyed its defunct weather satellite in a polar orbit. The test raised concern internationally, mainly because China attempted to conceal the test and the harmful impact on the environment caused by the resulting debris. The ASAT test itself was not in contravention of Article IV of the Outer Space Treaty.[49] Likewise, intercontinental ballistic missiles (ICBMs) also fall outside the scope of the prohibition, as they are launched from one place and hit the target elsewhere on the Earth via outer space, but are neither 'placed in orbit' nor 'installed or stationed in outer space'.[50] There are also no rules with regard to intercepting missiles from outer space. Space-based military applications beyond the defensive scope such as ASATs, ICBMs, and missile defence are not specifically banned and remain a controversial issue under the current regime.

In reference to defensive space applications, historically satellites have been extensively utilized for military purposes. In fact, 75% of the 2,200 satellites launched from 1958 to 1984 are allegedly military satellites.[51] The use of outer space for military purposes, which has constituted a substantial part of space activities up until now, is within the scope of legitimate space activities, provided that states comply with the provisions under international law, including the UN Charter. The partial authorization of military uses of outer space also invokes the need for a more comprehensive legal regime governing military activities, which are often outside the scope of specific regulations, at both international and national levels. This is particularly relevant, since remote sensing satellites can serve a dual-use purpose (e.g., both military reconnaissance and civilian remote sensing). Commercial activities are not directly mentioned in the Outer Space Treaty. It was not that such activities were intended to be excluded; rather, in 1967, space activities were state-based and reserved solely for a few space powers, so extensive commercial activities were simply not foreseen at that time. Although INTELSAT was created at the beginning of the 1960s, applying military systems to civil and commercial uses such as reconnaissance by civil remote sensing satellites or employing navigation systems for public use, was hardly conceivable.[52] Yet the use of outer space appears to have

---

[48] See S. Aoki, *Japanese Space Strategy*, Keio University Press, Tokyo, 2006, 159. In fact, US and Soviet Union have tested anti-satellite weapons the past. See E.S. Waldrop, Weaponization of Outer Space, XXIX *Annals of Air and Space Law* (2004), 333.

[49] See in general K. Nair, China's Test: A Demonstrated Need for Legal Reform, 33 *Journal of Space Law* (2007), 191–194.

[50] See S. Aoki, *Japanese Space Strategy*, Keio University Press, Tokyo, 2006, 131.

[51] See S. Aoki, *Japanese Space Strategy*, Keio University Press, Tokyo, 2006, 131. Amongst the military satellites, 40% of all military satellites are said to be reconnaissance satellites.

[52] M. Benkö & K.U. Schrogl, Article 1 of Outer Space Treaty Reconsidered after 30 years – 'Free Use of Outer Space' vs. 'Space Benefits', *Outlook on Space Law over the*

included the notion of exploitation during the Outer Space Treaty's formulation process. The *ad hoc* Committee for space treaties in 1959 had employed the term 'exploitation'[53] and, during the discussion to clarify the meaning of 'use', the representative of France suggested that 'use' was equivalent to 'exploitation'.[54] Furthermore, commercial satellite communications activities utilising the geostationary orbit were widely known then, and no objection or exception to those activities was raised either in the text of the Outer Space Treaty or during its negotiation.[55] Also, a comparison with the Antarctic Treaty and legal status of Antarctica supports the conclusion that commercial activities in outer space can be considered consistent with its legal status.[56] Commercial activities being within the scope of the Outer Space Treaty is supported by a more recent legal instrument, the Declaration on International Cooperation,[57] which incorporates specifically the concepts of commercial activities. This declaration has a concrete provision on contractual terms, stating that: "Contractual terms in such cooperative ventures should be fair and reasonable, and they should be in full compliance with the legitimate rights and interests of the parties concerned as, for example, with intellectual property rights."[58]

Today, commercial space activities can be seen in various arenas. Examples include, but are not limited to, commercial launchers such as Sea Launch and Space X, Satellite TV, car navigation systems using GNSS, and the increasing number of companies dedicated to space business such as Virgin Galactic, which specializes in space tourism. The applicability of the Outer Space Treaty to commercial activities is not much questioned, but gaps and discrepancies do exist between the provisions of a treaty intended primarily for state-based space activities and the current situation, which includes remote sensing from space that has been conducted more and more by non-state based operations, thus presenting critical issues not addressed sufficiently under the present regime.

---

next 30 years, G. Lafferanderie, and Crowther, D. (Eds), Kluwer Law International, Hague, 1997, 68.

[53] C.Q. Christol, *Past, Present and Future of Space Law*, Kluwer Law and Taxation Publishers, Deventer, 1991, 67.

[54] C.Q. Christol, *Past, Present and Future of Space Law*, Kluwer Law and Taxation Publishers, Deventer, 1991, 68.

[55] See C.Q. Christol, *Past, Present and Future of Space Law*, Kluwer Law and Taxation Publishers, Deventer, 1991, 68.

[56] See G.H. Reynolds & R.P. Merges, *Outer Space Problems of Law and Policy*, Second Edition, Westview Press, Colorado, 1997, 42.

[57] Declaration on International Cooperation in the Exploration and Use of Outer Space for the Benefit and in the Interest of All States, Taking into Particular Account the Needs of Developing Countries. Adopted on 13 Dec 1996 (resolution 51/122).

[58] See Principle 2.

As to subjects for conducting space activities and implications, many different actors participate in space activities, including states, international organizations, and private entities. Only the subjects of international law – states, international intergovernmental organizations, and the other traditionally recognised entities – can conclude treaties under international law. It is evident that states are directly bound by the Outer Space Treaty but the question arises as to how non-state actors are covered.

Article XIII of the Outer Space Treaty clarifies the Treaty's applicability to an individual state even when activities are conducted within the framework of an intergovernmental organization.[59] Consequently, state parties will not be exempt from complying with the Outer Space Treaty just because the particular space activity was conducted by the international organization to which they belong. Article XIV only specifies a state as a subject for being a party to the treaty.[60] Whilst an international organization cannot be held as a party to the Outer Space Treaty, its language assigns consequential rights and obligations: Article VI states that "(...) responsibility for compliance with this Treaty shall be borne both by the international organization and by the State Party to the Treaty (...)".[61] For example, responsibility for an activity of ESA led by the United Kingdom rests with both ESA and the United Kingdom. Still, the question arises as to the responsibility of an IGO when the participating state is not a party to the Outer Space Treaty.

Private entities by themselves are not subject to international law; hence they are neither given rights nor obligations directly under international law. Under the Outer Space Treaty, however, private entities are bound by it through their own state parties under Article VI. Article VI serves as a bridge between the international regime and the national regime as it imposes international responsibility upon state parties for conducting space activities, regardless of whether they are governmental or private in nature.[62] States may thus be held responsible for space activities under their jurisdiction, including those of their national space agencies, private entities,[63] and international organizations in which the state parties participate for any part of the activities of the organization. This provision was a result of a compromise between the United States, who supported the provision, and the Soviet Union, who believed in excluding private

---

[59] See Article XIII of the Outer Space Treaty.

[60] Article XIV of the Outer Space Treaty states that "This Treaty shall be open to all States for signature." Exclusions of international intergovernmental organisations from the treaty provisions arise from the Soviet's view that States alone may be a party to the treaties.

[61] Article VI of the Outer Space Treaty.

[62] Article VI of the Outer Space Treaty stipulates that "State parties shall bear international responsibility for national activities in outer space, whether such activities are carried on by governmental agencies or by non-governmental entities."

[63] This includes entities that have a nationality of that State as well as the foreign owned entity operating inside the territory of that State.

entities from engaging in space activities.[64] Actually, it is common that national laws contain a licensing regime for conducting space activities in general, and with respect to launching operations in particular; some deal specifically with remote sensing activities, which will be examined later in this Chapter.

Another important characteristic of international space law is the applicability of general international law to space activities. The Outer Space Treaty establishes the applicability of international law to the regime in outer space;[65] therefore, international legislation, the United Nations Charter, international customary law, as well as other general principles of law are applicable, complementing the Outer Space Treaty. This Article III provision needs further analysis. At first glance, there seems to be a certain amount of contradiction. Since the Outer Space Treaty establishes that neither the concept of sovereignty nor that of territory exist in outer space, it appears inappropriate to apply international law on the basis of state sovereignty to exploration of space and to those activities that take place wholly within outer space.[66] It is, therefore, much more logical to conclude that Article III of the Outer Space Treaty applies to Earth-oriented activities such as: 1) activities from space to the Earth; 2) activities within outer space, as far as these directly affect the relations among states on the Earth; 3) activities from the Earth or its airspace to outer space; and 4) activities from the Earth or its airspace via outer space back to the Earth or its airspace.[67] Major Earth-oriented activities such as launching a space object to outer space, GNSS, and remote sensing of the Earth from space, specifically and directly affect the relations between/amongst states on Earth. Such activities are subject to application of international law, and are also in need of more specific regulations.

2.2.3. *Overview of the Regime Governing Use and Exploration of Outer Space*

This section addresses the elements critical to an understanding of the outer space legal regime. It discusses the basic principles governing outer space, as well as the obligations associated with conducting space activities, which are endorsed by the Outer Space Treaty and the other relevant documents in the *corpus juris spatialis*. The basic principles underlie the concept of outer space

---

[64] See Y. Sneifer, The Implications of National Security Safeguards on the Commercialization of Remote Sensing Imagery, 19 *Seattle University Law Review* (1996), 549.

[65] See Article III of the Outer Space Treaty.

[66] See H. Wassenbergh, *Principles of Outer Space Law in Hindsight*, Martinus Nijhoff, Publishers, Dordrecht-Boson-London, 1991, 17.

[67] See H. Wassenbergh, *Principles of Outer Space Law in Hindsight*, Martinus Nijhoff, Publishers, Dordrecht-Boson-London, 1991, 17.

and set forth the norms for conducting space activities. The section then discusses the specific obligations involved in outer space: state responsibility in conducting space activities, liability for damage caused by space objects and registration of space objects.

2.2.3.1. *The Freedom of Outer Space*

The freedom of space activities was clearly established by the Outer Space Treaty, which states: "Outer Space, including the Moon and other celestial bodies, shall be free for exploration and use by all States (...) and there shall, be free access to all areas of celestial bodies."[68] Freedom of space means that space is open and accessible to all and freedom to conduct space activities is secured to all states, including those not parties to the Outer Space Treaty; hence, no state can impede any part of space activities carried out by other state.

Freedom cannot be detached from the notion of sovereignty. Freedom to conduct certain activities outside the state's own national territories rests on the foundation that either sovereignty of a state does not apply in the area (e.g., the high seas) or permission is granted by the sovereign state to other states that wish to conduct a particular activity inside the territory of that sovereign state (e.g., right to fly over the foreign state through commercial carriage).[69] As mentioned earlier, space law derives from other branches of international law, and the legal status of outer space open for free exploration and use originates in the principle established by the law governing the seas and Antarctica.[70] The 1982 United Nations Convention on the Law of the Sea[71] is worthy of further analysis.

The issue that is of particular relevance to outer space is the legal status of high seas established by the Law of the Sea. Article 87 of the Law of the Sea states that: "The high seas are open to all States."[72] The high seas may be used freely by the ships of all nations; several different types of freedom are recognized; amongst others, freedom of navigation, freedom of fishing, freedom to lay submarine cables and pipelines, and freedom to fly over the high seas.[73]

---

[68] Article I of the Outer Space Treaty.

[69] This includes the extensive practice in international air transport that certain freedoms such as freedom to fly over a sovereign State are granted in the form of agreements to exchange such rights associated with air space through bilateral agreements.

[70] The freedom of exploration and use can be regarded as an extension of the earlier freedom established under the Antarctic Treaty. See G.H. Reynolds & R.P. Merges, *Outer Space Problems of Law and Policy,* Second Edition, Westview Press, Colorado, 1997, 41.

[71] 1982 United Nations Convention on Law of the Sea, 13 *UST* 2312. [Hereafter: Law of the Sea Convention]

[72] Article 87 of the Law of the Sea Convention.

[73] See Article 87 of the Law of the Sea Convention.

The Law of the Sea Convention generally recognizes that the legal status of the air space above the high seas is analogous to that of the high seas proper, that is, *res omnium communis*, meaning that no state may exercise sovereign rights over the high seas or the airspace over the high seas.[74]

One can argue that freedoms associated with the high seas could be applicable to outer space. As mentioned earlier, the Outer Space Treaty provides for four freedoms in outer space: the freedoms of exploration, use, access, and scientific investigation.[75] It can be suggested that freedom of navigation in the high seas is analogous to freedom of exploration in outer space; freedom to lay submarine cables and pipelines is analogous to freedom of use of outer space; and overflight right is analogous to 'access' to outer space. An exception would be the freedom of fishing. This freedom does not appear to fit into outer space activity unless an aspect of mining is considered. Freedom to fish can be comparable to retrieving samples from celestial bodies as stipulated under the Moon Agreement;[76] but, theoretically, the right to collect samples is reserved solely for parties to the Moon Agreement, which has not been ratified by many spacefaring nations, since such right is not specifically referred to under the Outer Space Treaty.

The four freedoms of outer space allow states to use or explore outer space freely using manned or unmanned spacecrafts for a variety of applications, both to enhance life on Earth and to conduct scientific investigations insofar as they are in accordance with the provisions of the Outer Space Treaty.

Space applications such as remote sensing, navigation, and telecommunications demonstrate the free use of outer space by operating in certain orbits.[77] Similarly, satellite remote sensing advances the free use of outer space, since, in contrast to aerial remote sensing, it permits the states to conduct information gathering without intruding upon territorial sovereignty. For instance, one can

---

[74] See G.H. Reynolds & R.P. Merges, *Outer Space Problems of Law and Policy,* Second Edition, Westview Press, Colorado, 1997, 40.

[75] Article 1 of the Outer Space Treaty stipulates that "Outer Space, including the Moon and other celestial bodies, shall be free for exploration and use by all States (...) and there shall, be free access to all areas of celestial bodies. There shall be freedom of scientific investigation (...)" See H. Desaussure & N. Jasentuliyana (Eds.), The Freedom of Outer Space and their Maritime Antecedents, *Space Law Development and Scope*, Praeger Publishers, Connecticut, 1992, 5.

[76] Article 6.2 of the Moon Agreement states that "in carrying out scientific investigations and in furtherance of the provisions of this Agreement, the States Parties shall have the right to collect on and remove from the Moon samples of its mineral and other substances".

[77] Major orbits include 1) Low Earth Orbit (LEO), an orbit between 200–5,500 km above the Earth suitable for earth observation, 2) Medium Earth Orbit (MEO) between 10,000–20,000 km used for GNSS, and 3) Geostationary Orbit (GO) at 36,000 km suited for telecommunication and weather satellites.

find the practice of securing the freedom of use at the geostationary orbit. Since the geostationary orbit is regarded as a limited natural resource and is overcrowded with myriad satellites, the failure of a state to launch its satellite(s) to the designated slot within a certain time period would be seen by other states as hampering their use of this space resource. The International Telecommunication Union (ITU) regulates the allocation of communications frequencies and occupancy of orbital positions (often referred as 'slots') around the geostationary orbit.[78] A state that fails to launch a satellite within an agreed period would be deprived of its right of use of frequency and access to the slot, as it would be hampering the opportunity for other space community members to use this limited resource.

The launching and re-entry of spacecraft is based on the freedom of access to space. The idea of freedom of access needs to be treated with caution, since – in contrast to an activity which takes place solely in outer space – initial access to space is an Earth-oriented activity as any space object is launched from the ground of the Earth; hence, it is subject to international law. At present, a considerable number of states need to depend upon the use of foreign launch sites, so the absolute freedom of access to space is limited in one way or another. Indeed, launching and re-entry of space objects that take place from a foreign state affect the legal rights and interest of that state. A state with a launch site may deny a request from other states to launch their satellite for any reason whatsoever.

Since the launch and re-entry of a spacecraft involves passing through airspace, it is useful to refer to the international law governing airspace. The regime applicable in airspace, as stipulated in Article I of the Chicago Convention, is that "every State has complete and exclusive sovereignty over the airspace above its territory."[79] Therefore, passage through territorial airspace must be authorized either through a contractual arrangement or by expressed permission.[80] If a space object is launched from a foreign territory, whether it is a state or commercial launch service, the launching takes place with the consent, or at least with the knowledge, of the state where the launch takes place. In that sense,

[78] Article 33 of the ITU Convention states that "In using the frequency bands for space-radio services members shall bear in mind that radio frequencies and the geostationary orbit are limited natural resources, that they must be used efficiently and economically so that countries (…) may have equitable access to both in conformity (…) to their needs and technical facilities at their disposal".

[79] Article I of the Chicago Convention.

[80] See G.H. Reynolds & R.P. Merges, *Outer Space Problems of Law and Policy*, Second Edition, Westview Press, Colorado, 1997, 36. States have had to negotiate on a bilateral basis to exchange different types of freedom, a set of aviation rights granting a State's carriers the privileges to enter and land in another country's airspace, such as flying to and from the foreign territory and putting down carriage for commercial purposes.

authorization is granted to the state prior to the launch, which, in a sense, follows the air law regime.

On the other hand, the re-entry of a space object via foreign airspace back to the launch site of its own state is much more complex. It is much less obvious, and probably much less formally authorized. Unlike air law, in which a right of passage through the foreign airspace is pre-arranged for that particular purpose (e.g., the right to discharge commercial carriage in a foreign state), the return of a space object from outer space to the Earth sometimes inevitably accompanies passage through foreign airspace without prior legal arrangement. Considering the circumstances, one can draw a closer analogy with the concept of innocent passage, which grants foreign ships a right of passage through another state's territorial water under the 1982 Law of the Seas.[81] The passage is construed as 'innocent' if the ship is (a) "traversing that sea without entering internal waters or calling at a roadstead or port facility outside internal waters; or, (b) proceeding to or from internal waters or a call at such roadstead or port facility insofar as the passage is continuous and expeditious."[82] Both launch and re-entry are sufficiently analogous to the circumstances in which right of innocent passage is granted, as the pass through the foreign airspace is only temporary and on the way to a final destination, specifically outer space or the home state. In fact, precedents do exist such as the Soviet spacecraft flying over the airspace of adjacent states in its return to the Soviet Union; or the space shuttle Challenger on 13 October 1984 crossing the US-Canadian air border at an altitude of approximately 68 km on its way to landing in US territory.[83] Whilst for these two cases it is not certain as to whether the states were informed either prior to or after the launch with respect to their overflight, there is a precedent of prior notification of re-entry: In March 1990, during the flight stage of the Atlantis, the United States informed Russia, as a courtesy, of the planned descent of a spacecraft through the airspace of Soviet Union territory a few hours before the overflight took place.[84]

Whilst these precedents do not deny the possibility of the right of a spacecraft to pass through foreign airspace after re-entry, it should be noted that there is no sufficient support for a customary rule for the right of innocent passage specific to spacecraft.[85] The issue of the right of free passage for a state's spacecraft through the airspace of other states is of growing importance. In the

---

[81] Article 21 of Law of the Sea Convention.

[82] Article 16 of Law of the Sea Convention.

[83] See V.S. Vereschetin, Custom as a Source of International Law of Outer Space, 13-1 *Journal of Space Law* (1985), 119.

[84] See A.D. Terekhov, Passage of Space Objects through Foreign Airspace: International Custom?, 25-1 *Journal of Space Law* (1997), 4.

[85] See A.D. Terekhov, Passage of Space Objects through Foreign Airspace: International Custom?, 25-1 *Journal of Space Law* (1997), 5.

future, with the advent of commercial space tourism, re-entry of a manned vehicle through foreign airspace is likely to take place more frequently.

The principle of freedom establishes the rule that states do not have to obtain permissions for engaging in space activities and cannot prevent other states from conducting such activities, insofar as they do not interfere with the sovereignty of other states. In theory, states not party to the Outer Space Treaty may be engaged in space activities. Although it can be construed that the right to conduct space activities also establishes the legality of sensing from space, the Outer Space Treaty does not set forth under what conditions states are free to conduct remote sensing from space; therefore, freedom of space established by Article I of the Outer Space Treaty itself cannot substitute for a concrete code of conduct for remote sensing activities.

2.2.3.2. *The Principle of Non-Appropriation*

Another major principle is the principle of non- appropriation. Article II of the Outer Space Treaty states that: "Outer Space, including the Moon and other celestial bodies, is not subject to national appropriation by claim of sovereignty, by means of use or occupation, or any other means."[86] This means that sovereignty does not extend to outer space and no state can claim outer space or any celestial body as its territory and exercise sovereignty rights therein.[87] Article II precludes any individual state from obtaining legal rights (e.g., ownership rights) over outer space or any parts thereof. Consequently, the status of outer space is not *res nullius* or *terra nullius* – unclaimed land open to acquisition by any state – but would qualify as *terra communis* – land jointly managed by all states – or res extra commercium – areas outside state's sovereignty.[88] This means that only a community of all states acting collectively can regulate legally any activity in outer space.[89]

Incorporating the principle of non-appropriation of outer space was welcomed by the super powers because, at the time of formulation of the Outer Space Treaty, both of them were unsure which one would land on the moon

[86] For the principle of non-appropriation, see further S. Gorove, *Studies in Space law: its challenges and prospects,* Nijhoff Publisher, Leyden, 1977, 79–84. See also C.Q. Christol, Article 2 of 1967 Principles Treaty Revisited, 9 *Annals of Air and Space Law* (1984), 217–221.

[87] See M. Lachs, *The Law of Outer Space: an experience in contemporary law making,* Sijthoff Publisher, Leiden 1972, 43.

[88] See F.G. von der Dunk, The Moon Agreement and the Prospect of Commercial Exploitation of Lunar Resources, 32 *Annals of Air and Space Law* (2007), 97.

[89] See F.G. von der Dunk, The Moon Agreement and the Prospect of Commercial Exploitation of Lunar Resources, 32 *Annals of Air and Space Law* (2007), 97.

first.[90] As seen earlier, the inclusion of the non-appropriation principle met with unanimous approval in General Assembly Resolution 1721 A (XVI) in 1961.[91] Indeed, the principle articulated in Article II was the main impetus for the existence of the Outer Space Treaty.[92] From this one can infer that the formulation of the Outer Space Treaty was political in nature.

In recent years, the application of the principle of non-appropriation has become a subject for debate with the emergence of companies with a business model taking advantage of the status of celestial bodies. Some companies, such as Lunar Embassy, have virtually sold land of celestial bodies on the Internet, claiming that they own them. Lunar Embassy's argument was based upon the provision's language "(…) is not subject to *national* (emphasis added) appropriation (…)", which it interpreted as constituting a loophole in a literal sense. What is not given to a state is given neither to natural or juridical persons;[93] therefore, one can argue that any entity claiming the ownership of celestial bodies has no legal standing. The principle is likely to be challenged more as commercial exploitation of celestial bodies advances.

2.2.3.3. *The Principle of Cooperation and Mutual Assistance*

As mentioned previously, the historical development of a legal framework for space activities came through negotiation and consensus within the UN COPUOS, which served as a forum for cooperation. The principle of cooperation constitutes one of the central principles governing the regime of outer space.[94] Whilst most provisions referring to international cooperation in the Outer Space Treaty are recommendatory in nature,[95] certain Articles do create an obligation to cooperate on specific matters. One example is the obligation to cooperate with regards to the protection of the environments of outer space and the Earth.

Article IX of the Outer Space Treaty obligates party states to undertake international consultation in case of a potential threat of environmental harm. Article IX states that: "States Parties to the Treaty shall pursue studies of outer space, including the moon and other celestial bodies, and conduct exploration of them so as to avoid their harmful contamination and also adverse changes in

---

90 At first, the United States was against this principle; however, driven by the fear that the Soviet Union might be the first to land on the Moon, the US called for a conclusion of a treaty. See B. Cheng, The 1967 Outer Space Treaty: Thirtieth Anniversary, XXIII-4/5 *Air and Space Law* (October 1998), 160.

91 UN Doc. A/4987, 20 December, 1961.

92 See B. Cheng, The 1967 Outer Space Treaty: Thirtieth Anniversary, XXIII-4/5 *Air and Space Law* (October 1998), 160.

93 See W. Jenks, *Space Law*, Stevens Publisher, London, 1965, 201.

94 Article III, IX, X, XI of the Outer Space Treaty.

95 See, e.g., Article I and Article III of Outer Space Treaty.

the environment of the Earth resulting from the introduction of extraterrestrial matter and, where necessary, shall adopt appropriate measures for this purpose."[96] In addition, Article IX requires states to undertake prior consultation in case of potential harm: "If a State Party to the Treaty has reason to believe that an activity or experiment planned by it or its nationals in outer space, including the Moon and other celestial bodies, would cause potentially harmful interference with activities of other State parties in the peaceful use of outer space, it shall undertake appropriate international consultations before proceeding with any such activity or experiment."[97] As mentioned earlier, intentionally destroying a satellite in the manner of the 2007 ASAT test conducted by China is in contravention of this principle that requires prior consultations for activities that impact others. The fact that no prior warning was given by China resulted in criticism from international society. Furthermore, with the February 2009 in-orbit crash of two full sized satellites – a first-time incident – the ongoing issues concerning space debris came under renewed scrutiny. The more space activities that are undertaken, the more concerns over the impact on the environment will grow. It is highly conceivable that obligations associated with space debris will be elaborated into more concrete guidelines; or there may be separate legislation devoted to the issue addressed under Article IX.

Another obligation concerns the cooperation in disclosing the details of space activities undertaken. Article XI stipulates that: "In order to promote international cooperation … Signatories are obliged to inform the United Nations Secretary General, as well as the public, and the international scientific community, to the greatest extent feasible and practicable, of the nature, conduct, locations, and results of space activities." One can contend that the language of this provision is too soft to impose any concrete obligation to disseminate the information concerning space activities so as to share it with the relevant communities. The phrase, "to the greatest extent feasible and practicable" reduces any obligation that may derive from this Article to a level purely within the discretion of the space power concerned.[98] The interpretations of, and the positions with respect to, this provision are not uniform. The Soviet delegation adhered to the position that reporting of activities in outer space generally should be on a voluntary basis.[99] The United States has interpreted this obligation to require the widest practical dissemination of data and information obtained from the civil space program to provide improved space

[96] Article IX of the Outer Space Treaty.

[97] Article IX of the Outer Space Treaty.

[98] See B. Cheng, *Studies in International Space Law,* Clarenton Press, Oxford, 1999, 403–404.

[99] See G.H. Reynolds & R.P. Merges, *Outer Space Problems of Law and Policy,* Second Edition, Westview Press, Colorado, 1997, 74.

services to itself and to the rest of the world.[100] ESA promotes the view that member states shall share and exchange information amongst themselves, but they are not required to do so if they believe it would be against their own national interests.[101] This implies that the national interest prevails over the obligation to share information; also, it appears to support the earlier conclusion that no obligation for cooperation is established, so sharing of information related to space activities is not required.

The recommendation to share information on space activities has significant implications for remote sensing activities, as the 'result of space activities' infers the inclusion of data and information derived from satellite remote sensing. 'Informing' is an active obligation and, if applied to remote sensing, it could become an automatic obligation to disclose every time new remote sensing data is gathered. One can see the implementation of this provision in reality through the practice of US governmental entities such as NASA and NOAA making their data available in archives through the Internet at marginal cost, but at an international level that practice can only occur if such a commitment has been agreed upon – as in the case of the International Charter on Space and Major Disaster (hereafter: Disaster Charter) where data and maps are created to assist disaster relief and humanitarian operations within these frameworks via the internet, and which are free of charge to download.[102]

In practice, it appears that the sharing of all data and information from remote sensing is not required of states, otherwise it would be difficult to explain the presence of multilateral data exchange principles,[103] Memoranda of Understanding (MOUs), or other forms of agreements specifying authorisation of mutual access to data and/or assistance to provide data.

With regard to the form of cooperation in space activities in general, when states cooperate with other states on specific matters, they tend to enter into treaties, bilateral and multilateral agreements, or soft forms of agreement such as MOUs.[104] Historically, there has been extensive cooperation in various forms in the field of space; international cooperation has resulted in the establishment of numerous international organisations through conventions such as the International Telecommunications Satellite Organisation (INTELSAT), the Council on International Cooperation in the Study and Utilisation of Outer Space (INTERCOSMOS), the International Organisation of Space Communications (INTERSPUTNIK), and the European Space Agency (ESA).

---

[100] See J.I. Gabrynowicz, Defining Data Availability for Commercial Remote Sensing, 23 *Annals of Air and Space Law* (1998), 95.

[101] Article III of the Convention for the establishment of a European Space Agency.

[102] See http://www.disasterscharter.org/. Last accessed on 10 April 2010.

[103] Examples include the Committee on Earth Observation Satellites (CEOS) Principles and WMO Resolutions.

[104] For MOU in general, see A. Aust, *Modern Treaty Law and Practice*, Cambridge University Press, Cambridge, 2000, 26–46.

In the conventions establishing the aforementioned organisations, concrete terms of cooperation are provided.

In contrast to the multilateral conventions establishing international organizations, cooperation in the field of satellite remote sensing is commonly seen in voluntary forms of cooperation; that is, neither strictly binding nor completely lacking in legal significance, such as exchange of letters and MOUs. Cooperation has been particularly extensive in the field of developing remote sensing satellites, as well as in acquiring and processing data from satellites. Ground station arrangements are often based on MOUs.[105] Successful emergence and operation of a framework for cooperation such as the Disaster Charter to be discussed later, is not based upon legal obligations born of treaty-based law, but on a voluntary effort amongst the willing partners. The international collaborative effort under the Disaster Charter framework is based on voluntary cooperation and good endeavours.[106] Within such programmes, the beneficiaries can receive assistance through access to critical information derived from a combination of satellites. The partners' will to set up such mechanisms signifies the success of voluntary forms of cooperation in the field of satellite remote sensing.

2.2.3.4. *State Responsibility in Conducting Space Activities*

Whilst the rules addressed in previous sections were fairly general, some provisions of the Outer Space Treaty establish explicit obligations. They endorse concrete rules on responsibility, liability, and control of a space object, and state how the obligations are implemented.

The Outer Space Treaty articulates the important rule that states are free to conduct space activities, but are obligated to be internationally responsible. Gorove contends that "it seems that when we speak of responsibility, we are

[105] For instance, the United States of America and India: Memorandum of Understanding concerning remote sensing and acquisition of satellite data, 31 January 1975; and the Memorandum of Understanding between the European Space Agency and the United States National Aeronautics and Space Administration concerning ERS-1 data acquisition at Fairbanks, 14 January 1986. Landsat ground station agreements have been in the form of MOUs such as Memoranda of Understanding between the National Space Development Agency of Japan and the United States Department of Commerce National Oceanic and Atmospheric Administration, July 5 1983. For a discussion of Landsat arrangement, see further P.A. Salin, Landsat Contracts Signed by US Agencies with Foreign Ground Stations: Commercial Remote Sensing from NASA Scientific Experiments to EOSAT Private Endeavours, 41 *German Journal of Air and Space Law* (1992), 165–175.

[106] Article 3.1 of the Disaster Charter states that "The parties shall develop their cooperation on a voluntary basis, no funds being exchanged between them" as well as Article VI stipulating: "The parties shall use their best endeavours in the conduct of this cooperation, which shall proceed on the following basis."

dealing primarily with obligations imposed on people and institutions who are supposed to carry out certain activities or are accountable in given situations though not necessarily in the form of compensation for damages."[107] By mandating international responsibility to the state party for its national activities in outer space, the Outer Space Treaty denies anarchy in outer space and establishes a regime where space activities are controlled by state parties.

State parties have responsibility over their national space activities, whether conducted by governmental agencies or non-governmental entities.[108] This particular provision imposing a direct responsibility on states for national activities conducted by non-governmental entities is unique under international law.

Since a state has responsibility for national activities including those conducted by private persons, it is particularly important to clearly define the scope of 'national activities'. They can be interpreted as those activities over which states can exercise legal control, that is, activities that take place under their jurisdiction, including: activities carried out by nationals of a state; activities which take place within a territory of that state; and activities which are under the jurisdiction of a state through a registration of a space object (quasi-territorial jurisdiction);[109] and possibly by other arrangement. The ambit of 'national' space activities can be relatively broad.

Article VI further states that "The activities of non-governmental entities in outer space, including the Moon and other celestial bodies, shall require authorization and continuing supervision by the appropriate State Party to the Treaty."[110] As a consequence of this responsibility, the appropriate state needs to have a control regime with respect to the activities by non-governmental entities within its jurisdiction.

The definition of 'appropriate state' is absent from the entire Outer Space Treaty, and requires further analysis. The fact that there is no specific definition allows flexibility to extend the interpretation of 'appropriate state' to states concerned, instead of limiting that interpretation to 'launching state' or 'state of registry'. In principle, any state that exercises control over said activities – whether that control be national, territorial, quasi territorial, or designated specifically by agreement – qualifies as an 'appropriate state'. Since the term 'appropriate state' is in a singular form, only one state can exercise jurisdiction, even if more than one state may be involved in the activity in question.[111]

---

[107] S. Gorove, *Developments in Space Law,* Martinus Nijhoff Publishers, Dordrecht, 1991, 224.

[108] See further A. Kerrest, Remarks on the Responsibility and Liability for Damages Caused by Private Activity in Outer Space, *Proceedings of the Fortieth Colloquium on the Law of Outer Space,* IISL, AIAA, Washington, 1997, 134–140.

[109] See F.G von der Dunk, *Private Enterprise and Public Interest in the European Spacescape,* International Institute of Air and Space Law, Leiden, 1998, 18–19.

[110] Article VI of the Outer Space Treaty.

[111] See F.G. von der Dunk, *Private Enterprise and Public Interest in the European Spacescape,* International Institute of Air and Space Law, Leiden, 1998, 21.

The fact that 'appropriate state' is put in a singular form has some implications. It would clarify where the responsibility for non-governmental activities lies, and would allow the discretion of those states concerned as to which one will actually exercise jurisdiction, in case more than one state party is involved in the activity in question. In the meantime, authorization and supervision by a single appropriate state would also avoid overlapping jurisdictions and promote effective control over a particular space activity undertaken.

The extent of the concept of international responsibility is ambiguous. It is not clear as to whether the responsibility extends beyond compliance with the Outer Space Treaty to a wider scope of international and national law as well as to contracts.[112] Such scope of responsibility needs to be defined more precisely in the national context.

A state's responsibility in conducting space activities could result in international liability for damage caused to other states. Article VII of the Outer Space Treaty states that: "Each State Party to the Treaty that launches or procures the launching of an object into outer space, including the Moon and other celestial bodies, and each State Party from whose territory or facility an object is launched, is internationally liable for damage to another State Party."[113] This provision is elaborated in the 1972 Liability Convention.

2.2.3.5. *Liability for Damage Caused by Space Objects*

Liability involves the legal consequences of a particular action or omission that result in damage,[114] that is, a loss or injury to person or property. The 1972 Liability Convention addresses the legal consequences for a state when either its launch or operation of a spacecraft in orbit leads to damage to another state, and provides a civil remedy. The Convention establishes a victim-oriented international scheme for the compensation of damage resulting from launching activities that provides the maximum protection for potential victims. The Liability Convention establishes the fundamental rules regarding international liability for damage caused by space objects so that different circumstances of such accidents are satisfactorily addressed.

Overall, the Liability Convention sufficiently foresees different settings and clarifies what measures of protection are available to potential victims under differing circumstances. First, it makes clear that the damage caused within a launching state is an internal matter and sets such situations outside the scope of the Liability Convention. The Convention only applies to damage done to foreign territories as well as to foreign nationals not participating in launch

---

[112] See B. Cheng, The 1967 Outer Space Treaty: Thirtieth Anniversary, XXIII-4/5 *Air and Space Law* (October 1998), 156.

[113] Article VII of the Outer Space Treaty.

[114] S. Gorove, *Developments in Space Law,* Martinus Nijhoff Publishers, Dordrecht, 1991, 223–224.

and/or operation of a satellite.[115] For example, the Liability Convention did not apply to the instance of the disintegration of the space shuttle Columbia on 1 February 2003, and the consequent precipitation of debris over US territory. The Liability Convention predominantly relies on a state-based compensation; however, it does not exclude the possibility of individual compensation as an alternative as Article XI(2) provides that "Nothing in this Convention shall prevent a State, or natural or juridical persons it might represent from pursuing a claim in courts or administrative tribunals or agencies of a launching state."[116] This gives victims opportunities to seek recovery in a civil suit without having to rely on the government; however, claims for the same damage cannot be presented in parallel on a state-to-state and individual basis.[117]

Closer examination of the mechanism for state-based compensation reveals that measures for dispute settlement are available at two levels: first, through diplomatic negotiation; second, through settlement by a Claims Commission. The former measure needs to be exhausted before application of the latter, within the time limit of one year.[118] If the dispute is not settled by diplomatic negotiation within this timeframe, a Claims Commission consisting of three members appointed by the claimant state, launching state, and both states respectively[119] is established. The Commission shall give its decision or award, and the decision of the Commission shall be final and binding if the parties have so agreed; otherwise the Commission shall render a final and recommendatory award, which the parties shall consider in good faith.[120]

Pursuant to Article VII of the Outer Space Treaty, the Liability Convention establishes an inseparable link between the launching state and damage caused by its space object. The Liability Convention adopts two distinct bases of liability, depending on the location where damage is caused. It imposes absolute liability on a launching state in case of damage on Earth: "A launching State shall be absolutely liable to pay compensation for damage caused by its space object on the surface of the Earth or to aircraft in flight."[121] It imposes fault-based liability in outer space: "In the event of damage being caused elsewhere than on the surface of the Earth to a space object of one launching State or to

---

115 Article VII of the Liability Convention.

116 Article XI (2) of the Liability Convention,

117 Article XI (2) of the Liability Convention states that: "A state shall not, however, be entitled to present a claim under this Convention in respect of the same damage for which a claim is being pursued in the courts or administrative tribunals or agencies of a launching state or under another international agreement which is binding on the States concerned."

118 Article XIV of the Liability Convention.

119 Article XIV of the Liability Convention. For the details of the Claims Commission, see Article XV.

120 Article XVIII of the Liability Convention

121 See Article II of the Liability Convention.

persons or property on board such a space object by a space object of another launching State."[122] There is no ceiling on the amount of compensation under the Liability Convention and this adds to the financial burden imposed on launching states.

When the damage takes place on the surface of the Earth, it is not necessary to prove fault; the only required element of proof is causation; that is, the damage was actually caused by the space object. Therefore, just by being qualified as a launching state, that state would potentially be held liable in case of damage on the Earth's surface. The only circumstance in which exoneration from liability is possible is when "the damage has resulted from gross negligence or an act or omission done with intent to cause damage on the part of a claimant State or natural or juridical persons it represents."[123] If a private entity is grossly negligent, the government may be exonerated from international liability over such an act by that entity. On the other hand, the Convention does not refer to exoneration in case the damage is caused by the fault of a third party or is due to *force majeure* (e.g., natural disasters).[124]

For damage caused either elsewhere than the Earth's surface or aircraft in flight, a launching state is liable only if it is at fault for damage caused in outer space (e.g., collision of/with a satellite).[125] In contrast to damage on the surface of the Earth, the claimant launching state has to demonstrate that the other launching state was at fault in causing such damage in outer space. If the collision was because of an inevitable accident with no fault on either side, no liability is invoked.

As discussed above, the point of departure for determining compensation is the identification of a launching state. For a close analysis of the terms of the Liability Convention, the critical requirement is the definition of 'launching state'. "The term launching State means (i) a State which launches or procures the launching of a space object; and (ii) a State from whose territory or facility a space object is launched."[126] Whilst a 'State that launches a space object' is fairly straightforward, a 'State which procures the launching' is not. The term is not defined concretely and is open to different interpretations. It is logical to regard a state with jurisdiction over the entity requesting launch services conducted from the foreign territories as a 'State which procures the launching'. The term may also be interpreted as including those states whose jurisdiction covers the activities of entities in manufacturing and arranging the launch of a satellite on behalf of its customer. Through such differences in interpretation,

---

[122] Article III of the Liability Convention.

[123] Article VI (1) of the Liability Convention.

[124] See A. Kerrest, M. Benkö and K.U. Schrogl (Eds.), Liability for Damage Caused by Space Activities, *Space Law: Current Problems and Perspectives for Future Regulation*, Eleven International Publishing, 2005, 100.

[125] Article II of the Liability Convention.

[126] Article I(c) of the Liability Convention.

the term, 'launching state' could offer potential victims more than one choice of targets to from which to seek recovery. The other side of the coin is, of course, that there is more potential risk for liability to all states that are in some way involved in launching operations either as a customer or service provider.

The Liability Convention holds launching states jointly and severally liable if more than one launching state is responsible for damage.[127] Given the four types of launching states coupled with the open definition of a 'State that procures launching', it is very common to have more than one launching state; it is increasingly so nowadays, with the growing number of commercial satellite manufacturers and launcher services that are offering customers around the globe the opportunities to get their satellites manufactured and launched outside their home states.

By providing for joint and several liability, the Liability Convention aims to implement proper allocation of costs as well as to minimise the burden of a potential claimant. This procedural measure can particularly be characterized as 'victim oriented' since an injured state reserves its right to seek the entire compensation from any or all of the launching states[128] without requiring them to identify which launching state(s) are liable for what portions of the damage. In case one launching state paid the compensation to a victim state, that state may request reimbursement from other responsible launching states.[129] In this fashion, the Liability Convention has established a mechanism to provide remedies promptly to victims and leaves the rest of the business of settling claims to the launching states concerned.

In case launching states are jointly liable for damage, the foremost question concerns how liability is apportioned between/amongst them. Article IV states that "(...) the burden of compensation for the damage shall be apportioned between the first two States in accordance with the extent to which they were at fault; if the extent of the fault of each of these States cannot be established, the burden of compensation shall be apportioned equally between them."[130] This provision, however, appears to be rather inadequate to address cases where more than two states are responsible for damage and the degrees of contribution to the damage by each party do not clearly show which states are the 'first two States'. Furthermore, the Liability Convention does not describe how the launching state that has given compensation may get reimbursement from the other states that share liability.[131] Since harmonized regulations for

[127] Article V of the Liability Convention. Article V (2) stipulates that burden of compensation for damage shall be apportioned.

[128] See Article V(2) of the Liability Convention.

[129] See Article V(2) of the Liability Convention.

[130] Article IV(2) of the Liability Convention.

[131] See A. Kerrest, M. Benkö and K.U. Schrogl (Eds.), Liability for Damage Caused by Space Activities, *Space Law: Current Problems and Perspectives for Future Regulation*, Eleven International Publishing, 2005, 101.

reimbursement are absent at present, and in case specific agreements arranged amongst launching states concerned are also absent, it is conceivable that one launching state will bring actions against another to shift the responsibility for the loss.

In terms of application of the Liability Convention to satellite remote sensing, the matter of highest relevance is perhaps the interpretation of 'damage caused by its space object'. Damage is defined as "loss of life, personal injury or other impairment of health or loss of or damage to property."[132] The Liability Convention does not define a space object fully, but clarifies that it "(...) includes the component parts of a space object as well as its launch vehicle and parts thereof."[133] This can be interpreted to cover manned vehicles such as the Space Shuttle; unmanned vehicles; and payloads for activities such as telecommunication, Earth observation, and navigation.

The common interpretation of 'damage caused by its space object' is that it would only cover direct damage.[134] It is logical to interpret that 'damage' refers to the physical impact, since it specifies a location where damage is caused. As a result, consequential damage associated with remote sensing is not likely to be covered under the Liability Convention. Yet such damage may be incurred from a particular action or through decisions taken on the basis of data. The ambiguities identified in the issues of responsibility and liability also exists with respect to control of a space object.

2.2.3.6. *Registration of Space Objects*

As has been discussed, identification of a launching state is a critical starting point for settlement of international disputes over damage caused by a space object. Nevertheless, with growing complexities of business configurations, it may not always be simple to assign blame to a single state. In this regard, registration of a space object by one state performs an essential function to identify a single state that officially takes charge of the space object. Pursuant to Article VIII of Outer Space Treaty, which establishes a rule that a state that registers a launched space object retains jurisdiction and control over such object, the 1975 Registration Convention elaborates a mechanism of control of a launched space object.

Article II of the Registration Convention establishes that a single launching state registers the space object and, in case of more than one launching state, they shall jointly determine which one registers the space object.[135] This makes

[132] Article I (a) of the Liability Convention.

[133] Article I (d) of the Liability Convention.

[134] See I. Diederiks-Verschoor, *An Introduction to Space Law*, Kluwer Publisher, Hague, 1999, 41.

[135] See Article II (3) of the Registration Convention.

clear the indisputable rule; that is, only one state of registry exists for a launched space object. The rule to document the single state of registry may involve a selection process of the 'principal' launching state amongst several states that fall within the scope of broadly defined 'launching states' with distinct forms and/ or degree of participation/involvement in a launch. A determined state of registry is then registered both at a national and international level and the international registry is maintained by the Secretary General of the United Nations.[136]

The question arises as to how a state of registry is determined. Then, the discussion comes back to the analysis of different type of launching states. Amongst the four definitions of 'launching state'–that is, a state that: launches the space object, procures the launching of a space object, from whose territory the space object is launched, or from whose facility a space object is launched[137] – which one is most suited to be the state of registry?

State practice appears to reflect that the state of registry is determined less on the basis of the location of launch of a space object than on who procures the launch. For instance, the United States would not register space objects just because they were being launched from US territory, but only if they were owned or controlled by "[the] State's private or governmental entities."[138] Likewise, France registers its national satellites irrespective of the state from which they were launched.[139] From these examples, one can recognize the possibility that a state operating commercial launch services from its territory, although construed as a launching state, does not necessarily become the state of registry just because an arrangement was made to launch a satellite from its territory.

The Registration Convention imposes upon the state of registry the duty to control the registered space object. The status most commonly assumed by the Registration Convention is that in which a single state qualifies as the launching state, the state of registry, and the state that retains control of a launched satellite. The Convention nevertheless indicates the possibility that jurisdiction and control may be exercised by a launching state other than the state of registry through certain arrangements.[140] In case a specific arrangement creates the situation that the launching state and state retaining control of a

---

[136] Article III (1) of the Registration Convention.

[137] Article I (c) of the Liability Convention.

[138] United States Department of State. United States criteria for the registration of space objects November 2002, Cf. A. Kerrest, M. Benkö and K.U. Schrogl (Eds.), Liability for Damage Caused by Space Activities, *Space Law: Current Problems and Perspectives for Future Regulation*, Eleven International Publishing, 2005, 94.

[139] See Y. Lee, Registration of space objects: ESA member states' practice, 22 *Space Policy* (2006), 44.

[140] Article II (2) of the Registration Convention states that "(…) without prejudice to appropriate agreements concluded or to be concluded among the launching States jurisdiction and control over the space object and over any personnel thereof."

launched satellite are different, the links established through registration amongst a launching state, the state of registry, and the state that actually controls the space object become much less effective, if not invalid; also confusion may arise as to where responsibility lies.[141]

Finally, the extent of control by the state of registry is not clear from the provisions of the Registration Convention. It is uncertain as to what type of control should be exercised. With respect to satellite remote sensing, it is particularly important to establish the extent of control, such as the control on the dissemination of data from the satellite, and whether control extends to potential damage arising from the data.

In summary, the Outer Space Treaty together with the Liability Convention and Registration Convention spells out the basic legal norms in conducting space activities and serves as an overall framework for space applications, including satellite remote sensing. It is, however, both ambiguous and too limited to address sufficiently rules for satellite remote sensing. It is therefore necessary to look further at the regime specifically dealing with satellite remote sensing to see whether rules for remote sensing could be better addressed under specialized legislation, and to clarify the rights and obligations that are currently established in association with remote sensing activities.

2.2.4. *Specific Law for Remote Sensing*

As introduced briefly, the body of law specifically dealing with remote sensing was adopted by consensus in the form of the UN General Assembly Resolution in 1986. Christol contends that the UN Remote Sensing Principles gave formal approval for states to engage in information gathering activities from space as international customary law,[142] that is, practices and beliefs that are so vital and intrinsic a part of a social and economic system that they are treated as if they were laws. Indeed, the UN Remote Sensing Principles are a re-establishment of legitimacy for data collection from space, which was already an accepted practice inferred under the Outer Space Treaty and, for the first time, establishing concrete rules with regard to data access and availability at an international level.

2.2.4.1. *Drafting History of the UN Remote Sensing Principles*

Following the initial considerations to codify principles at the United Nations in the late 1960s, it took more than fifteen years for states to come to a consensus in 1986 on the general principles regarding remote sensing. The lengthy

---

[141] See S. Aoki, *Japanese Space Strategy*, Keio University Press, Tokyo, 2006, 79.

[142] C.Q. Christol, The 1986 Remote Sensing Principles: Emerging or Existing Law? *Proceedings of the Thirtieth Colloquium on the Law of Outer Space*, IISL, AIAA, 1987, 134–138.

drafting history is attributed to difficulties in achieving a consensus amongst various nations with different positions and conflicting interests. The stances adopted by states were fundamentally split between those COPUOS members representing developed nations with sensing capabilities on the one hand and the developing nations that lacked such technology on the other.[143] The former position was taken by the United States and many Western states, who insisted upon the freedom of acquiring and imparting information as a human right; the latter position was held by those states amongst the developing and Soviet-led socialist countries, who insisted upon the principle of state sovereignty over natural resources as well as the information concerning such resources."[144]

The disparate positions stemmed from the interpretation of the term sovereignty,[145] and location where the sovereignty in question was stressed. The first question was whether freedom of sensing was permissible. The location of a target for sensing, in which the sovereign right of states was exercised, was stressed by developing nations;[146] the developed nations stressed the location from which sensing takes place, where sovereignty does not extend. The second question was whether sovereignty extends to acquired information. Whilst developing nations claimed their sovereign right over acquired information relating to natural resources in their territory, developed nations argued otherwise. The conflicting positions seemed to be impossible to reconcile at first. Indeed, the drafting history of the UN Remote Sensing Principles had been a difficult process of compromise between the dissenting views, which progressed gradually over a period of time.

The initial examination of the subject by the UN COPUOS started in 1968 in Vienna.[147] Until the text of what is now the UN Remote Sensing Principles was adopted, a series of proposals by different nations were presented. [148] The major proposals to be examined in this Chapter are as follows:

---

[143] See C.M. Hayward, Remote Sensing: Terrestrial Laws for Celestial Activities, 8 *Boston University International Law Journal* (1989), 170.

[144] See S. Gorove, *Developments in Space Law,* Martinus Nijhoff Publishers, Dordrecht, 1991, 295.

[145] See M. Benkö, W. de Graaff & G.C.M. Reijnen, *Space Law in the United Nations,* Martinus Nijhoff Publishers, Dordrecht, 1985, 19.

[146] The line of argument was: 1) sovereignty of States prohibits interference in the territory of another State; 2) effects of remote sensing bear upon a State's internal affairs; thereby 3) affects the sovereign right of the State; and 4) such activity cannot be carried out without consent.

[147] See E.R.C. Van Bogaert, *Aspects of Space Law*, Kluwer Law and Taxation Publishers, Deventer, 1986, 240.

[148] Other proposals and working papers submitted include India, the Soviet Union, France, Venezuela, Greece, Austria, Romania, and Uruguay.

1. 1970 Proposal by Argentina[149]
2. 1973 Proposal by the Soviet Union co-sponsored by France[150]
3. 1974 Joint draft by the Soviet Union and France[151]
4. 1974 Joint Draft by Argentina and Brazil[152]
5. 1975 Proposal by the United States[153]
6. 1981 Proposal by Mexico[154]
7. 1982 Proposal by Brazil[155]
8. 1984 Proposal by Chile[156]

The content of the proposals, which were subject to negotiation, can be categorized into three critical elements: conditions for sensing, conditions for access and distribution of data, and international responsibility regarding conducting remote sensing activities.

The negotiation was put forward mainly in two stages: first, debates on conditions for sensing – such as prior consent, limit on location, and veto right; and second, debates on rules concerning the resulting information – such as priority access and control on dissemination of data. The detailed examination of drafting history can be roughly divided into three periods: 1) the era prior to the 1978 Moscow Convention on the Transfer and Use of Data of Remote Sensing of the earth from Space;[157] 2) the Moscow Convention and International Satellite Monitoring Agency; and 3) post Moscow Convention towards the 1986 UN Remote Sensing Principles. The debate regarding the freedom of sensing took place primarily during period 1 and the debate regarding access and dissemination characterising priority access and control on the dissemination of data took place in periods 2 and 3. It is important to look at how divergent were the views and positions presented, how they met opposition from the other side, and how gradual convergence to agreement resulted in each period.

The initial formulation phase of the UN Remote Sensing Principles began with a 1970 proposal by Argentina, which claimed that prior consent was required from the sensed state before another state conducted remote sensing. This position was based on the contention that, since certain UN General Assembly Resolutions provide that states have permanent sovereignty over their

---

149 UN Doc. A/AC.105/C.2/L.73, 26 (1970).
150 UN Doc. A/AC/105/111 (1973).
151 UN Doc. A/AC.105/C.2/L.99 (1974).
152 UN Doc. A/AC/1/1047, Oct 15, 1974.
153 UN Doc. A/AC.105/C.2/L.103 (1975).
154 UN Doc. A/AC.105/288 (1981).
155 WG/RS (1982)/WP.11 (1982).
156 WG/RS (1984)/WP.12 (1984).
157 UN Doc. A/33/162, 29 June, 1978. [Hereafter: Moscow Convention]

own natural resources, information on the these assets was also included in the concept of sovereignty.[158]

The 1974 Brazil and Argentina draft proposal again insisted on prior consent for sensing any foreign territory and rigid control over sensing natural resources – a proposal so extreme that it would have denied the remote sensing practice that had been conducted by space powers at that time, and would have made remote sensing activities practically impossible to conduct. That proposal was not acceptable to space-faring nations. The United States strongly opposed this position for legal, technical, and practical reasons, contending that it was legally inconsistent with the provisions of the Outer Space Treaty and that the legality of sensing should depend upon the site of the sensing device rather than the location of the sensed object.[159] Furthermore, the United States argued, from a technological standpoint it is not possible for remote sensing equipment to distinguish international borders; and besides, it would be financial impracticable to develop such capability.

The United States maintained its stance for freedom of sensing, adhering to the Open Skies Policy and recognizing the commercial potential of remote sensing data.[160] The United States did not mention the sovereignty issues in its proposal, a working paper entitled *Legal Implications of the Earth from Space.*[161] The United States contended that the disclosure of useful information derived from sensing would contribute to the "(…) preservation of environment and effective management and control by states of their natural resources."[162] Freedom of sensing was proposed by only a few states then.

The Soviet Union and France made a joint proposal in 1974. This proposal also required prior consent and the legality of sensing was limited in its location, suggesting that freedom of sensing be permitted only in the areas outside the national jurisdiction of any state.[163] The proposal also introduced the notion of control over the dissemination of data and suggested a requirement of prior consent from sensed states.

---

[158] See G.H. Reynolds & R.P. Merges, *Outer Space Problems of Law and Policy,* Second Edition, Westview Press, Colorado, 1997, 198. See UN General Assembly Resolution 1803 (XVII) on Permanent Sovereignty over Natural Resources, 14 December 1962, and Resolution 3171 (XXVIII) on Permanent Sovereignty over Natural Resources adopted by the UN General Assembly at its 28th session in 1973.

[159] See C.M. Hayward, Remote Sensing: Terrestrial Laws for Celestial Activities, 8 *Boston University International Law Journal* (1989), 174.

[160] See C.M. Hayward, Remote Sensing: Terrestrial Laws for Celestial Activities, 8 *Boston University International Law Journal* (1989), 174.

[161] UN Doc A/AC.105/C.2/L.103 (1975).

[162] See C.Q. Christol, Remote Sensing and International Law, 5 *Annals of Air and Space Law* (1980), 395.

[163] See G.M. Danilenko, Principles relating to Remote Sensing of the Earth from Space: Territorial Sphere of Application, *Proceedings of the Thirtieth Colloquium on the Law of Outer Space*, 1987, 292.

Under the French-Soviet proposal, data was divided into two types.[164] Data acquired with a spatial resolution greater than 50m would be classified as 'global' and would not be subjected to restrictions imposed for sensing and dissemination; data of higher than 50 m resolution would be classified as 'local' and would be subject to control.[165] This proposal was based on the view that sovereignty of sensed states extended to acquired information by sensing states, rather than any grant of superior rights to sensing states to have access to critical information on natural resources in the territory of sensed states.

The only exception to the application of prior consent and the right of veto in the French–Soviet proposal was distribution for the purpose of disaster management and environmental protection.[166] The draft stipulated that no veto rights would be exercised in the event of a natural disaster or phenomena detrimental to the environment; and the sensing state could distribute data without the knowledge or approval of the sensed state(s).[167] Because it called for 1) classification of data by ground resolution with consequential dissemination policies, and 2) giving preferential treatment for data availability for public benefit, the proposal can be regarded as an attempt to establish concrete data policies hitherto not adopted for inclusion in the final document.

The 1974 Soviet-French proposal led to the 1978 Moscow Convention on the Transfer and Use of Data of Remote Sensing of the Earth from Space.[168] This convention set forth the detailed rules with respect to the dissemination of data, which can be seen as a data policy. The convention was signed in Moscow by Cuba, Czechoslovakia, the German Democratic Republic, Hungary, Mongolia, Poland, Romania, and the Soviet Union.[169] At the time of this convention, the argument was shifting from the freedom of sensing from space to the control of resulting information and attempts to regulate the access and

---

164 UN Doc A/AC.105/195 (1977).

165 See H. Desaussure, Remote Sensing Satellite Regulation by National and International Law, 15 *Rutgers Computer & Technology Law Journal* (1989), 356.

166 See H. Desaussure, Remote Sensing by Satellite: What future for an international regime, 71-4 *American Journal of International Law* (1977), 720.

167 See SubSection 5(c) of the Joint USSR-France Draft on Remote Sensing.

168 UN Doc. A/33/162, 29 June, 1978. See further G. Zhukov, Une expérience historique: la convention de moscou de 1978 sur le transfert de l'utilisation des données de téléobservation de la terre a partir de l'espace, *Droit télédétection et environnement: actes du colloque international: le droit face aux techniques de télédétections par satellite au service de l'environnement*, Strasbourg, 2-3-4 juin 1993, Sides, Antony, 1994, 189–194. See also G. Gál, Legal principles of remote sensing in the Moscow Convention of 1978, *International space law miscellaned: liber amicorum honouring Professor Dr Andrzej Gorbiel in his 65th anniversary*, E. Palyga (Eds.), Andrzej Frycz Modrzewski Foundation 1995, 91–96.

169 See M. Benkö, Principles on Remote Sensing of the Earth from Outer Space as adopted in United Nations resolution 41/65 of the 3rd December 1986, *Space: Legal and Commercial Issues 1987*, International Bar Association, London, 1987, 252.

dissemination of data. The control over dissemination specifically imposed in Article IV of the Moscow Convention stipulates that "A Contracting Party in possession of initial data of the remote sensing of the Earth from outer space, with a better than 50 metres ground resolution on the terrain, relating to the territory of another Contracting Party, shall not disclose or make them available to anyone except with an explicit consent thereto of the Contracting Party."[170]

Article IV of the Moscow Convention therefore required prior consent from sensed states for dissemination of initial data and derived information concerning natural resources of the sensed states,[171] and also established international responsibility for use of remote sensing data.[172] The intention was clear: Article IV attempted to protect the sovereign rights and interests of sensed states through control of dissemination of the information. Prior consent for the dissemination of data infers veto rights by sensed states, and it can be construed that if a party considers the dissemination of the data as detrimental to them, they can hamper the dissemination by exercising their right of veto.

The Soviet Union was specifically opposed to the free dissemination of data and rejected the idea that remote sensing data should be freely purchased, as that would constitute a free market in remote sensing data. At the time of the convention, the Soviet Union also proposed that the sensing state should have international responsibility for any serious damage caused by the dissemination of the data;[173] however, such proposal was not taken into account.

The Moscow Convention was not ratified by any further states apart from the first signatories of the socialist nations and failed to become a globally accepted legislation on remote sensing. Over time, it was clear that the convention had been discarded. This was a natural outcome, as communism came to an end in the 1990s; the former Eastern-bloc nations – which comprised the majority of the parties to the convention – have joined the western systems since then. Today, an increasing number of states that were strictly the subjects of sensing activities back in 1986 now have their own space capabilities.

It should be noted that in 1978 – during the first special session of the UN General Assembly devoted to disarmament, France proposed establishment of the International Satellite Monitoring Agency (ISMA), a specialized agency of the UN, responsible for collecting, processing, and disseminating data and information gathered by satellites.[174] The proven capability of satellite remote sensing to verify the bilateral agreements such as the SALT and ABM treaties led to this proposal, which would also apply such capability to multilateral

---

170 Article VI of the Moscow Convention.

171 See Article V of the Moscow Convention.

172 See Article VI of the Moscow Convention.

173 See A.L. Brigham, Natural Resources: View from Above, 55-2 *Journal of Air Law and Commerce* (1989), 529.

174 UN Doc. A/S-10/AC.1/7, 1 June, 1978.

agreements. The main functions for ISMA were to be the monitoring of the implementation of multilateral arms control and disarmament agreements and the investigation of crisis situations.[175] The ISMA was to be established in three phases: 1) it would operate an international data processing and interpretation facility and disseminate resulting information to member states where data would be acquired from states that operate surveillance satellites; 2) it would establish ground receiving stations, possibly working with national satellites; and 3) it would establish its own space segment as a supplement to national systems.[176]

In response to the submission for a vote on the ISMA to the United Nations, the United States and the Soviet Union abstained and the General Assembly created a committee of experts[177] to conduct a feasibility study.[178] The feasibility study concluded that establishing the ISMA was technically feasible and not in conflict with international law. The ISMA did not come into existence, however, because both superpowers were against its establishment. The Soviet Union argued that allowing an international agency to gain control of military or strategic information would have violated the principle of national sovereignty, whilst the United States contended that such an agency would have permitted states to engage in espionage or activities that were inconsistent with national security or national policies.[179] The failure to establish the ISMA indicates that, even amongst the developed nations with remote sensing capabilities, their interests were not uniform.

The Moscow Convention did not resolve the fundamental problem associated with achieving consensus for a set of principles governing remote sensing; thus, during the period from 1976 to 1981 no progress was made toward an agreement on the principles of state sovereignty and prior consent.[180]

With the 1981 Mexico proposal,[181] meaningful debate toward agreement resumed. Mexico proposed priority access as well as prior notification.

---

[175] Department of Disarmament Affairs Report of Secretary General United Nations, The Implications of Establishing an International Satellite Monitoring Agency, New York, 1983, 23.

[176] See B. Jasani, International Satellite Monitoring Agency- Has the time come for its establishment?, London, 2003. Available on the Internet at http://esarda2.jrc.it/events/other_meetings/inmm/2003-esarda-inmm-Como/2-paper%20pdf/2-10-jasani-040130.pdf. Last accessed on 20 April 2010.

[177] None of them were Soviet or US experts.

[178] See K.L. Tanner, *Eyes in the Sky: The Technology, Politics, and Law of Remote Sensing,* Unpublished thesis presented to Princeton University, 1987, 96.

[179] See K.L. Tanner, *Eyes in the Sky: The Technology, Politics, and Law of Remote Sensing,* Unpublished thesis presented to Princeton University, 1987, 96.

[180] M. Benkö, W. de Graaff & G.C.M. Reijnen, *Space Law in the United Nations,* Martinus Nijhoff Publishers, Dordrecht, 1985, 21.

[181] UN Doc. A/AC.105/288 (1981).

The phrase 'prior notification' instead of 'prior consent' was a reflection of some attitudes of compromise shown by developing states,[182] as 'prior notification' – in contrast to 'prior consent' – does not imply veto rights. The proposal also required that sensing states provide preliminary information and final results, whilst prohibiting dissemination of information without the prior consent of sensed states.[183] The debate was shifting more towards the conditions necessary for access and dissemination, together with the responsibility of states to carry out remote sensing activities.

The 1982 Brazilian working paper provided a compromise solution by way of a restriction of state responsibility as well as data availability.[184] It advocated for priority access and international responsibility on dissemination, and it proposed priority access by sensed states as: a "Sensed State shall have timely and non-discriminatory access to primary data obtained by remote sensing of the Earth from outer space concerning territory under its jurisdiction before access is granted to any third party."[185] The Brazilian proposal suggested that sensing states should be held internationally responsible for dissemination of data that adversely affect the interests of sensed states.

Both socialist nations and developing countries were ready to agree to this proposal;[186] however, it proved unacceptable to sensing states as it would have meant that state responsibility extended to monitoring data dissemination by private entities conducting remote sensing.[187]

Specific responsibility for the use of remote sensing data was also proposed in a working paper by Chile in 1984.[188] Principle XI states that "States shall bear international responsibility for improper or discriminatory use of analysed information resulting from remote sensing."[189] This language can be construed to mean that the misuse of data that could be detrimental to the interests of sensed states is prohibited.

---

[182] See K.L. Tanner, *Eyes in the Sky: The Technology, Politics, and Law of Remote Sensing,* Unpublished thesis presented to Princeton University, 1987, 73.

[183] See H. Feder, The sky's limit? Evaluating the International Law of Remote Sensing, 23-2 *NYU Journal of International Law and Politics* (1991), 610.

[184] H.L. Van Traa-Engelman, *Commercial Utilization of Outer Space,* S.I Publisher, Utrecht, 1989, 240.

[185] M. Benkö, W. de Graaff & G.C.M. Reijnen, *Space Law in the United Nations,* Martinus Nijhoff Publishers, Dordrecht, 1985, 26.

[186] See W. Hampe & R. Mueller, The International Legal framework for Conducting Remote Sensing Activities *Proceedings of the Thirtieth Colloquium on the Law of Outer Space,* IISL, AIAA, 1987, 318–321.

[187] M. Benkö, The Principles on Remote Sensing of the Earth from Outer Space as adopted in United Nations Resolution 41/65 of the 3rd December 1986, Paper presented at the Section on Business Law's 9th Conference, International Bar Association, London, 1987, 255.

[188] WG/RS (1984)/WP.12.

[189] Principle XI of the Remote Sensing Principles.

The emergence of civilian-operated remote sensing satellites made the developed states adhere even more to their position for freedom of sensing with unregulated dissemination of data. The United States began operating Landsat in 1972, and its position on remote sensing reflected a clear intention to vitalise the Landsat system and its data dissemination. The United States intended to commercialize the remote sensing system through the 1984 US Land Remote Sensing Commercialization Act.[190] France initially supported the Soviet view for requiring prior consent, but shifted to follow the US position in anticipation of the development of the French SPOT – the first non-US civilian satellite – in 1986.[191]

The final UN Remote Sensing Principles mainly reflect the US view and are clearly the product of a high degree of compromise by the developing states. None of the proposals from developing nations with regard to the rights of sensed states and obligations of sensing states were incorporated in the way that these sensed states originally wished. The major contention of developing nations – including, but not limited to, prior consent for sensing and dissemination of data, privileges in accessing data by sensed states, concrete data policies, and responsibility for dissemination of data – were discarded; only ambiguous phrases on international responsibility and the respect of the interests of sensed states remained, and these were incorporated into the final UN Remote Sensing Principles. It was a victory for the proponents of sensing,[192] represented by the United States, which managed to codify most of its contention that the freedom of sensing and the unregulated dissemination of data were consistent with the Open Skies Policy. The only matter that was not incorporated into the Remote Sensing Principles was the US view regarding intellectual property rights. The United States has claimed that once collected data is transformed into product though processing and analysis, it becomes the intellectual property of its owner and thus subject to copyright law.[193] This view is now incorporated in data policies of individual entities and widely practiced, but this practice invokes questions as to the applicability of such data policies to different types of products.

The UN Remote Sensing Principles were adopted after a long period of difficult negotiations. In the end, the Principles adopted turn out to be

---

[190] Land Remote Sensing Commercialization Act of 1984, Pub. L. No. 98–365, 98 Stat. 451 (1984) This act was replaced by the Land Remote Sensing Policy Act of 1992. For a more detailed discussion on the legislation, see section II of this Chapter.

[191] See H. Feder, The sky's limit? Evaluating the International Law of Remote Sensing, 23-2 *NYU Journal of International Law and Politics* (1991), 611.

[192] It is interesting to contrast with the drafting history of DBS, which ended with the victory of the receiving side. For a detailed analysis of the DBS Principles drafting history, see M.N. Taishoff, *State Responsibility and the Direct Broadcasting*, Frances Printer Publishers, London, 1987, 143–145.

[193] Copyright Act, Pub. L. 94–553, 90 Stat. 2451 (1978).

ambiguous; they do not impose specific obligations, and they contain significant loopholes. The resulting Principles reiterate general concepts of international law or formalize existing customary practice.[194] Even after the formulation of the Remote Sensing Principles, dissatisfaction – particularly amongst the developing countries – remained undiminished. For example, the Latin American nations' claim for more regulated legislation came to a concrete proposal in 2003. Argentina, Brazil, and Mexico submitted a working paper to the UN COPUOS on a new agenda item calling for an international convention for remote sensing.[195]

2.2.4.2. *The Legal Value of the UN Remote Sensing Principles*

The legal value of the UN Remote Sensing Principles is often debated. In examining their legal value, it is necessary to approach the issue from two standpoints: the analysis of UN General Assembly Resolutions amongst international law instruments, and the actual practice associated with remote sensing.

The UN Remote Sensing Principles are in the form of a UN General Assembly Resolution. It is generally perceived that a UN General Assembly Resolution is not binding as a document in itself, as it is clearly distinguished from a binding instrument such as a treaty or a convention.[196] A binding instrument is a law made, executed, and ratified by governments.[197] In fact, the Assembly is considered to be an advisory body with binding decisions only on issues of threats to international peace and security as adopted by its Security Council.[198] Hence, it is understood that the UN General Assembly Resolution itself is recommendatory in nature – as a recommendation made to the member states of the United Nations in accordance with the UN Charter.[199]

Although commonly perceived as recommendatory, different types of UN General Assembly Resolutions have varying degrees of value. Particularly, a distinction should be made between a declaration and mere principles. Kopal compares the 1963 Declaration of Legal Principles to the UN Remote Sensing Principles; he contends that only the former is called a 'declaration' and

---

[194] See H.L. van Traa-Engelman, *Commercial Utilization of Outer Space*, S.I Publisher, Utrecht, 1989, 21.

[195] UN Doc. A/AC.105/C2/L.244 (2003), Why is an international convention on remote sensing of the Earth from outer space necessary?

[196] See B. Cheng, *Studies in International Space Law,* Clarenton Press, Oxford, 1999, 132–136. For instance, Canada distinguishes among 'merely a General Assembly Resolution', 'declaration,' and 'binding treaty or convention'.

[197] UN Doc. A/AC.105/C.2/SR.23 (25.4.63), 4.

[198] See M. Shaw, *International Law,* Fifth Edition, Cambridge University Press, Cambridge, 2003, 108.

[199] See Article 10 of the United Nations Charter.

declared principles qualify as legal.[200] In addition, the 1963 Declaration opened a treaty-making process that lead to the conclusion of legally binding instruments such as the Outer Space Treaty. In that respect, the UN Remote Sensing Principles were not designated as a declaration, nor did they open a treaty-making process; therefore, they do not go beyond the status of a recommendation.

Even non-declaration types of General Assembly Resolutions (such as the UN Remote Sensing Principles) address the general code of conduct for all UN member states, or even for all states around the globe; hence, they possess not only political meaning but moral weight.[201] A certain view exists that the significance and weight is increased for a General Assembly Resolution that is unanimously adopted; this means that no states found shortcomings in the resolution and did not vote against it to block its adoption.[202] Indeed, delegates of states such as the United Kingdom and India distinguish between ordinary General Assembly Resolutions and those that are unanimously accepted,[203] and India even contends that unanimously accepted resolutions are accepted as part of international law.[204] Since the UN Remote Sensing Principles document is accepted unanimously, it is considered authoritative – if not universally recognized as a binding document.

Even if the UN Remote Sensing Principles document does not gain sufficient support for it to be binding in nature, its content can be considered to reflect an element of international customary law.[205] Article 38 of the International Court of Justice (ICJ) Statute recognizes international custom as evidence of a general practice accepted as law.[206] The element of international custom requires *opinio juris* and state practice, and the UN Remote Sensing Principles satisfy both. First of all, as previously discussed, large numbers of proposals from states including Argentina, the Soviet Union, Brazil, France, Chile, Mexico, Venezuela, and Austria were taken into account in the drafting process prior to the adoption of the Resolution. It can be considered, therefore,

---

[200] See V. Kopal, The Role of United Nations Declarations of Principles in the Progressive Development of Space Law, 16 *Journal of Space Law* (1988), 18.

[201] See V. Kopal, The Role of United Nations Declarations of Principles in the Progressive Development of Space Law, 16 *Journal of Space Law* (1988), 18.

[202] See F.G. von der Dunk, *Private Enterprise and Public Interest in the European Spacescape*, International Institute of Air and Space Law, Leiden, 1998, 100.

[203] A/AC.105/C.2/SR.17 (17.4.63), 9.

[204] A/AC.105/C.2/SR.22 (24.6.63), 10.

[205] See B. Cheng, *Studies in International Space Law,* Clarenton Press, Oxford, 1999, 136. See also S. Marchisio, Remote Sensing for Sustainable Development in international law, *Outlook on Space Law over the next 30 years – Essays Published for the 30th Anniversary of the Outer Space Treaty*, G. Lafferanderie and D. Crowther (Eds.), Kluwer Law International, Hague-London-Boston, 1997, 338–340.

[206] Article 38(b) of the ICJ Statute.

that the Resolution is a final synthesis of different countries' views on remote sensing from space.

The UN Remote Sensing Principles have been serving as general guidelines for activities in the field of remote sensing from outer space and no disputes have arisen thus far with regard to their application.[207] As far as freedom of sensing is concerned, the Open Skies Policy is a clear indication of *opinio juris* between the United States and Russia since 1960, and have been accepted later by many other states.

The UN Remote Sensing Principles are incorporated into continuing state practice. Over two decades, the UN Remote Sensing Principles have been widely followed by various states and entities. With regard to the most fundamental of these Principles, namely, the principle of sensing from space without prior consent, states have sought no permission and such practice has not encountered protests from the states whose interests were affected.[208] Non-discriminatory access to data is also realized worldwide through an extensive availability of satellite data and derived products via the Internet. These Principles are viewed as a codification of customary legal principles that are binding on nations, pursuant to years of practical implementation,[209] as nations continue specifically to refer to the Principles in domestic, bilateral, and multilateral legal documents.[210] The Principles are actually the only source of legislation devoted to satellite remote sensing at an international level, and other relevant legislation or principles tend also to follow them. For instance, the phrase 'non-discriminatory basis' is used in international data exchange

---

[207] See A.D. Terekhov, UN General Assembly Resolutions and Outer Space Law, *Proceedings of the Fortieth Colloquium on the Law of Outer Space*, IISL, AIAA, 1997, 98–99.

[208] See V.S. Vereschetin & G.M. Danielenko, Custom as a Source of International Law of Outer Space, 13-1 *Journal of Space Law* (1985), 121.

[209] Cf. J.I. Gabrynowicz, Defining data availability for commercial remote sensing systems, XXIII *Annals of Air and Space Law* (1998), 95. See American Society for Photogrammetry and Remote Sensing and the American Bar Association, Earth Observation Systems Legal Considerations for the 90's, Chicago: American Society for Photogrammetry and Remote Sensing & the American Bar Association, 1990, 127.

[210] See J.I. Gabrynowicz, Expanding Global Remote Sensing Services, *Proceedings of Workshop on Space Law in the Twenty-first Century,* ST/SPACE/2, United Nations, New York, 2000, 101. UN Remote Sensing Principles are referred to in: Commercialization Act, 15 USC 4201–4292 (1984); Land Remote Sensing Policy Act, 15 USC 9 5601–5642 (1992); RADARSAT Data Policy, Document Number: RS , CA – PR0004. Sec 10.1 b, (Canadian Space Agency), July 13, 1994, at 11.; ESA Envisat Data Policy, ESA/PB-EO (97) rev.3, Paris, (European Space Agency), 19 Feb. 98; Principles of the Provision of ERS Data to Users, ESA/PB-EO (90)57, rev. 6, Paris, 9 May 1994, (European Space Agency, Earth Observation Programme Board), Sec. 2 General Principles, 2.1 Legal Principles, para.2, at 2.

instruments such as the WMO Resolution,[211] as well as in data policies of space agencies such as NASA and ESA, which will be discussed in detail in Chapter 5. However, in the absence of confirmation given by the ICJ on its customary value, the binding value of UN General Assembly Resolutions has to be treated with caution.

2.2.4.3. *The Scope of the UN Remote Sensing Principles*

To understand the scope of the UN Remote Sensing Principles, it is necessary to look at the precise applications, activities, and types of data to which they apply. Principle I provides the various definitions and scope of the Principles, as discussed in detail below.

Unlike the Outer Space Treaty, which allows for certain military uses of outer space and is open to commercial activities, the scope of the UN Remote Sensing Principles pertains to civil remote sensing for limited applications – specifically natural resource management, land use, and protection of the environment.[212] From this provision, it can be interpreted that military applications fall outside the scope of the Principles, thus excluding satellites used for reconnaissance or intelligence purposes. A question remains as to whether or not dual use remote sensing satellites, which serve for both civil and military purposes such as Global Monitoring for Environment and Security (GMES), are covered under these Principles. Indeed, GMES is intended to monitor the Earth's environment as well as to serve for security purposes; thus, the programme is partially covered under the Principles. Furthermore, it is difficult to construe that commercial applications other than the three purposes stated above are covered under the UN Remote Sensing Principles. A particular issue may be raised with respect to applying the Remote Sensing Principles to high resolution satellites, which are nowadays a critical source of information in various decision-making activities.[213] These satellites have the capability to reveal very sensitive information and the application of the UN Remote Sensing Principles to remote sensing activities of such a nature would affect the concerned parties extensively.

Today there are many other applications apart from natural resource management, land use, and protection of environment in civil and commercial remote sensing – regardless of their humanitarian, commercial, or military nature – to name but a few: disaster management, registry of land, archaeology, urban planning, map making, agriculture, and the variety of verification purposes already mentioned in the Introduction. The use of data for these

[211] World Meteorological Organization Resolution 40 (Cg-XII)

[212] Principle I of the UN Remote Sensing Principles.

[213] See further M.M. Fomtchenko & A.S. Movlyav, High Resolution Remote Sensing: New Aspects and Problems, *Proceedings of the Thirty-Ninth Colloquium on the Law of Outer Space*, IISL, AIAA, 1996, 85–93.

applications may not fall within the scope of the UN Remote Sensing Principles and issues associated with these applications are not addressed under the present regime, which needs improvement with regard to this point.

The UN Remote Sensing Principles give clear definitions with regard to the extent of remote sensing activities: 1) the operation of remote sensing space systems; 2) the operation of primary data collection and storage stations; 3) the activities in processing; 4) interpreting; and 5) disseminating the processed data. Although data handling activities and data distribution are covered under the scope of the Principles, the use, as well as the application of remote sensing information is not covered.

It is significant to point out that data handling activities and dissemination, which take place on the Earth, are within the scope of remote sensing activities defined under the UN Remote Sensing Principles. Indeed, only the first and second activities listed above take place in space; the rest are conducted on the Earth. Long before the formulation of the Remote Sensing Principles, W.P. Heere stated that Earth resources are a terrestrial matter, and thus should fall under terrestrial law and not space law; likewise, Franco Fiorio suggested that a distinction should be made between remote sensing activity itself and the use made of the resulting picture.[214] Contrary to such opinions, the Principles do not distinguish between data collection taking place in outer space and data handling operations taking place on the Earth. The Principles apply even if states are involved in only terrestrial operations such as data processing, interpretation, value-adding activities, or just information extraction, as long as satellites are the original source. Although remote sensing takes place in outer space where there is no sovereignty, the data are collected for potential use on Earth, where state sovereignty prevails over territory and airspace. The international community needs to recognize that the situs of remote sensing activities lie on the Earth, where the information acquired is transformed into useful media.[215] The transformation to useful knowledge, as well as use of such knowledge, is the part that should be regulated and the current legal framework needs improvement in this regard.

It is also important to examine the coverage of the UN Remote Sensing Principles with regard to activities conducted by private entities. The UN Remote Sensing Principles provide nothing more than is provided by the Outer Space Treaty. They are silent as to private entities except for Principle XIV, which makes a link between state activities and those of private entities, recalling Article VI of the Outer Space Treaty. This means that private entities are

---

[214] See G.C.M. Reijnen, Remote Sensing Satellites and Legality, *Legal Implication of Remote Sensing From Outer Space,* N. Matte and H. Desassure (Eds.), Leyden, Sijthoff, 1976, 22.

[215] See C.M. Hayward, Remote Sensing: Terrestrial Laws for Celestial Activities, 8 *Boston University International Law Journal* (1989), 182.

bound by the regulations imposed by the states whose jurisdiction covers such entities but neither Article VI of the Outer Space Treaty nor Principle XIV of the UN Remote Sensing Principles give details as to how non-governmental entities should be authorized and supervised. Either the particular national jurisdiction exercises control through a licensing regime, or such activities are left unregulated.

As mentioned in the Introduction, remote sensing data are classified in three types: primary data, processed data, and analyzed information.[216] In reality, there are many other terms apart from the three definitions used, such as: raw data, unenhanced data, derived product, and value-added product. The elaboration of what constitutes analyzed information is of particular importance from the standpoint of intellectual property rights, as it addresses the critical question of whether or not the information includes content that has been modified to the extent that it does not retain the original pattern and, as such, may not then be subject to protection under intellectual property laws. The coordination of these definitions with actual practice, as well as elaboration, may be necessary.

2.2.4.4. *Contents of the UN Remote Sensing Principles*

As has been discussed, the UN Remote Sensing Principles were the codification of the interests of the spacefaring nations in safeguarding their right to gather information from space. A number of propositions on the rights of sensed states and their obligations associated with sensing arose during the formulation process, but were not incorporated. The Remote Sensing Principles proposed a general framework with respect to data collection, data availability, and a state's responsibility for conducting remote sensing activities, all of which allowed for flexible interpretation. The resulting Principles have established new rules whilst repeating the provisions that were already in place. This section analyzes the content through key elements newly established in the UN Remote Sensing Principles, and reiteration of principles established by the Outer Space Treaty.

The newly established Principles clearly set forth the fundamental norms and rules regarding data collection but are limited to the issues of access, distribution, and use of data. The key elements the UN Remote Sensing Principles can be divided into three categories: freedom of remote sensing, respect of rights and interests of sensed states, and data rules.

Principle IV is a core provision and one of the two most critical Principles of all fifteen. Principle IV stipulates that "Remote sensing activities shall be conducted in accordance with Article I of the Treaty on Principles Governing the Activities of States in the Exploration and Use of Outer Space"[217] which

[216] Principle I of the UN Remote Sensing Principles.

[217] Principle IV of the UN Remote Sensing Principles.

establishes the free use of outer space. Although not explicitly stated, Principle IV establishes the right of states in sensing the entire globe from outer space using satellites. Accordingly, this provision established the data collection rule as: 1) there is no restriction on areas for observation based on geography or objects for observation including natural resources or sensing sea surface beyond national jurisdiction; 2) no prior consent, consultation, or notification is required before sensing and, therefore, no veto right is granted to sensed states for their territories being sensed;[218] and, 3) no conditions are imposed for sensing capabilities in terms of spatial and temporal resolution, or type of sensors such as radar or optical.[219]

All states are given the right of data collection from outer space. Later in Principle IV, the sovereign rights of sensed states are respected, in that: "These activities shall be conducted on the basis of respect for the principle of sovereignty of all States, and peoples over their own wealth and natural resources (…). Such activities shall not be conducted in a manner detrimental to the legitimate rights and interests of the sensed State."[220] As seen earlier, the sovereignty of sensed states was the major issue of contention during the negotiation process before formulation. Yet both sentences are ambiguous and fail to clarify what the principles of sovereignty and the legitimate rights of sensed states actually entail.[221] The obligation not to conduct remote sensing activities against the legitimate rights and interests of sensed states appears to refer only to the conduct of remote sensing itself and not to what may happen after the dissemination of data.[222] The interest of sensed states is most commonly affected by the use and misuse of the resulting data, but the Principles impose neither specific obligations on sensing states (such as the obligation to be liable for damage suffered by sensed states for misuse of data), nor clarify the specific rights of sensed states (such as control of access and dissemination for data concerning natural resources). Those who are concerned that data gathered from outer space might be used or misused to their detriment by others – either by the sensing state or

---

[218] See F.G. von der Dunk, United Nations Principles on Remote Sensing and the User, *Earth Observation Data Policy and Europe*, R. Harris (Eds.), A.A. Balkema Publishers, Lisse, 2002, 33.

[219] See A. Ito, Improvements to the Legal Regime for the Effective Use of Satellite Remote Sensing Data for Disaster Management and Protection of the Environment, 34–1 *Journal of Space Law* (2008), 49.

[220] Principle IV of the UN Remote Sensing Principles.

[221] For instance, it is not clear as to whether either property rights of sensed states or priority access to the data are implied from the provision. See C.M. Hayward, Remote Sensing: Terrestrial Laws for Celestial Activities, 8 *Boston University International Law Journal* (1989), 176.

[222] See G. Catalano Sgrosso, Prevention and Management of Natural Disasters, *Outlook on Space Law over the next 30 years,* Kluwer Law International, 1997, 299.

by third parties – cannot confidently rely on the provision of the UN Remote Sensing Principles.[223]

Principle XII addresses data rules, establishing the rule regarding data access and dissemination, and thus determining the general availability of remote sensing data. Principle XII stipulates that: "As soon as the primary data and the processed data concerning the territory under its jurisdiction are produced, the sensed state shall have access to them on a *non-discriminatory basis* and on *reasonable cost terms*" (emphasis added).[224] There are several important implications of this provision. It can easily be inferred from the provision that prior consent before dissemination is not required. Neither veto rights nor control on information for dissemination are granted to sensed states; hence, theoretically, sensed states can neither choose the states to which their territorial information is disclosed nor determine the type of data to disseminate, such as data of high ground resolution.

Principle XII means that sensing states have an obligation to provide the data to sensed states under the same conditions as they provide the data to other interested states. Indeed, sensed states do not have an exclusive, free, or preferential right of access to the data.[225] No privilege is given to sensed states in terms of timing or cost to access to the data. Sensed states are not granted the right of access to the data prior to the sensing states – or any other state – which may affect their interests extensively. Likewise, no financial privilege is given to sensed states, such as data provision either free of charge or at a lower cost than for other states. The provision 'reasonable cost terms' is ambiguous and open to different interpretations, as it is not defined under the UN Remote Sensing Principles.[226] It could mean marginal cost (which is more specifically phrased on a domestic level as the data policy set by different data generators),[227] or it could mean commercial market price if it is considered to be 'reasonable' for a particular image.

Non-discriminatory access within Principle XII generates an obligation to make imagery available to the sensed states; however, whether sensing states actually make the data available hinges on their good faith. There are no means

---

[223] See B. Cheng, *Studies in International Space Law,* Clarenton Press, Oxford, 1999, 596. For the issues of sovereignty over natural resources, see D.S. Myers, Remote Sensing and National Sovereignty over Natural Resources: Assessment of Mexican View, 14-1 *California Western International Law Journal* (1984), 22.

[224] Principle XII of the UN Remote Sensing Principles.

[225] See F.G. von der Dunk, United Nations Principles on Remote Sensing and the User, *Earth Observation Data Policy and Europe*, R. Harris (Eds.), A.A. Balkema Publishers, Lisse, 2002, 33.

[226] Principle XII of the UN Remote Sensing Principles.

[227] A more refined term equivalent for reasonable cost terms include the Landsat data "cost of fulfilling user request" and for ESA's ERS and ENVISAT, the price will be set "at or near cost of reproduction."

to force sensing states to release data to sensed states in case the sensing state does not follow this rule. It is practically impossible for sensed states to keep track of the precise time and location of observation of their territories by all the satellites. Whether data is actually accessed by the sensed state is entirely at the discretion of the data generators and the interested state's ability to pay for the data. Therefore, this Principle can be construed as a 'passive' obligation to disseminate data to any other states if requested.

Principle XII is based upon the presumption that a distinction is made only on the basis of sensing states and other states, and not on consideration of other factors such as the purpose for which the data is used. Principles X and XI oblige the disclosure of information for environmental protection and disaster management, but they do not stipulate the conditions for such disclosure.[228] In reality, certain data use – such as for disaster relief – and certain types of users under governmental space programmes receive preferential treatment, including a price waiver. The practical questions of under what conditions free provision of data is allowed, and the subsequent implications arising are not addressed anywhere.

Another issue to be addressed is the availability of analysed information derived from remotely sensed data. There is a loophole in Principle XII, and the provision becomes even more complicated regarding this analysed information, specifically, "The Sensed States shall also have access to the available analysed information concerning the territory under its jurisdiction in the possession of any state participating in the remote sensing activities on the same basis and terms."[229] Strictly speaking, this may exclude cases where the analysed information is *not* in possession of any state. The obligation to provide analysed information may not extend to individuals or juridical persons[230] engaged in analysing information, such as a processing wholesaler and/or information extractor. This is again a loophole and results in a gap between the implied position and the reality of current practice. The mechanisms surrounding data availability should be phrased differently in the context of data policy.

The other principles endorsed by the UN Remote Sensing Principles fundamentally repeat the provisions established under the Outer Space Treaty. Yet, some reiterated principles are tailored to situations specifically associated with remote sensing from space and worthy of attention. This section reviews the issues of application of international law, international cooperation, and responsibility and liability.

---

[228] Principles X and XI of the UN Remote Sensing Principles.

[229] Principle XII of the UN Remote Sensing Principles.

[230] The interpretation that private entities are outside the scope of such obligation is supported by Principle XIV on state responsibility. See H.L. van Traa-Engelman, *Commercial Utilization of Outer Space*, SI Publisher, Utrecht, 1989, 256.

Principles II and III fundamentally reiterate the provisions of the Outer Space Treaty, stipulating that remote sensing activities are to be conducted for the benefit and interests of all countries and in accordance with international law. Application of general international law means that remote sensing activities are carried out under the rubric of the UN Charter and the Outer Space Treaty.[231]

Several principles refer to the issue of strengthening international cooperation, which basically repeat the norm established by the Outer Space Treaty, but adapted to remote sensing activities. Principle V and Principle XIII share the common objective to promote international cooperation through remote sensing activities. Principle V states that "States carrying out remote sensing activities shall promote international cooperation in these activities. To this end, they shall make available to other States opportunities for participation therein."[232] This principle is slightly further elaborated in Principle XIII, which states that "To promote and intensify international cooperation, especially with regard to the needs of developing countries, a State carrying out remote sensing of the Earth from space shall, upon request, enter into consultations with a State whose territory is sensed in order to make available opportunities for participation and enhance the mutual benefits to derived therefrom."[233] These Principles, however, do not impose any obligation for price waivers or mutual exchange of data.[234]

Principles VI, VII, and VIII provide concrete guidance on how to promote international cooperation. Principle VI encourages states to conclude agreements for the establishment and operation of data collecting and storage stations and processing and interpretation facilities, and likewise, Principle VII encourages remote sensing operating states to give technical assistance to other interested states. Principle VIII invokes the role of UN agencies and the relevant agencies in promoting cooperation and technical assistance. Principle IX calls for the transparency of remote sensing activities, requiring the sensing states to report their activities to the UN Secretary General, and encourages the dissemination of information to those states affected by the activity; however, nothing substantial is added to the rights of sensed communities.

Principles X and XI affirm that states must warn of danger in cases of harm to the environment or impending disasters by rapidly and freely disclosing the relevant information.[235] These two concepts had been supported from the

231 See H. Feder, The sky's limit? Evaluating the International Law of Remote Sensing, 23-2 *NYU Journal of International Law and Politics* (1991), 613.

232 Principle V of the UN Remote Sensing Principles.

233 Principle VIII of the UN Remote Sensing Principles.

234 See G. Catalano Sgrosso, Prevention and Management of Natural Disasters, *Outlook on Space Law over the next 30 years,* Kluwer Law International, 1997, 322.

235 Principle X states that: "States participating in remote sensing activities that have identified information in their possession that is capable of averting any phenomenon

initial phase of negotiation on the formulations of the UN Remote Sensing Principles; indeed, remote sensing has been used extensively for assessing damage resulting from natural disasters. For example, when flooding occurred around the city of Saint Louis, Missouri, in the United States, the ERS-1 satellite assessed the extent of the flooded area and its effect on cultivation.[236] The Disaster Charter, originally started as an agreement between CNES and ESA, is a now viable example of extensive international cooperation making satellite data and derived information available for any state affected by disaster.

Finally, Principle XIV of the UN Remote Sensing Principles that deals with international responsibility also needs to be examined.[237] Principle XIV makes reference to Article VI of the Outer Space Treaty and stipulates that: "States operating remote sensing bear international responsibility for their activities."[238] It is possible to argue that the state is internationally responsible for remote sensing activities, including the dissemination of data, but not for what happens afterwards as a result of any activity by the state or a third party.[239] Furthermore, a loophole exists in Principle XIV where it states that: "*States operating remote sensing satellites* (emphasis added) bear international responsibility for their activities (…) whether such activities are carried out by governmental or non-governmental entities".[240] Although government entities are not the only entities carrying out remote sensing activities, the language in Principle XIV raises questions as to its coverage for all potential parties who may be held responsible. Therefore, the provision may be construed literally – that responsibility is imposed only on states operating remote sensing satellites.

Strictly speaking, the UN Remote Sensing Principles do not impose responsibilities on either those states or entities that are not operating satellites but who are actually receiving images and/or engaged in data handling activities

harmful to the Earth's natural environment shall disclose such information to States concerned." Principle XI states that: "States participating in remote sensing activities that have identified processed data and analysed information on their possession that may be useful to States affected by natural disasters, or likely to be affected by impending natural disasters, shall transmit such data and information to States concerned as promptly as possible."

[236] See S. Courteix, Toward the legal Recognition of a new method of proof for the defence of the Environment, *Proceedings of the Thirty-Seventh Colloquium on the Law of Outer Space,* IISL, AIAA, 1994, 226.

[237] The Principle XIV language dealing with responsibility was based on the Mexican and Brazilian proposal. See B.L. Deekshatulu, V.R. Venkataraman and M.G. Chandrasekhar, Overview of Remote Sensing, *Recent Trends in International Space Law and Policy,* Lancers Books, New Delhi, 1997, 225.

[238] Principle XIV of the UN Remote Sensing Principles.

[239] See S. Gorove, *Developments in Space Law,* Martinus Nijhoff Publishers, Dordrecht, 1991, 308.

[240] Principle XIV of the Remote Sensing Principles.

as well as distribution, or on the third parties who use the data. In fact, there are many states and entities that do not operate satellites, but rather house value-added resellers where the users should also be held responsible for the misuse of data. This can be seen as a significant loophole and the gap needs to be filled.

Overall, the UN Remote Sensing Principles succeed in codifying basic norms to govern sensing from space and derived operations. They clarify the conditions for sensing as well as data dissemination, and establish the general responsibilities expected of sensing states. The rights and obligations provided by the Remote Sensing Principles are: 1) the right of any state to sense anywhere on the earth from space without prior consent; 2) the obligation of sensing states to provide data to sensed states on a non-discriminatory basis; 3) the right of sensed states to access data on a non-discriminatory basis; 4) the obligation not to infringe the right of sensed states in sensing; and 5) the obligation to bear international responsibility in conducting remote sensing.

On the other hand, the UN Remote Sensing Principles are ambiguous and inadequate because they do not address some critical issues. There are gaps, shortcomings, and loopholes, which result in the Principles not being entirely effective.

There are major shortcomings in that the rights and obligation of concerned parties in remote sensing are not properly addressed in the UN Remote Sensing Principles. They are ambiguous with regard to responsibility and the liability associated with remote sensing data and its use in terms of consequential damage. Another shortcoming is that issues with respect to data availability and accessibility are not sufficiently addressed. Principle XII is the only provision related to the practical issues of data availability and pricing policy, but it does not serve as a concrete guideline for the variety of policies on satellite remote sensing data that exist today. With regard to data accuracy and data validation, the Principles are completely silent. The present legal framework is inadequate in terms of data policy, criteria for data accuracy, and guidelines for data authentication – all of which affect widespread use of data for applications such as disaster management and protection of the environment.

It should be reemphasized that the content of the UN Remote Sensing Principles centres on the rights and obligation of sensing states and the interests of sensed states, which highlight a gap and discrepancy with reality. The Principles reflect the politics and technologies of the 1980s, rather than the present day; they were sensible back then, since the most important issue at that time was to reconcile the interests of sensed states and sensing states. Only a few sensing states existed, and the rest of the world was composed of sensed states. Over the two decades since the Principles were adopted, increasing numbers of nations that were once standing on the 'sensed' side, now possess their own remote sensing capabilities in orbit. Consequently, the debate has shifted more to considering the rights and obligations of data providers, data recipients, and third parties. The debate that is now of importance focuses on issues related to data policy, data accuracy, and authentication as well as the responsibility associated with the provision and use of data. The question is whether these issues that are not sufficiently regulated internationally are covered in the national context. In order to make the analysis complete, it is necessary to look at the national measures related to space activities and remote sensing.

2.3. National Laws and Policies

As noted earlier, the *corpus juris spatialis* provides the general framework for member states to regulate national space activities through more specific rules.[241] National laws are the concrete implementation of Article VI of the Outer Space Treaty, which requires the authorization and continuing supervision by the appropriate state party of the national activities conducted by its non-governmental entities.[242] The provision, in reality, requires national licensing and regulatory legislation;[243] in fact, domestic legislation devoted to space activities has been endorsed in several nations.

2.3.1. *National Legislation on Space Activities*

Currently, states that possess national space legislation relating to space activities include: the United States,[244] Sweden,[245] the United Kingdom,[246] Russia,[247]

---

[241] See A. Kerrest & M. Benkö and K. Schrogl (Eds.), Liability for Damage Caused by Space Activities, *Space Law: Current Problems and Perspectives for Future Regulation*, Eleven International Publishing, 2005, 107.

[242] See A. Kerrest & M. Benkö and K. Schrogl (Eds.), Liability for Damage Caused by Space Activities, *Space Law : Current Problems and Perspectives for Future Regulation*, Eleven International Publishing, 2005, 107.

[243] See P.P.C. Haanappel, *The Law and Policy of Air Space and Outer Space*, Kluwer Publisher, Hague, 2003, 57

[244] National Aeronautics and Space Act, Public Law 85–568, 85th Congress, H.R. 12575, 29 July 1958, Commercial Space Launch Act, Public Law 98–575, 98th Congress, H.R. 3942, 30 October 1984, Land Remote Sensing Commercialization Act, Public Law 98–365, 98th Congress, H.R. 5155, 17 July 1984, Land Remote Sensing Policy Act 1992, Public Law 102–555, 102nd Congress, H.R. 6133, 28 October 1992.

[245] Act on Space Activities 1982: 963, 18 November 1982. [Hereafter: Swedish Space Act].

[246] Outer Space Act, July 18, 1986, 1986 Chapter 38. [Hereafter: UK Outer Space Act]

[247] Law of Russian Federation on Space Activity, August 20, 1993, Presidential Edict No. 185. [Hereafter: Russian Space Act]

South Africa,[248] Ukraine,[249] Hong Kong,[250] Australia,[251] Belgium,[252] Korea,[253]the Netherlands,[254] Japan,[255] and France.[256] The fundamental role of national legislation is to provide a regulatory means for a government to authorize actors under its jurisdiction to engage in space activities and to exercise control over them. By and large, rules established by national legislation comprise the concrete obligations under Articles IV, VII, and VIII of the Outer Space Treaty – namely, measures for authorization, continuing supervision, sanctions, responsibility, and liability, as well as for exercising jurisdiction and control through the registration of space objects. It is common that authorization is given through a permit or license to conduct space activities, particularly regarding launch operations; the government supervises continuously to ensure that licensed entities are in compliance with the requirements established under such licenses. This section examines the scope of government authorization for, and continuing supervision of, space activities carried out by domestic actors.

2.3.1.1. *Scope*

The scope of space activities covered under domestic laws of different countries varies by legislation, but primarily extends to one or a combination of the following: 1) launching of a space object and/or operation of a launch facility; 2) operation of a space object in orbit; and 3) conducting any other activities in outer space. Domestic law is less likely to apply to simply manufacturing a satellite or receiving signal from a satellite at the ground station.[257]

Generally speaking, national space legislation applies to space activities carried out by nationals, and those space activities that take place from its

[248] Space Affairs Act, September 6, 1993. [Hereafter: South African Space Act]

[249] Ordinance of the Supreme Soviet of Ukraine on Space Activity Law of Ukraine, 15 November 1996. [Hereafter: Ukrainian Space Act]

[250] Outer Space Ordinance of 1997, 13 June 1997.

[251] Space Activities Act, December 21, 1998, No. 123. [Hereafter: Australian Space Act]

[252] The Law of 17 September 2005 (Belgian Official Journal of 16 November 2005) on the Activities of Launching, Flight Operation or Guidance of Space Objects.

[253] Space Exploration Promotion Act, 1 December, 2005. [Hereafter: Korean Space Exploration Promotion Act]

[254] Rules Concerning Space Activities and the Establishment of a Registry of Space Objects, 13 June 2006. [Dutch Space Activities Act]

[255] Basic Law for Space Activities adopted 13 May 2008.

[256] French Law on Space Operations, Act no. 2008–518 du 3 juin 2008 [Hereafter: French Space Operations Act]

[257] An exception to this is, e.g., the Dutch legislation that applies to activities that are not conducted in outer space for preparation of space activities. See Article 3 of the Dutch Space Activities Act.

territory – including activities conducted by foreign nationals. A noteworthy example that is distinctive from others is the United States. In contrast to other states, the United States requires licenses from foreign persons even if they conduct space activities from outside US territory or areas outside its jurisdiction if the controlling interest is held by either US nationals or entities incorporated under US law.[258] In addition, foreign nationals need to obtain a license in cases where there is an agreement between the United States and a foreign state stating that the United States exercises its jurisdiction over a specific activity such as launching operation.[259] In this way, the US Commercial Space Launch Act illustrates how the terms 'national activities' and 'appropriate state party' can be broadly interpreted in the national context.

2.3.1.2. *Authorization*

Authorization is required for entities wishing to carry out space activities if required under domestic law of the state whose jurisdiction covers the activities of the entities in question. Although not always stated explicitly, in accord with the 'appropriate state party' language of the Outer Space Treaty, authorization appears to be required from a single jurisdiction even if an activity may fall under multiple jurisdictions.[260] For instance, the UK Outer Space Act states that a license is *not* required "for activities in respect of which it is certified by Order in Council that arrangements have been made between the United Kingdom and another country to secure compliance with the international obligations of the United Kingdom."[261] In Australia, with regards to overseas launches, space legislation states that overseas launch requires an overseas launch certificate if "a space object is launched from a launch facility located outside Australia and the launch is not authorized by an overseas launch certificate held by any person; and an Australian national is a responsible party for the launch."[262]

Whilst there are several elements required for authorization, this section focuses primarily on the discussion of liability and financial aspects, specifically a regime of reimbursement, a regime surrounding third party loss, and a regime surrounding inter-party liability.

Given the high risk of space activities, financial capability of a licensee – particularly in terms of preparedness for liability – is one of the essential elements needed for government authorization. Since the responsibility for national space activities falls upon states, states seek certain security in the event of accident or negligent failures. Accordingly, a state has an imperative to

[258] See Section 6(a)(2)(3) of the US Commercial Space Launch Act.
[259] Section 6(a)(3)(B) of the US Commercial Space Launch Act.
[260] Article VI of the Outer Space Treaty.
[261] Section 3(2)(b) of the UK Outer Space Act.
[262] Section 12 of the Australian Space Act.

properly arrange the measures for reimbursement between the state and entities that are subject to its jurisdiction and control. Most of the time, domestic legislation – as endorsed by the United Kingdom,[263] the Netherlands,[264] Sweden,[265] and France[266]– requires indemnification for the loss caused by the licensee to the government when the respective state is held internationally liable under the scope of the Liability Convention. More often than not, failure occurs during a launching phase, so an agreement between a government and a launch operator is not uncommon. For instance, with regard to a launch by an Ariane rocket, it is agreed that Arianespace would be required to reimburse the French government up to a maximum of € 60 million.[267] This was the practice for years in Ariane operations and the approach seen in the agreement was codified into the French Space Operations Act, which states that in case the government paid compensation for damage, it may then present a claim for indemnification against the operator that caused the damage for which France was held internationally liable – to the extent that the government has not already benefited from the insurance or financial guarantees of the operator up to the amount of the compensation. As commercial launch services emerge and grow in an increasing number of states, such an arrangement or domestic law referring to the rule is deemed essential and likely to proliferate in the future.

As a reasonable measure to prepare for risks, some states such as Australia, Russia, Ukraine, Korea, and the Netherlands make a compulsory insurance regime part of their licensing requirements. Most compulsory insurance is coverage for a third party loss; thereby, a state secures *de facto* a measure of protection in case of failures. The United States has a detailed liability regime applicable to launch operations. The Commercial Space Launch Act requires the licensees either to purchase insurance or demonstrate financial capability so as to cover the potential third party loss. The maximum amount of obligation a licensee owes for the third party loss is US$ 500 million, and US$ 100 million with respect to the loss of or damage incurred to government property.[268] On the other hand, the government would pay up to US$ 1.5 billion for damages that exceed US$ 500 million insofar as the damage was not due to wilful misconduct of the licensee.[269] The French Space Operations Act takes a similar approach

[263] See Section 10(1) of the UK Outer Space Act.

[264] See Chapter 4 Section 12 (1) of the Dutch Space Activities Act.

[265] See Section 6 of the Swedish Space Act.

[266] See Article 14 of the French Space Operations Act.

[267] See Section III(1)h of the Declaration by certain European Governments on the Launchers Exploitations Phase of Ariane, Vega, and Soyuz from the Guiana Space Centre, Paris 30 2007.

[268] See Section 16(a)(1)(B) of the US Commercial Space Launch Act.

[269] See Section 16(4)(b)(1) of the US Commercial Space Launch Act states that "The secretary shall not provide for payment of any part of such claims for which the death,

by obliging licensees to obtain insurance or another financial guarantee[270] whilst placing a ceiling on the maximum amount that operators need to pay for third party loss during the launch phase.[271] Claims by the government to operators for indemnification are limited to that amount: operators can benefit from the state guarantee if the amount exceeds € 60 million.[272] This state guarantee – that is, reimbursement of indemnifications by the government if liability exceeds € 60 million – is also available in the cases where operators are sued in any court.[273]

Inter-party liability is less addressed within a legislative context; rather, it is tailored to each contractual agreement. Some legislation, however, does address and refer specifically to this matter, such as the US Commercial Space Launch Act and the French Space Operations Act. The Commercial Space Launch Act contains a reciprocal waiver of liability regime that applies to all the parties involved in a launch, including the government, any of its agencies, licensee, contractors, subcontractor, customers, and their contractors and subcontractors.[274] Likewise, the French regime holds inter-party waiver of liability in case of damage to a person taking part in the space operation or the production of space object(s) except in the case of wilful misconduct.[275]

Even if it is not established in legislation, the stance taken by the parties engaged in launch activities is to waive liability amongst them. For instance, ESA agrees to waive all claims for compensation against France as far as these claims result from launch operations at the Centre Spatial Gyuanais.[276] This minimizes the risk of inter-party claims. The practical tendency in the exercise of inter-party waiver of liability logically follows the rule of the Liability Convention, which does not offer protection for damage incurred to persons participating in the launch.

For the most part, the government's intention is to minimize the potential cost incurred in the instances of its international liability resulting from irresponsible operations of national actors, as well as in the instances of loss incurred

---

bodily injury, or loss of or damage to property has resulted from wilful misconduct by the licensee or transferee".

[270] See Article 6 of the French Space Operations Act.

[271] See Articles 14 and 16 of the French Space Operations Act.

[272] See Section III(1)h of the Declaration by certain European Governments on the Launchers Exploitations Phase of Ariane, Vega, and Soyuz from the Guiana Space Centre, Paris 30 2007.

[273] See Article 15 of the French Space Operations Act.

[274] Section 16 (C)(D) of the US Commercial Space Launch Act amended in 1988.

[275] Article 20 of the French Space Operations Act.

[276] Article 13(1), Centre Spatial Guyanais. Cf. F.G. von der Dunk, *Private Enterprise and Public Interest in the European Spacescape*, International Institute of Air and Space Law, Leiden, 1998, 160.

amongst the concerned parties. These measures help the concerned parties to quantify properly the involved risk.

2.3.1.3. *Continuing Supervision*

Continuing supervision is more complex and difficult to exercise. The term infers the obligation to 'keep an eye' on the authorized person after the initial authorization. Continuing supervision can well be associated with the monitoring and control of the licensed entity and launched space object that are subject to the jurisdiction of a particular state. This part reviews the measures for monitoring compliance with licensing conditions, sanction, and control of a space object, as related to registration.

Overall, national legislation attempts to control licensees through monitoring their compliance for license requirements and, in case of non-compliance, suspend or revoke the license. In South Africa,[277] the United Kingdom,[278] and Sweden,[279] licenses may be withheld if the licensing requirement is breached. Conditions for suspension differ in each state, but they all perform a function of continuous supervision and regulation of the licensed activity. In the case of South Korea, the license is withheld if the launch does not take place within one year from the scheduled launch date.[280] This may be a stringent condition, as launch delays are frequent.

Continuing supervision of national space activities and control of a space object can also be enforced through punitive measures. Some states specifically provide measures for sanctions in case of breach of the provisions of the legislation. The type of sanction imposed is usually fine or imprisonment, or both. For instance, the Swedish Space Act establishes that persons conducting space activities without a license, whether wilfully or through negligence, are fined or sentenced to imprisonment for up to one year.[281] The US Commercial Space Launch Act holds any person convicted of committing a prohibited act liable for civil penalties of not more than US$ 100,000 per day for each violation.[282] A fine imposed in such a manner generates great incentives to offenders to cease the prohibited act immediately. Sanctions serve as a powerful incentive for actors to refrain from engaging in space activities without authorization from the appropriate jurisdiction, as well as for licensees not to breach licensing provisions.

Continuing supervision can be exercised through control over a space object based on the registration of that space object established by Article VIII of the

[277] See Section 13(2)(a) of the South Africa Space Act.
[278] See Section 6(2)(a) of the UK Outer Space Act.
[279] See Section 4 of the Swedish Space Act.
[280] See Article 13 of the Korean Space Exploration Promotion Act.
[281] See Section 5 of the Swedish Space Act.
[282] See Section 19(a) of the US Commercial Space Launch Act.

Outer Space Treaty. Several nations including Ukraine,[283] Russia,[284] Australia,[285] the Netherlands,[286] and Korea[287] require the registration of space objects and have established a national registry in accordance with the Registration Convention. Pursuant to the Registration Convention that obliges state parties to register domestically, national space legislation establishes a mechanism of state registry with more specific rules as to how the domestic registration of a space object takes place. There are two major rules with regard to registration established under the Registration Convention: control of a satellite to be exercised effectively by a state of registry – that is, usually the launching state – and prohibition of double registration; both are incorporated into national space legislation.

In reality, with the increasing lifespan of satellites, conceptual separation between the launching and operational phase of a satellite is feasible;[288] the gap is growing between the law and practice. The situation is even more complicated recently with the cases of lease or transfer of ownership in orbit, including sale of a spacecraft after it is placed in orbit. Such complexities could give rise to at least five different possibilities for the status of a satellite in terms of its link with a launching state, state of registry, and controlling state, as shown in the table below:

Table 2.1: Possible Status of Satellites with Different Types of Configurations

Launching State	State of Registry	Controlling State
A	A	A
A	A	B
A	B	B
A	B	A
A	B	C

*A signifies as state A, B signifies as state B and C signifies as state C.

The first case A-A-A is the orthodox situation that is assumed under the Registration Convention – A single launching state registers and controls the launched space object. The second case, A-A-B, would be best explained by a situation of a lease of a satellite. State A launched, and registered the launched

283 See Article 13 of the Ukrainian Space Act.

284 Article 17 of the Russian Space Act.

285 See Article 76 of the Australian Space Act.

286 See Chapter III Section 11 of the Dutch Space Activities Act.

287 See Article 8 of the Korea Space Exploration Promotion Act.

288 See F.G. von der Dunk, *Private Enterprise and Public Interest in the European Spacescape*, International Institute of Air and Space Law, Leiden, 1998, 61.

space object, but later leased it to state B; therefore, state B controls the space object, even though state A still owns it. The A-B-B case could be illustrated by the situation that would involve certain transfer of ownership after the launch. The foremost example is the sale/purchase of a satellite in orbit. State A launched the space object and later transferred its ownership to state B, which re-registers and controls the object. For the A-B-A case, one can envisage a scenario as follows: State B, a developing state, bought a satellite in orbit from state A, a developed state. Later, from a technical reason, state B becomes no longer able to control the space object; state A agrees to resume the operation for the benefit of state B. The last A-B-C case could be explained by the combination of a transfer of ownership of a satellite from state A to state B, who then leases it to state C.

The above described situations are not at all hypothetical. In reality, there have been cases where an original state of registry does not exercise jurisdiction and control over a space object later in the operational phase due to its change of ownership. Such examples concern four communication satellites returned from the United Kingdom to China in accordance with the return of Hong Kong.[289] Following the return of Hong Kong on 1 July 1997, the United Kingdom informed the United Nations on 27 March 1998 that it was no longer the state of registry and then requested an amendment to the UN registry as well as removal of those satellites from its national registry.[290] Even though a launching state is no longer linked with a state of registry, this case is relatively simple because jurisdiction and control is transferred along with the change of ownership so that only one state of registry exists for those satellites in orbit. This case illustrates the A-B-B scenario in the table above.

A more complex situation exists when jurisdiction and control may be exercised by more than one state. An example of such a case concerns a broadcasting satellite BSB-1A. BSB-1A was originally registered by the United Kingdom in 1990, but purchased in orbit by Sweden in 1996.[291] In contrast to the case discussed above, although Sweden registered BSB-1A nationally as Sirius-1 in 1999 and notified the United Nations with specific mention of the United States as the launching State, the United Kingdom did not remove BSB-1A from its registry.[292] Therefore, BSB-1A is 'doubly' registered, and this creates a

[289] They are AsiaSat-1, AsiaSat-2, APSTAR-I and APSTAR-IA which were launched from Xichang, China in 1990, 1995, 1994 and 1996 respectively and registered in the UK national registry. Cf. Y. Lee, Registration of space objects: ESA member states' practice; 22 *Space Policy* (2006), 46.

[290] See Y. Lee, Registration of space objects: ESA member states' practice, 22 *Space Policy* (2006), 46.

[291] UNGA Doc. A/AC/105/768. See Y. Lee, Registration of space objects: ESA member states' practice, 22 *Space Policy* (2006), 47.

[292] UN Secretariat Doc. ST/SG/SER.E/377 (24 October 2000).

situation that is not envisaged by the Registration Convention: the launching state is the United States; the state of registry is both the United Kingdom and Sweden; and the controlling state is Sweden. This situation is a slightly complicated version of the A-B-C case; that is, the state of registry is really two states, B/C rather than just state B. The cases like BSB-1A diminish the effectiveness of the control regime established by the Registration Convention.

Not only was the BSB-1A type of situation not addressed under the Registration Convention, there is no national legislation at present that adequately fills the existing gap – particularly with regard to either leasing or the transfer of ownership of a space object and the subsequent impact on the jurisdiction and control over the space object. The ongoing and future practice of lease, sale, and transfer of ownership of a satellite may require adaptation of the rule in accordance with the practice at a domestic level.

Overall, there is no national legislation addressing how the responsibility associated with a space object are divided between the concerned parties when the state of registry and the state that actually controls the space object are different. Such cases may be increasingly common in the future, and there is a need for further clarification.

Finally, national legislation is inadequate to address how control by the state of registry over its space object should take place. The obligation to exercise jurisdiction and control by the state of registry is particularly relevant to remote sensing since information acquired by the satellite affects the interests of both the state and the international community. Increasingly, states with remote sensing capabilities are adopting laws and/or policies to control the data collection and subsequent dissemination. The next section reviews those regulations relevant to remote sensing and analyses the overall tendencies and implications for remote sensing activities.

2.3.2. *National Regulations Associated with Remote Sensing*

National regulations addressing space activities in general are written to implement concrete measures for continuous supervision of the actors under a government's jurisdiction. A similar role is performed by national regulations on satellite remote sensing. There are two main purposes for governmental control of remote sensing activities imposed on commercial remote sensing satellite systems: 1) to protect national security by denying or restricting access to information; and 2) to guarantee that the government has priority access to the system's capabilities during times of crisis.[293] The measures taken to achieve these

---

[293] See D. Clarke, Access Control of Remote Sensing Satellites, J.F. Keeley & R. Hueber (Eds.), *Commercial Satellite Imagery and UN Peacekeeping*, Ashgate Publisher, Aldershot, 2003, 173.

goals can be divided into restrictions on data collection and/or data dissemination. This section first examines those states that have legislation specialising in remote sensing, and then reviews other states that have adopted certain policies with regard to data collection and dissemination.

Although some provisions of national space legislation addressing launching and general space activities have some implications on satellite remote sensing operations, generally speaking, existing domestic legislation on space activities does not address properly the rules applicable to remote sensing. Corresponding to the emergence of private actors and commercial remote sensing satellites, the United States, Canada, Germany, and France have established specific regulations/legislation devoted to remote sensing from space. It is notable that amongst these, the legislation of Canada and Germany addresses solely the matters regarding satellite remote sensing, whilst neither country possesses domestic legislation addressing overall space operations. The existing legislation on remote sensing authorizes and controls the actors.

2.3.2.1. *The United States*

As stated earlier, the United States has enacted several pieces of space legislation, one of which is dedicated to remote sensing activities. The legislation on remote sensing has evolved in accordance with the trends surrounding remote sensing from space, particularly reflecting advances in technology and market development as well as changing policies. Compared to legislation from other countries, US legislation addresses practical matters such as data policies as well as measures of protection to maintain the competitiveness of domestic actors.

Looking at the historical evolution of US laws and policies associated with remote sensing, the major rationale for making both law and policy reflects the government's striving to maintain the leadership in this field whilst safeguarding national security and foreign policy interests. Starting its operation in 1972 under the auspices of NASA, by the late 1970s the United States started to consider privatizing the Landsat system. That attempt was codified into the Land Remote Sensing Commercialisation Act of 1984.[294] This Act was adopted to commercialise the Landsat programme by transferring remote sensing operations to the private sector. The Earth Observation Satellite Corporation (EOSAT), a partnership of Hughes Aircraft and Radio Corporation of America (RCA), was selected as the executor of the project to commercialize Landsat under a ten year contract. The original plan was to delegate a private sector operation of the Landsat system that would become commercially viable through data sales and then to develop the successive Landsat series on a private-sector-driven basis. Commercialisation did not progress as planned

[294] The Land Remote Sensing Commercialization Act, Public Law 98–365, 98th Congress, H.R. 5155, 17 July 1984.

because of the sharp increase in data price which resulted in a sharp drop in data needs. Coupled with the 1986 emergence of the powerful competitor, SPOT, which benefited from funding by the French government – and the significant market share attained by SPOT soon afterwards – the United States relinquished the attempt at commercialization.

Having experienced the unsuccessful attempt to privatize Landsat, at least in the short term, the US government was compelled to return to the regime to support the Landsat system. The Land Remote Sensing Commercialisation Act of 1984 was then supplanted by the Land Remote Sending Policy Act in 1992.[295] The 1992 Land Remote Sensing Policy Act gave the operation of Landsat back to the public sector under the Department of Defense (DOD) and NASA, whilst leaving the licensing authority under the Secretary of Commerce through NOAA.[296]

In the longer term, United States had not renounced its plan for commercialization.[297] The strategy was to let the public recognize the value of the Landsat data. In the Land Remote Sensing Policy Act, availability and affordability of Landsat data was set up to diffuse its utility. Cost hurdles based on the market price were lifted and a standard for pricing of unenhanced data was established. The Act states that "to maximize the value of the Landsat program to the American Public, unenhanced Landsat 4 through 6 data should be made available at a minimum to US agencies, to global environmental change researchers, and to other researchers who are financially supported by the US government at cost of fulfilling user requests, and unenhanced Landsat 7 data should be made available to all users at the cost of fulfilling user requests."[298] Furthermore, US government and affiliated users could access the unenhanced data at a reduced rate.

---

[295] The Land Remote Sensing Policy Act 1992, Public Law 102–555, 102nd Congress, H.R. 6133, 28 October 1992. See further C.C. Joyner & D.R. Miller, Selling satellites: the commercialization of LANDSAT, 26 *Harvard International Law Journal*, 63–102, See also J.I. Gabrynowicz, The Promise and Problems of the Land Remote Sensing Policy Act of 1992, 9 *Space Policy* (Nov. 1993), 319–328.

[296] Sections 101, 102 and Section 201–205 of the Land Remote Sensing Policy Act.

[297] Section 5601 (6) of the Land Remote Sensing Policy Act states that "Full commercialization of the Landsat program cannot be achieved within the foreseeable future, and thus should not serve as the near-term goal of national policy on land remote sensing; however, commercialization of land remote sensing should remain a long-term goal of United States policy".

[298] Section 5601 (13) of the Land Remote Sensing Policy Act. Cost of fulfilling user requests means that "incremental costs associated with providing product generation, reproduction, and distribution of unenhanced data in response to user requests and shall not include any acquisition, amortization or depreciation of capital assets originally paid for by the United States Government or other costs not specifically attributable to fulfilling user requests."

In an attempt to stimulate the development of the commercial market, exclusive rights for Landsat data sale were no longer granted to a single entity. Under the Land Remote Sensing Policy Act, a data policy for Landsat 7 was set to allow competition within the private sector for distribution of unenhanced data and value added services. This provision assumes the possibility of competition amongst different licensed entities that provide unenhanced data and, particularly, value added services at a non-uniform price. The Act emphasizes that the license application should not be rejected on the basis of protection of existing licensees from competition,[299] thus permitting data collection and sale of Landsat data worldwide.[300]

The Land Remote Sensing Policy Act addresses the issue of licensing for the operation of private remote sensing systems. The Act anticipated that commercial players would operate remote sensing systems other than the Landsat and undertake subsequent data dissemination. The Act required persons under US jurisdiction and control to obtain a license to operate private remote sensing systems. Conditions of operation were imposed on licensees carrying out the operation in a manner to preserve US national security, to make unenhanced data available to the government of any country, and to inform the Secretary of Commerce on both the characteristics of the system and any intentions of concluding agreements with foreign states.[301] The Act contains an enforcement mechanism similar to that in the aforementioned Commercial Space Launch Act, but the penalty is much lower – US$10,000 per day per violation.[302]

The 1992 Land Remote Sensing Policy Act remains as the principal legislation governing US remote sensing activities today and serves as an overall framework, updated, complemented, or implemented concretely by relevant policies and regulations, namely: Presidential Decision Directive 23 (PDD-23); the US Policy on Foreign Access to Remote Sensing Capability;[303] the Kyl-Bingaman Amendment to the 1997 National Defense Authorization Act; the 2003 Commercial Remote Sensing Policy; and the 2006 Regulation on Licensing of Private Land Remote Sensing Space Systems. The following chronological progression of law and policy developments serves to capture the essential transitions in the field of remote sensing in the United States.

In 1994 PDD-23 was announced, recognizing the emergence of non-US remote sensing capability and the need to ensure continued US industrial

[299] See Section 5621(d) of the Land Remote Sensing Policy Act.

[300] See generally, P.A. Salin, Landsat Contracts Signed by US Agencies with Foreign Ground Stations: Commercial Remote Sensing from NASA Scientific Experiments to EOSAT Private Endeavours, 41 *German Journal of Air and Space Law* (1992), 165–175.

[301] See Section 5622(b) of the Land Remote Sensing Policy Act.

[302] See Section 5658 of the Land Remote Sensing Policy Act.

[303] Presidential Decision Directive 23, US Policy on Foreign Access to Remote Sensing Capability, 9 March 1994. [Hereafter: PDD-23].

competitiveness in the global arena. The objective of PDD-23 is roughly threefold: 1) to relax restrictions placed on data sale; 2) to clarify the operational regime of Landsat; and 3) to specify national security measures as a licensing requirement. As to the first objective, PDD-23 addressed the question of which agency develops and operates the Landsat system and markets the resulting data[304] by establishing an operational regime for Landsat that delegates different operations to three agencies: the operation of satellites is conducted by NOAA; archiving and distribution is conducted by the US Geological Survey under the Department of the Interior; and the development of satellites is conducted by NASA.

As to the second objective, in order to allow the national actors to compete internationally, PDD-23 – for the first time – allowed US entities to sell satellite images up to one metre ground resolution.[305] PDD-23 relaxed the restrictions, encouraged the sale of high resolution data services from US providers, and even permitted the export of turnkey systems and sensitive components on a case-by-case basis, subject to export controls and requirements for government-to-government agreements.[306]

As to the last objective, a concrete measure was taken toward ensuring US national security and foreign policy interests. An explicit provision to justify the interruption of operation, which is known as 'shutter control', was introduced: "[W]hen national security or international obligations and/or foreign policy may be compromised as defined by the Secretary of Defense or the Secretary of State, respectively, the Secretary of Commerce may after consultation with the appropriate agencies, require the licensee to limit data collection and or distribution by the system to the extent necessitated by the given situation."[307] This provision remains today to be invoked and enforced if necessary.

An earlier provision of the Land Remote Sensing Policy Act on a licensee's obligation to preserve US national security and foreign policy interests was implemented concretely with the Kyl-Bingaman Amendment to the 1997 National Defense Authorization Act. The amendment prohibits licensees from disseminating imagery of Israel that is more detailed or precise than what is

---

[304] See A. Florini & Y.A. Dehqanzada, No More Secrets?: Policy Implications of Commercial Remote Sensing Satellites, Carnegie Paper no. 1, Carnegie, Washington, July 1999, 6.

[305] See R.J. Aamoth, Use of Remote Sensing Imagery by the News Media, *Heaven and Earth: Civilian Uses of Near-Earth Space*, Martinus Nijhoff Publishers, Hague, 1997, 150.

[306] See VI. Foreign Access to US Commercial Remote Sensing Capabilities. See also P. Hays, *Spacepower for a New Millennium*, McGraw Hill Companies, New York, 2000, 13.

[307] Presidential Decision Directive 23, US Policy on Foreign Access to Remote Sensing Capability, 9 March 1994. [Hereafter: PDD-23].

available from foreign commercial sources.[308] The section on prohibition of collection and release of detailed satellite imagery relating to Israel states that "A department or agency of the United States may issue a license for the collection or dissemination by a non-Federal entity of satellite imagery with respect to Israel only if such imagery is no more detailed or precise than satellite imagery of Israel that is available from commercial sources."[309]

At present, limitation of spatial resolution is implemented through a two-tiered policy. Electro-optical imagery of 0.5 meters can be marketed generally, whereas imagery higher than 0.25 meters can only be disseminated with specific authorization to recipients individually authorized by the US government.[310] For the latter case, before such authorization is granted to a US firm, the State Department must first obtain assurances from the foreign government that the imagery will not be disseminated further without authorization.[311] It is worth noting the policy for implementation of *de facto* shutter control. This policy is known as 'buy to deny' because all the data for specific locations was bought up by the government to hamper the access to data by any other users for a certain period of time. The notable example is the invocation of this policy in 2001 when US Federal agencies essentially restricted access to satellite imagery over Afghanistan by acquiring exclusive rights to all IKONOS satellite imagery.[312]

Realizing that remote sensing was not yet commercially viable, the United States recognized the need to actively support its national remote sensing actors and to foster the industry. In 2003, the US government issued a Commercial Remote Sensing Policy that supersedes the earlier PDD-23.[313] The objective of the 2003 Commercial Remote Sensing Policy was to enhance the competitiveness of national actors by freeing restrictions as well as supporting the national remote sensing industry.

The 2003 Commercial Remote Sensing Policy further realizes the deregulation of the earlier PDD-23. The policy has taken away the restrictions on spatial, spectral, and radiometric resolution and encourages US companies to build

[308] Kyl-Bingaman Amendment to the 1997 National Defense Authorization Act, 15 U.S.C. 5621.

[309] Public Law 104–201 div. A, Title X, Sec1064, September 23, 1996, 110 Stat. 2653.

[310] See J.C. Kessler, Leadership in the Remote Sensing Satellite Industry, US Policy and Foreign Competition, 8.

[311] See J.C. Kessler, Leadership in the Remote Sensing Satellite Industry, US Policy and Foreign Competition, 8.

[312] See Crowsey Incorporated, *Legal Assistant Guide to Legal Applications for geospatial information*, June 10, 2002, 28.

[313] Fact sheet: US Commercial Remote Sensing Policy of April 25, 2003. [Hereafter: 2003 Commercial Remote Sensing Policy]

and operate commercial remote sensing systems, inferring the favourable review of license applications.[314]

The key feature of the 2003 Commercial Remote Sensing Policy is the enhancement of governmental reliance on US commercial remote sensing capabilities. The policy states that the US government will "rely to the maximum practical extent on US commercial remote sensing space capabilities for filling imagery and geospatial needs."[315] The policy also states that the US government should fill the gap only "if fulfilling their data needs cannot be effectively, affordably, and reliably satisfied by commercial providers."[316] This confirms the government's intention to become a major customer for data from US private remote sensing systems.

In order to implement the policy concretely, the 2003 Commercial Remote Sensing Policy calls for the establishment of the National Imagery and Mapping Agency (NIMA)[317] that is responsible for acquisition and dissemination of data for national security and foreign policy requirements to enhance the governmental reliance on US commercial remote sensing capabilities.[318] The 2003 Commercial Remote Sensing Policy is clearly aimed at encouraging operations of domestic commercial remote sensing actors through deregulation and the 'data buy' approach.

To effectively implement the 1992 Land Remote Sensing Policy Act on private actors, the United States adopted the Licensing of Private Land Remote Sensing Space Systems in 2006 (Remote Sensing Licensing Regulations).[319] This latest licensing regulation endorses detailed procedures for licensing terms and conditions, review procedures, and monitoring compliance pursuant to the provisions related to the US Remote Sensing Licensing Regulations under the Land Remote Sensing Policy Act.

In addition to the first time authorization to operate remote sensing systems, any new actions are closely monitored. The regulation requires a license holder to file an application in case of amendment to the license as well as to notify when a license holder intends to enter into an agreement with any foreign nation or foreign entity. Moreover, the regulation requires licensees to undergo

---

[314] See W. Von Kries, The US Commercial Remote Sensing Policy of April 28, 2003: Some Comments, 52–4 *German Journal of Air and Space Law* (2003), 555.

[315] Section II of the 2003 Commercial Remote Sensing Policy.

[316] Section II of the 2003 Commercial Remote Sensing Policy.

[317] As of 2004, the name of the agency was changed to the National Geospatial Intelligence Agency.

[318] See Section V on United States Government Use of Commercial Remote Sensing Capabilities.

[319] Federal Register Vol. 71, No. 79, April 25, 2006, 15 CFR Part 960, Licensing of Private Land Remote Sensing Systems; Final Rule. [Hereafter: the 2006 Remote Sensing Licensing Regulations].

an annual audit, as well as to record their data for one year and report on the data acquisition and dissemination related to Israel.

Audits are undertaken annually to review compliance with licensing conditions and changes associated with the licensee. The licensees are obligated to report at different time frames: 12 months before launch; six month prior to launch; when the system is in operation; and every three months afterwards.[320] Specifically, the regulations implement shutter control measures on a regular basis: four times a year, licensees are obliged to report on estimated resolution of all images collected and disseminated on the state of Israel.[321] The Licensing Regulations require licensees to keep all records and data from the previous 12 months, including tasking information and associated metadata.[322] Under the requirement, the government is able to ascertain the time and locations of images collected from private remote sensing systems. This close monitoring mechanism enables the systematic and regulated control on different satellites operated by different entities.

Another notable provision of the Remote Sensing Licensing Regulations establishes data policies that take satellite funding mechanism into account. The Regulations distinguish three different situations: 1) remote sensing systems that are not funded by the US government;[323] 2) systems that are partially funded by the US government;[324] and 3) systems fully or significantly funded by the US government.[325] As to the non-government- funded systems, a licensee is allowed to provide access to its unenhanced data at reasonable commercial terms and conditions, subject to the requirement of providing data to the government of any sensed state. As to the systems that are partially funded by the US government, it will be determined "whether the interest of the United States in promoting widespread availability of remote sensing data on reasonable cost terms and conditions requires that some or all of the unenhanced data from the system be made available on a non-discriminatory basis in

---

[320] See Sub Part A Section II of the 2006 Remote Sensing Licensing Regulations.

[321] Section V(2) of the 2006 Remote Sensing Licensing Regulations.

[322] Section VI of the 2006 Remote Sensing Licensing Regulations. Provision 2 states that "metadata includes date/time of collection, image number, imager used, imager corner points in latitude/longitude, inertial position, scan duration and azimuth."

[323] See §960.12(b) of the 2006 Remote Sensing Licensing Regulations that states that "the US Government has not funded and will not fund either directly or indirectly, any of the development, fabrication, launch, or operations costs of a licensed system."

[324] See §960.12(c) of the 2006 Remote Sensing Licensing Regulations that states that the "US Government has or will directly fund all or substantial part of the development, fabrication, launch, or operation costs of a licensed system."

[325] See §960.12(a) of the 2006 Remote Sensing Licensing Regulations that states that the "US Government has funded some of the development, fabrication, launch, or operation costs of a licensed system."

accordance with the Act."[326] For fully or significantly funded systems, a licensee is required to make all data available on a non-discriminatory basis.

2.3.2.2. *Canada*

Canada, like the United States, has comprehensive domestic law relating to remote sensing – specifically, an Act governing the Operation of Remote Sensing Space Systems (hereafter: the Remote Sensing Space Systems Act)[327] complemented by the Remote Sensing Space Systems Regulations[328] that implemented the detailed rules of the Act. Whilst the law was adopted in 2005, steps to formulate certain rules on satellite remote sensing date back to the late 1990s. The rationale behind the formulation of rules lies in the planned change in the operational regime of a Canadian radar satellite system, RADARSAT, which was first launched in 1995. Following the success of RADARSAT-1 – owned and operated by the Canadian government – Canada decided to increase the private sector involvement in its remote sensing missions starting with RADARSAT-2. Since RADARSAT-2 was funded by the private sector, the government of Canada was forced to consider the need for legally established controls on the collection and dissemination of data, which were not required in the case of the earlier RADARSAT-1 system.[329] The first legal instrument for establishing rules on satellite remote sensing was the conclusion of a bilateral agreement between Canada and the United Stated, the Agreement between the Government of Canada and the Government of the United States of America Concerning the Operation of Commercial Remote Sensing Satellite Systems, in 2000.[330]

The Remote Sensing Space Systems Act regulates the overall activities of remote sensing actors through licensing and continuous supervision using measures of enforcement of regulations such as inspections by the authorities, and imposition of sanctions.

The scope of the Act is broad: it applies to public, private, and public-private systems; this means that the Act not only applies to commercial satellites, but also covers satellites of the Canadian Space Agency (CSA) as well as other governmental satellites. The coverage for 'public' systems exists because commercial

---

[326] See §960.12(c) of the 2006 Remote Sensing Licensing Regulations.

[327] Remote Sensing Space Systems Act, S.C. 2005, c. 45

[328] Remote Sensing Space Systems Regulations, SOR/2007-66.

[329] See D. Clarke, Access Control of Remote Sensing Satellites, J.F. Keeley & Rob Hueber (Eds.), *Commercial Satellite Imagery and UN Peacekeeping,* Ashgate Publisher, Aldershot, 2003, 174–175.

[330] Agreement between the Government of Canada and the Government of the United States of America concerning the Operation of Commercial Remote Sensing Satellite Systems, U.S.- Can., June 16, 2000, 2000 Can. T.S. No. 2000/14.

operations under the framework of CSA and the government remain subject to the data distribution controls set by the Minister of Foreign Affairs.[331] The intended use of a satellite is irrelevant to the application of the Act;[332] hence, the Act could apply to dual use satellites. The fact that the law applies to public systems is distinctive, as most domestic legislation is intended to regulate only private and public-private systems.

The wide scope of this Act is evident in other aspects as well. A 'remote sensing satellite' is defined broadly as "a satellite that is capable of sensing the surface of the Earth through the use of electromagnetic waves,"[333] and satellites that are subject to control include satellites with optical, radar, thermal infrared, multi-spectral, and other types of sensors.[334] Application of the Act is not limited to high resolution satellites; furthermore, it applies to operations from inside and outside Canada. As to the operation outside Canada, the law is applicable to certain persons such as 1) Canadian citizens; 2) permanent residents; 3) Canadian corporations; and 4) classes of persons as maybe specified in the Regulations.[335]

Whilst the Act holds virtually every possible Canadian remote sensing system within its scope, the Minister of Foreign Affairs is authorized to exempt any persons, satellite systems, or data, on an individual or class basis from any or all aspects of the licensing regime.[336] This exemption would avoid the situation where the Canadian government unnecessarily exercises jurisdiction over operations that are primarily in the hands of other states.[337] With respect to the operation of remote sensing satellite systems by CSA or the Department of Defense, the Governor in Council may issue an order to adapt the provision of this Act or the regulations.[338] It is anticipated, therefore, that the operation of a purely governmental satellite could be excluded from the application of the Act at any time.[339] In this way, Canada takes the approach to first cover all potential remote sensing activities that may fall under Canadian jurisdiction, and can later make exemptions to the Act on a case-specific basis.

---

[331] See B.W. Mann, First Licence issued under Canada's Remote Sensing Satellite Legislation, 34 *Journal of Space Law* (2008), 73.

[332] See B.W. Mann, First Licence issued under Canada's Remote Sensing Satellite Legislation, 34 *Journal of Space Law* (2008), 72.

[333] Section 2 of the Remote Sensing Space Systems Act.

[334] See Section 2 of the Remote Sensing Space Systems Act.

[335] See Section 6 of the Remote Sensing Space Systems Act.

[336] See Section 4(3) of the Remote Sensing Space Systems Act.

[337] See B.W. Mann, First Licence issued under Canada's Remote Sensing Satellite Legislation, 34 *Journal of Space Law* (2008), 72.

[338] See Section 4(2) of the Remote Sensing Space Systems Act.

[339] See B.W. Mann, First Licence issued under Canada's Remote Sensing Satellite Legislation, 34 *Journal of Space Law* (2008), 73.

The Remote Sensing Space Systems Act establishes a licensing regime to authorize persons under the jurisdiction of Canada to carry out remote sensing activities. Licenses are granted by the Ministry of Foreign Affairs and can be amended, suspended for a period of up to 90 days, or cancelled at the discretion of the government.[340] Conditions include licensee compliance with restrictions that may be placed upon communication of raw data from the licensee to any persons other than the licensee, as well as on the provision of remote sensing products from the licensee to any persons other than the licensee.[341]

A large portion of the Act is devoted to the obligations of the licensee and rights of government in relation to managing the operation of the remote sensing system and control of subsequent data distribution. In fact, the government can issue orders in order to exercise governmental control of operations. One such order concerns the licensed person's obligation to follow measures deemed appropriate by the government upon the cancellation or suspension of a license considering the national security and system disposal plan.[342] Others concern interruption of service and priority access. For the orders described here, opportunity for making representation within a limited period of the receipt of the notice of an order is granted to licensees.[343]

The Act establishes the government's right to interrupt remote sensing services by actors subject to the jurisdiction and control of Canada. Both the Minister of Foreign Affairs (hereafter: the Minister) and Minister of Defence may order a licensee to restrict or interrupt any operation for a specified period if the Minister believes that continuation of that operation would harm the national interest of Canada.[344] The Act is silent as to the specific spatial resolution subject to the interruption of service by the government. As to the concrete measures of restriction, it is understood that a licensee is not required to stop data collection and distribution, but the distribution could be delayed, and the spatial resolution of data may be reduced as well.[345] Since such restrictive measures could affect commercial business operations, interruption of service is only invoked at a ministerial level and only in the rarest of circumstances.[346]

---

[340] See Section 10–13 of the Remote Sensing Space Systems Act.

[341] See Section 8(6) & (7) of the Remote Sensing Space Systems Act.

[342] See Section 13 of the Remote Sensing Space Systems Act.

[343] See Sections 13, 14, and 15 of the Remote Sensing Space Systems Act. Section 13 states that "the order may be effective immediately on notice to the person, but the Minister shall give to the person an opportunity – during a period of 15 days after the notice or any longer period that the Minister specifies – to make representations regarding it". Sections 14 and 15, respectively, refer to essentially equivalent provision seen in Section 13.

[344] See Section 14 of the Remote Sensing Space Systems Act.

[345] See T. Gillon, Regulating Remote Sensing Space Systems in Canada- New Legislation for A New Era, 34 *Journal of Space Law* (2008), 30.

[346] See T. Gillon, Regulating Remote Sensing Space Systems in Canada- New Legislation for A New Era, 34 *Journal of Space Law* (2008), 30.

The Canadian government retains the right of priority access to the data. Priority access to services means that the Act permits the Minister, the Minister of Defence, and the Minister of Public Safety and Emergency Preparedness to order a licensee to provide any remote sensing service to the government of Canada for the purpose of international relations and national defence interests of Canada.[347] More specifically, priority access to service is available to her Majesty, the Royal Canadian Mounted Police, and to the Canadian Intelligence Service.[348] The Act guarantees the Canadian government's right to priority access to data in case of emergency. Whilst it is anticipated that priority access will not be invoked regularly, it could be invoked more frequently than shutter control, since priority access can be invoked at the Deputy Minister level.[349]

Whilst the government may cover the cost associated with priority access, no right to compensation is granted to a licensee against the government for any financial losses resulting from actions taken in good faith by the government under this act.[350] Such actions include: amendment of a system disposal plan; amendment, suspension, or cancellation of a license; ordering licensees to take measures upon cancellation or suspension of a license; interruption of service; and priority access.[351] Hence, the government is not liable for the loss suffered by licensees unless the government intentionally caused such loss.

The Remote Sensing Space Systems Act establishes how continuing supervision by the governmental authority is exercised effectively through operations by inspectors and enforcement officers. The Act authorises inspectors designated by the Minister to enter and inspect, at any reasonable time, any place owned by or under the control of a licensee, a system participant, or any other person who may be carrying on a controlled activity in the operation of a remote sensing space system.[352] An inspector then could serve as an enforcement officer, and may issue a notice of violation in any case in which he believes there are reasonable grounds that a violation has been committed.[353]

The Act has a detailed enforcement mechanism. There are two levels of breach of the provisions of the Act: one is 'violation' and the other is 'offence'. Violation is not a criminal act and a licensee is entitled to make representation upon a notice of 'violation' within 30 days, whereas a licensee who commits an offence is liable on summary conviction with no reference to the opportunity to respond. Whilst a violation may concern the breach of any of 43 provisions of the Act and regulations,[354] an offence concerns only four provisions: 1) operation

---

[347] See Section 15 of the Remote Sensing Space Systems Act.

[348] See Section 15 of the Remote Sensing Space Systems Act.

[349] See T. Gillon, Regulating Remote Sensing Space Systems in Canada- New Legislation for A New Era, 34 *Journal of Space Law* (2008), 30.

[350] See Section 22 of the Canadian Act on Remote Sensing.

[351] See Section 22 of the Remote Sensing Space Systems Act.

[352] See Section 18 of the Remote Sensing Space Systems Act.

[353] See Section 25 of the Remote Sensing Space Systems Act.

[354] See Schedule 2 of the Remote Sensing Space Systems Regulations.

of satellite systems without a license; 2) letting a person outside Canada command the satellite; 3) not complying with the Minister's order to take measures upon suspension or cancellation of a license; and 4) not complying with the Minister's order with regards to interruption of service.[355]

Sanctions for breaches of the Act are set at two tiers. Any individual committing a violation will be fined up to $ 5,000; the fine is $ 25,000 for other cases. It should be noted that the amount is imposed per day and adds up as long as the violation continues.[356] This approach is equivalent to that taken by the US Land Remote Sensing Policy Act. By contrast, a person guilty of an offence can be fined up to $ 50,000, receive 18 month imprisonment, or both. In other cases, most notably, a juridical person committing an offence is fined up to $250,000.[357] Certain illegal acts are regarded as both violations and offences, but a person cannot be charged for the same act first as a violation and again as an offence.[358]

2.3.2.3. *Germany*

Germany, without a launch site in its territory, had no pressing need for developing domestic space legislation. This changed, however, with the development of high resolution remote sensing satellites and private sector involvement. In fact, the drafting for a law specific to remote sensing from space began during the implementation of the first large scale Public-Private Partnership (PPP) in the field of Earth observation, TerraSAR-X. Drafting for the German Act on Satellite Data Security, *Satellitendatensicherheitsgestz,* came into force on December 1, 2007.[359] The law established governmental control over collection and dissemination of data from high performance remote sensing satellites to minimize any threat to the security of Germany and the security of other nations.[360]

The German Act on Satellite Data Security applies to 1) the operation of high-grade remote sensing systems; and 2) the handling of data generated by a high-grade remote sensing system.[361] A 'high grade remote sensing system' is

---

[355] See Section 5, 16(1), 13, and 14 of the Remote Sensing Space Systems Act.

[356] Section 32 of the Remote Sensing Space Systems Act states that "a violation that is committed or continued on more than one day constitutes a separate violation for each day on which it is committed or continued."

[357] See Section 23 of the Remote Sensing Space Systems Act.

[358] Section 34(1) of states that "if it is possible to proceed with any act or omission as a violation and it is also possible to proceed with it as an offence, proceeding in one manner precludes proceeding in the other."

[359] Satellite Data Security Act Federal Gazette BGBI.2007 I. No.58, 28.11.2007.

[360] See B. Schmidt-Tedd & M. Kroymann, Current Status and Recent Developments in German Remote Sensing Law, 34-1 *Journal of Space Law* (2008), 7.

[361] See Section 1 and 2 of the German Act on Satellite Data Security.

defined as "a space-based transport or orbital system, including the ground segment, by means of which data about the Earth are generated, where its sensor is itself/sensors are themselves technically capable either alone or in combination with one or more other sensors of generating data with a particularly high information content within the meaning of Paragraph 2."[362] Information content must be determined according to 1) geometric resolution, 2) spectral coverage, 3) the number of spectral channels and the spectral resolution, 4) the radiometric resolution, and 5) the temporal resolution.[363] In addition, the information content of microwave sensors or radar sensors must also be determined according to the polarization characteristics and phase history.[364] The German Act on Satellite Data Security limits its scope to high grade remote sensing systems and appears to exclude its application to low resolution satellites. The Act applies to both the operation of high-grade earth remote sensing systems and to the handling of data generated from those systems by: a) German nationals – juridical persons or associations under German law; b) foreign juridical persons or associations of foreign nationals having their head office within the territory of Germany; and c) if the command to the system is given from the territory of the Federal Republic of Germany.[365]

It is notable that the German Act on Satellite Data Security distinguishes between remote sensing system operators and data distributors, and imposes different types of authorization, respectively. Whereas the Act requires a remote sensing system operator to obtain 'approval', it requires a data distributor to obtain a 'license.'[366] Considering that other legal systems tend to have only one type of license – usually, for a system operator – it is unique to hold two distinctive types of authorizations related to remote sensing actors. This corresponds well to the growing tendency of separation between the satellite operation phase and operations related to data exploitation and distribution.

The approved and licensed persons share common obligations such as the duty to notify the responsible authority for any changes after the authorization; the duty to provide information requested by the authority; and the duty to let the responsible authority inspect the operations and undertake examinations.[367]

---

362 Section 2(1), see also (4), of the German Act on Satellite Data Security. Paragraph 2 states that "The information content shall thereby be determined according to: 1. geometric resolution, 2. spectral coverage; 3. the number of spectral channels and the spectral resolution; 4. the radiometric resolution; and 5. the temporal resolution. The information content of microwave sensors or radar sensors shall also be determined according to 1. the polarization characteristics and 2. the phase history."

363 See Section 2(2) of the German Act on Satellite Data Security.

364 See Section 2(2) of the German Act on Satellite Data Security.

365 See Section 1(1) of the German Act on Satellite Data Security.

366 See Section 11(1) of the German Act on Satellite Data Security.

367 See Section 12 of the German Act on Satellite Data Security.

Furthermore, both remote sensing system operators and data providers have a duty to prioritize government requests in terms of generation and distribution of data.[368] These governmental rights will be restricted to rare cases of crisis, and transaction costs can be charged to the government at a market price.

Obligation for documentation is distinct for system operators and data distributors. A system operator is required to record: command sequences to control the sensor, orbital, or transport systems; time and path of the command sequences; and details of encryption processes, and they must maintain the record for at least five years.[369] For data distributors, obligations with respect to duty of documentation seem much more detailed and burdensome. All data requests need to be recorded, regardless of their sensitivity. The information data distributors need to record are, among others: requesting party (ies) and their usual places of residence; procedure for, and the result of, sensitivity checks; data generation order; reports on the processing sequences of the ground segment; metadata – including the location of observation, time of data generation, sensor operating mode, and data processing parameters; and invoices.[370] In short, a data distributor is obliged to record the precise detail of every transaction. Again, the records must be maintained for at least 5 years and need to be available for inspection.[371]

When a data distributor wishes to comply with the data request, a sensitivity check is required. The sensitivity of the data request is determined by considering the elements of the information content of data, the location observed, who purchases the data, the country of destination, and the time between data acquisition and the processing of the data request.[372] As to the information on the purchaser, the data provider must check the identity of the requesting party, as well as the names of the persons who necessarily come into contact with the data and their places of residence.[373] Some areas are excluded for almost all the customers, for instance, military operation zones.[374] This can be construed as ban on the data collection for customers rather than a ban on data dissemination.

The sensitivity check consists of a twofold process: first, to be performed by, and in the full responsibility of, the data distributor; second, to be conducted on a case-by-case basis by the Federal Office of Economics and Export Control.[375]

368 See B. Schmidt-Tedd & M. Kroymann, Current Status and Recent Developments in German Remote Sensing Law, 34-1 *Journal of Space Law* (2008), 111.

369 See Section 5 of the German Act on Satellite Data Security.

370 See Section 18 of the German Act on Satellite Data Security.

371 See Section 18 of the German Act on Satellite Data Security.

372 See Section 17 of the German Act on Satellite Data Security.

373 See Section 17 of the German Act on Satellite Data Security.

374 See B. Schmidt-Tedd & M. Kroymann, Current Status and Recent Developments in German Remote Sensing Law, 34-1 *Journal of Space Law* (2008), 113.

375 See B. Schmidt-Tedd & M. Kroymann, Current Status and Recent Developments in German Remote Sensing Law, 34-1 *Journal of Space Law* (2008),108–109.

As to the first stage, evaluation of the sensitivity of a data request is undertaken by the data distributor in accordance with a set procedure and clearly established criteria with no room for discretionary assessment.[376] If the data request is classified 'not sensitive', the data distributor is allowed to deliver the requested data without further approval from the government.

In case a data request is found to be sensitive as a result of the evaluation by the distributor, and yet the distributor wishes to proceed with the data request, the data distributor files an application for a permit to the Federal Office. The Federal Office then conducts a case by case review evaluating the risks involved in complying with the data request.[377] A permit is required to comply with sensitive data requests, and cases where data are to be distributed without data request are treated likewise.[378] The Federal Office issues a permit if the data request in question would not endanger the security of Germany.[379] In case compliance with the data request is found to be risky, a permit may be conditionally granted through measures such as reduced spatial resolution of data, time delay, reduced processing quality of the data, or the avoidance of certain target areas.[380] The flow for the evaluation of sensitivity and resulting action taken is depicted in the Figure 2.1. Although the government reserves its right to prohibit data distribution, approximately 99% of permit applications is usually granted;[381] therefore, the restriction imposed on data distribution is considerable but not at the level to seriously disrupt the commercial operation.

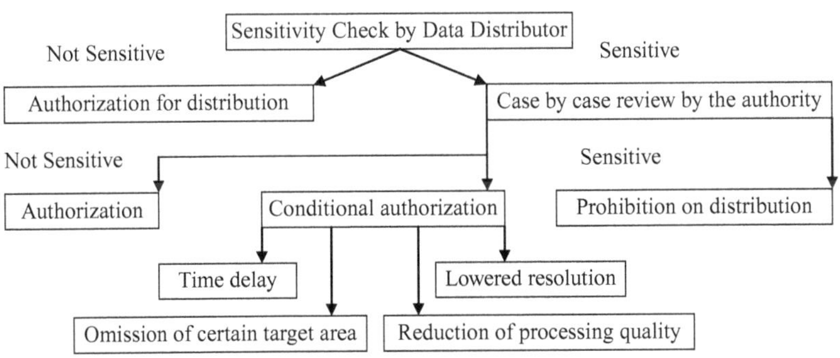

Figure 2.1: Flow of the Sensitivity Check and Resulting Actions

[376] See Section 17 of the German Act on Satellite Data Security.

[377] See Section 19 of the German Act on Satellite Data Security.

[378] See Section 19 of the German Act on Satellite Data Security.

[379] See W. Schneider, Background Information for the Act on Satellite Data Security, File no. VII B3, 2008, 11.

[380] See B. Schmidt-Tedd & M. Kroymann, Current Status and Recent Developments in German Remote Sensing Law, 34-1 *Journal of Space Law* (2008), 109.

[381] See B. Schmidt-Tedd & M. Kroymann, Current Status and Recent Developments in German Remote Sensing Law, 34-1 *Journal of Space Law* (2008), 113.

In addition, to avoid hampering the commercial transaction unnecessarily, the fundamental policy is to limit the duration of a case specific review by governmental authority to maximum one month.[382]

As is the case in Canada, the German government reserves its right of priority access to data in the event of an emergency. Data providers owe an obligation to prioritize in case of threat to German military and civilian forces in a foreign country or to foreign officers employed at German foreign embassies.[383] These rights will be restricted to rare occasions of crisis.[384]

Rules endorsed in the German Act on Satellite Data Security are enforced through sanctions. As to the form of sanctions, a three-tier penalty scheme is set depending on the type of offence. Offences can be categorized as either serious ones with heavier fines that may be replaced by sentence to imprisonment, or those that are less serious with lower fines. The highest fine of up to € 500,000 is imposed for offences such as: operating remote sensing system without permission; complying with a sensitive data request without permission; and distributing data without a permit.[385] A fine of up to € 50,000 is imposed for offences such as failing to undertake the sensitivity check and conducting the check not using the designated procedure or conducting it incompletely.[386] Any of the offences listed above in the first category may result in imprisonment up to five years instead of a fine.[387] Finally, for those offences categorized as least serious, a fine of up to € 25,000 is imposed. Such offences include failing to provide the information requested by the authority as well as failing to report any changes that occurred after the issuance of license.

2.3.2.4. *France*

Unlike many other states, France does not have a specialized law for remote sensing as such, but the French Space Operations Act addressing general spacecraft control includes Title VII (Title VII : Space-based Data) devoted to the issues of remote sensing operations and data dissemination control.[388] Remote sensing activities are addressed only by three articles – Articles 23 to 25 in Title VII of this law – and provisions are fairly general. The domestic French space law serves as an overall framework and is complemented by more detailed

[382] See B. Schmidt-Tedd & M. Kroymann, Current Status and Recent Developments in German Remote Sensing Law, 34-1 *Journal of Space Law* (2008), 109.

[383] See Section 21 of the German Act on Satellite Data Security.

[384] See B. Schmidt-Tedd & M. Kroymann, Current Status and Recent Developments in German Remote Sensing Law, 34-1 *Journal of Space Law* (2008), 112.

[385] Section 28 of the German Act on Satellite Data Security.

[386] Section 28 of the German Act on Satellite Data Security.

[387] See Section 29 of the German Act on Satellite Data Security.

[388] Title VII of the French Space Operations Act addresses the issues of space-based data.

regulations in the form of decrees. The decree of 6 June 2009 supporting the applications of measures provided by Title VII of the French Space Operations Act[389] effectuates the regulations of Title VII of French Space Operations Act, and articulates how rules are applied.

It is noteworthy that the part of French legislation addressing data control was developed separately by the Ministry of Defence, whilst the rest of the law was prepared by the Council of State,[390] signifying the critical impact data collection and dissemination may have on national security. The scope of the legislation is limited to the natural or juridical person responsible for the programming of a remote sensing satellite or responsible for the collection of space-based remote sensing data.[391] Therefore, mere data resellers and value added service providers that are not involved in the operation of the satellite or data collection are not covered under the legislation. Data from military spacecraft or data collected on behalf of the Ministry of Defence is excluded from the scope: it neither applies to activities of CNES nor to military applications.[392] Title VII applies to data from panchromatic optical sensors, multi-spectral optical sensors, stereoscopic sensors, infrared, and radar sensors.[393] A separate decree is to be adopted for further defining the data resolution, and the characteristics of precision.[394]

Remote sensing data exploitation is subject to prior declaration to the Administrative Authority.[395] It is clarified in the French Remote Sensing Decree, which states that the decree of 6 June 2009, which supports the applications of measures provided by Title VII of the French Space Operations Act, provides that the Administrative Authority is Secretary General for National Defence (SDGN).[396] A declaration needs to be made at least two months prior to the starting of operations.[397] In case the declaration is not in conformity with the regulations, or is incomplete, the authority asks the candidate to declare in a designated or complete manner.[398]

---

389 Decree 2009-640 of 9 June 2009 supporting the applications of measures provided by Title VII of French Space Operations Act. [Hereafter: French Decree on Remote Sensing]

390 See P. Achilleas, French Remote Sensing Law, 34-1 *Journal of Space Law* (2008), 6.

391 Article 1 of the French Space Operations Act.

392 Article 26 of the French Space Operations Act. It does not apply to data originated from military spacecraft or collected on behalf of the Ministry of Defence.

393 Article 1 of the French Decree on Remote Sensing.

394 Article 23 of the French Space Operations Act.

395 Article 23 of the French Space Operations Act.

396 Article 2 of the French Decree on Remote Sensing.

397 See Article 3 of the French Decree on Remote Sensing.

398 See Article 3 of the French Decree on Remote Sensing.

Governmental control is exercised by declaration instead of through a license. Particular attention should be paid to French law adopting a system of a declaration. A declaration, which is a weaker form of a license, is closer to a notification than to an authorization. A declaration is a one-sided operation and implies that such a declaration will not be rejected – unlike the case of a license, which normally requires an application that is then granted or rejected upon review by the authority. No conditions to declare are articulated under the French Space Operations Act.

Title VII of the French Space Operations Act refers to possible restrictions on remote sensing actors to safeguard the national interest. Article 25 makes it clear that the Administrative Authority shall ensure that the primary operator of space-based remote sensing data does not interfere with the fundamental interests of the nation. Concrete measures of restriction are not specified, but there are historic incidents of suspension of data for a certain period, such as when France exercised the control of data dissemination by suspending Spot Image's data dissemination to Iraq during the Gulf War. In fact, a mechanism of control by the French government on commercial distribution has been operational via GIRSPOT, an inter-ministerial working group that reports on the specific cases that trigger the restrictive measures.[399] GIRSPOT is composed of representatives of SDGN: the Ministry of Foreign Affairs, Ministry of Defence, Ministry of Space, and CNES.[400] Upon the recommendation made by GIRSPOT, the Prime Minister has the responsibility to impose limitation on Spot Image.[401] Since operations of GIRSPOT remain confidential and the legality of data control implemented in its framework was questioned, the need arose for a law to formally address the issue.[402] Incorporating rules associated with satellite remote sensing into French law on space operation will codify and implement the operational regime already in place and, at the same time, will address the issue of shutter control in a more official and transparent manner.

Types of restrictions considered to be implemented include: the immediate suspension on data distribution for a limited period of time; the obligation to delay the data distribution, and the permanent prohibition of data dissemination for certain locations.[403] It is envisaged that locations such as nuclear plants

[399] See P. Achilleas, French Remote Sensing Law, 34-1 *Journal of Space Law* (2008), 5.

[400] See P. Achilleas, French Remote Sensing Law, 34-1 *Journal of Space Law* (2008), 5.

[401] See P. Achilleas, French Remote Sensing Law, 34-1 *Journal of Space Law* (2008), 5.

[402] See P. Achilleas, French Remote Sensing Law, 34-1 *Journal of Space Law* (2008), 7.

[403] Article 2 of the French Decree on Remote Sensing.

in France, the location of French troops abroad (e.g., the Ivory Coast), or the location of allied troops abroad (e.g., US Army troops in Iraq) are most likely to be the targets for restrictions.[404]

Finally, the French Space Operations Act has a concrete enforcement mechanism imposing a € 200,000 fine against a person that distributes data without a declaration, as well as those who do not comply with data restriction regulations.[405] This is a civil penalty with no distinction between an individual, and other person (e.g., juridical person). One can conclude at this point that the French Space Operations Act is intended primarily to ensure governmental control of remote sensing activities and subsequent data distribution, and has little provision for fostering the industry.

Overall, the remote sensing laws that exist at the domestic level serve as a regulatory regime for non-governmental as well as public-private remote sensing operations. With the exception of Canada, the respective laws are not applicable to national and military operations. One can conclude that the existing national laws addressing remote sensing perform primarily two functions:

1) to provide a regulatory regime to authorize and supervise remote sensing operations, and enforce associated regulations through a licensing framework; and
2) to codify and ensure enforcement of the government's right to interrupt services, and to have priority access to data.

The laws reflect the government's intention to regulate remote sensing activities, and to exercise control to ensure that a licensed operation does not harm the government's interests. The laws, however, are not intended to safeguard the right of the potential users or that of other parties affected.

Generally speaking, national remote sensing laws are not intended to serve the purpose of promoting remote sensing activities. Only the United States takes an active approach to realize that goal. The United States has kept pace with changing policies and regulations in line with the market evolution in the field of EO, and has attempted to maintain the competitiveness of national actors by adopting measures to foster the industry. Such a tendency is found to a minimal extent in other states. Certain elements of consideration for not hampering commercial business can be seen, however, to the extent of limiting restrictive measures by the governmental authority to the benefit of licensees. Examples include limiting the exercise of shutter control to only rare occasions; adopting a review process that is not overly lengthy; and government's willingness to pay for the transaction costs associated with priority access. The next section reviews the policies associated with remote sensing and analyses the

[404] Articles 5–6 of the French Space Operations Act.
[405] Article 25 of the French Space Operations Act.

general tendencies found in the existing remote sensing laws that also apply to such policies.

Whilst not established as a law, an increasing number of states adopt certain policies with respect to data dissemination. Driven by the needs that have emerged with the launch or planned launch of high performance remote sensing satellites and consequent data dissemination, states have moved to adopt, or consider framing, policies that endorse data- specific rules worldwide.

It is of particular importance to gain an understanding of the existing remote sensing policies of various states and to analyze their feature and impact on the data access environment. In theory, certain data policies exist for all satellites of different states currently in operation. The following part will look only at the policies implemented and exercised by state governments that significantly affect the actors; the discussion will focus on a state-by-state review of the policies and practices of China, India, and Russia.

2.3.2.5. *China*

The milestone from the standpoint of China's Earth Observation policy was the launch of Beijing-1, a high resolution satellite of 4 m in 2005. Beijing-1 forms a part of the Disaster Monitoring Constellation (Hereafter: DMC) that consists of satellites from several states/entities, whereby 32 m data can be exchanged amongst the DMC partners. It is the policy of China not to provide its 4 m resolution data of Chinese territory from the Beijing-1 satellite to users outside China.[406] This practice is, in a way, *de facto* shutter control. Furthermore, the National Mapping Service has up-to-date maps available, but users who are not authorized by the Chinese government are only allowed to access up to 1: 1 million scale maps.[407] Restrictions placed on a map scale for dissemination is relevant to remote sensing, as maps are often based upon or derived from satellite remote sensing data. This may serve as an impediment to open access to the satellite-derived maps and serves as a restriction on data availability on a global basis.

2.3.2.6. *India*

India has a comprehensive EO data policy. The Remote Sensing Data Policy established in 2001,[408] fulfils a twofold purpose: 1) to establish a licensing mechanism; and 2) to establish guidelines to control the dissemination of remote sensing data.

---

[406] Communication to Martin Sweeting of SSTL on 22 December 2006.

[407] R. Harris, and R. Browning, *Data Policy Assessment for GMES Final Reports*, EVK2-CT-2002-80012-DPAG., University of London, London, 2004, 52.

[408] ISRO: EOS: POLICY-01: 2001, Remote Sensing Data Policy.

With regard to licensing, India's policy requires a license and /or permission from the ministry in charge – the Department of Space – to operate remote sensing systems as well as to acquire /distribute remote sensing data within the country.[409] The Department of Space assigns licensing operations to two entities: the National Remote Sensing Centre, and Antrix Corporation Limited, known as a commercial arm of the Department of Space. The National Remote Sensing Centre takes charge of the acquisition/distribution of remote sensing data within India[410] whilst Antrix Corporation handles granting license for acquisition/distribution of IRS data outside of India.[411]

With regard to guidelines for data dissemination, the policy is three-tier, depending on the resolution of data: 1) data lower than 5.8 m; 2) data between 5.8 m and 1 m; and 3) data higher than 1 m. The Remote Sensing Data Policy makes data lower than 5.8 m ground resolution available on a non-discriminatory basis and on an "as requested basis".[412] Data between 5.8 m and 1 m resolution "can be distributed to users after screening and ensuring that the sensitive areas are excluded."[413] Distribution of higher than 1 m differs by type of users. Whilst government users do not require further clearance, other users are subject to case-by -case review by an inter-agency High Resolution Image Clearance Committee (HRC), which makes decisions as to whether or not to grant such requests.[414] The distribution of 1 m resolution commercial imagery is kept under check through specific sales agreements containing non-disclosure clauses.[415]

The Remote Sensing Data Policy does not provide a definition of sensitive areas; however, the examples of certain sensitive areas can be illustrated using an agreement concluded by the Indian government with a foreign operator in the past. Under a July 2000 agreement between the government of India and Space Imaging – a foreign private entity – sensitive Indian installations such as military bases and airfield will be blotted out of IKONOS images before their distribution in India.[416] The conclusion of this agreement in 2000 signifies that a case-specific legal instrument was already adopted before the establishment of

---

409 A. Ghosh, Remote Sensing Data Policies. Available on the Internet at http://www.gisdevelopment.net/magazine/global/2007/july/54_1.htm. Last accessed on 17 April.

410 See Section 2 of the Remote Sensing Data Policy.

411 See Section 3 of the Remote Sensing Data Policy.

412 See Section 4 of the Remote Sensing Data Policy.

413 See Section 4 of the Remote Sensing Data Policy.

414 See Section 4 b (iii)(iv)of the Remote Sensing Data Policy.

415 See Section 4 b (v)of the Remote Sensing Data Policy.

416 K.S. Jayaraman, "India US Firm agree to Sale of 1 meter Imagery", *Space News*, 17 July 2000. Cf. Jakhu, R., International Law Governing the Acquisition and Dissemination of Satellite Imagery, *Commercial Satellite Imagery and UN Peacekeeping*, Ashgate, Aldershot, 2003, 28.

the Remote Sensing Data Policy that addresses guidelines for data dissemination in a general context. Furthermore, the National Remote Sensing Centre prohibits the dissemination of data concerning India's borders up to eighty kilometres inside the territory, including the states of Jammu and Kashmir.[417] Whilst the Remote Sensing Data Policy does not provide for sanction measures such as fines, it is clearly characterised by rigid governmental control, and is closer to a law than a policy.

2.3.2.7. *Russia*

Russia went through phases of relaxing the restrictions on the resolution of available data whilst retaining the right of data refusal in the interests of national security. It was not until 1987 that the commercial images of 5 m high resolution were authorised for dissemination, and it was 1992 when Russia authorized the dissemination of 2 m images.[418] Russia started to distribute high resolution images acquired by its KVR-1000 panchromatic camera, which was initially designed for intelligence applications. Whilst restrictions may be imposed on data from specific locations of military sensitivity, in response to the evolution in marketing of commercial images Russia lowered the restriction on resolution to 1 m in 2000[419] when the Russian Space Agency announced the commercial distribution of 1 m spatial resolution space images taken by its KOMETA satellite system. Russia currently operates Resurs DK1, and imagery from Resurs DK1 and other remote sensing satellites is marketed on a commercial basis by SOVZOND JSC, a private firm established in 1992 to exploit the data.[420]

Russia has placed restrictions on the map scale, just like China. The maps of a scale better than 1:200,000 are regarded by the state as confidential, and a combination of the Federal Security Service and Roskartografia, the National Mapping Agency of Russia, reviews and controls all cartography at map scales better than 1:200,000.[421] The KOMETA images are used to produce 1: 50,000 scale topographic maps that are currently prohibited for civil commercial use.[422]

---

[417] P. Achilleas, Data Policy and Security Issues, Communications to the 12th European Centre for Space Law (ECSL) Practitioners Forum, March 12, 2004, 4.

[418] For instance, Russian entities may be prohibited from selling the imagery of certain places under crisis such as Chechnya. See P. Achilleas, Data Policy and Security Issues, Communications to the 12th European Centre for Space Law (ECSL) Practitioners Forum, March 12, 2004, 4.

[419] P. Achilleas, Data Policy and Security Issues, Communications to the 12th European Centre for Space Law (ECSL) Practitioners Forum, March 12, 2004, 4.

[420] See A. Ghosh, Remote Sensing Data Policies, 14.

[421] See R. Harris & R. Browning, *Data Policy Assessment for GMES Final Reports*, EVK2-CT-2002-80012-DPAG., University of London, London, 2004, 52.

[422] A. Ghosh. Remote Sensing Policies. Available on the Internet at http://www.gisdevelopment.net/magazine/global/2007/july/54_1.htm. Last accessed 15 July 2010.

Again, this can serve as an impediment to access to data for public benefit purposes.

From the foregoing discussion, one can argue that 'control of the space object' if applied to remote sensing satellites (i.e., data collection and dissemination), is much more specifically defined and implemented in the national framework addressing satellite remote sensing. As far as the remote sensing policies discussed in this Chapter are concerned, the tendency found in the analysis of law is only applicable to the first component; that is, to perform the function of a regulatory framework to authorize and supervise the domestic actors, as well as to ensure the rights of the government to interrupt the remote sensing operations. With regard to the remote sensing policies exercised at a national level, their aim is primarily to safeguard the national interests and they reflect little to no element of fostering the domestic industry.

Remote sensing actors are, to a greater or lesser extent, subject to restrictions imposed by governments in the national context. As conditions are mandatory on the licensed remote sensing operators, actors are forced to adhere to the provisions under the domestic regime, despite their possible inconsistency with the international regime. Indeed, if a government decides to exercise interruption of service, limitation on access to the data from any entity that is under the jurisdiction of that government is inevitable.

National legislation and policies relevant to remote sensing that exist today in different countries largely serve as restrictions, rather than to encourage a global data use or to bridge gaps that exist in the present international regime. They only clarify what can and cannot be done by national actors within the particular jurisdictions, and do not adequately complement the international regime by covering the issues not addressed therein.

2.4. Conclusion

The current legal framework surrounding remote sensing neither promotes nor regulates remote sensing activities sufficiently. International rules do not adequately address the critical issues associated with supply and use of data, whilst national legislation largely serves as a means for a government to exercise control on the activities of actors under its jurisdiction.

The inadequacy of the current international legal regime closely reflects the highly political background for, and drafting histories of, the existing international laws governing outer space. The Outer Space Treaty was a compromise of views of the early space powers, largely motivated by the justification of their interests, such as the non-appropriation of outer space and the protection of 'spy' satellites; but the Treaty was not intended to generate practical rule-making to realize the stimulation and regulation of space activities. The UN Remote Sensing Principles codified the views of states with sensing capabilities, mainly reflecting the viewpoint of the United States, which won the negotiation with a high degree of compromise by the developing nations. It was not that concrete rules – such as stating the liabilities arising from the dissemination of data – were never considered. Quite the contrary: various proposals to formulate concrete rules were not incorporated because the states failed to reach consensus. In certain respects, the ambiguities in the current regime were inevitable or, to some extent, even generated deliberately.

With respect to existing national laws and policies, these were primarily motivated by interests of respective states, and are characterized as regulating national actors through governmental control largely with regard to national security and foreign policy interests. Whilst certain rules are intended to minimise the negative impact of restrictive governmental measures on commercial operations, they would only benefit domestic actors, and not the global communities affected by remote sensing activities at large.

The present international law regime on satellite remote sensing supports data applications for environmental protection and disaster management; however, the provisions dealing with data availability and pricing policy do not particularly promote the use of data for these applications. Furthermore, the current international legal regime fails to address important commercial and operational issues regarding use and application of data, such as data policy issues including intellectual property rights, guidelines for accuracy validation and data authentication, and liability issues. The uncertain legal environment currently pertaining to remote sensing from space restricts the widespread use and realization of the full benefits of this powerful tool.

The international legal regime is now at a turning point, as the environment surrounding remote sensing has changed since the formulation of the UN Remote Sensing Principles; this change has been particularly dramatically in the last few years. Today remote sensing has a more global presence, is used for a wide range of applications, and has further potential as a critical information tool. It is now time to regulate and promote remote sensing activities through an unambiguous legal framework. There is an urgent need for clarification of ambiguities and for studies of current practice, as well as a more comprehensive regime.

Chapter Three

Environmental Law

3.1. Introduction

The degradation of the Earth's environment is an imminent and serious problem that has been aggravated dramatically during the last few decades in every aspect. For instance, 2/3 of agricultural land has degraded, and 4% of the forest around the globe as well as 27% of coral reef has disappeared.[1] Particularly, the global warming of the Earth is an impending threat. According to the third assessment report of the International Panel on Climate Change, the global average surface temperature has risen since 1861 by 0.6 degrees centigrade and is estimated to increase further between 1.4 and 5.8 degrees over this next century.[2] The effects of global warming are detrimental to many nations, perhaps most dramatically to Tsuvalu, a Polynesian island nation, and the Maldives Islands off the coast of India that are under constant threat of sinking because of the rise in sea level.

Today there are a range of environmental treaties and a variety of legislation addressing diverse environmental issues. The degradation of the environment is outpacing the environmental protection achievable through the current environmental regime; hence, the present regime needs reinforcement. Satellite

[1] See K. Sagawa, *Global Warming and Responsibility of America*, Fujiwara Press, Tokyo, 2002, 44.

[2] See B. Muller, The Global Climate Change Regime: Taking Stock and Looking Ahead, *Yearbook of International Cooperation on Environment and Development 2002/2003*, Earthscan Publications, London, 2002, 27.

remote sensing is a powerful tool to monitor and assess the state of the Earth's environment, as well as to implement and enforce certain environmental legislation. Remote sensing has already been used for various environmental applications, such as monitoring oil spills and studying deforestation, as well as for the verification of environmental claims. Remote sensing can contribute greatly to offset the inadequacies of the environmental regime.

At present, the use of remote sensing data for environmental applications has its limitations in terms of both accessibility and reliability. Remote sensing data could be used more extensively for environmental applications if it were more easily accessible and its reliability ensured. In this Chapter, the shortcomings of the current legal framework of remote sensing described in Chapter 2 are further explored in terms of environmental applications to demonstrate the need for clear data policies to facilitate data use, as well as authenticity and validity of data for use as evidence.

The initial discussion provides a brief look at the background of environmental protection, both to follow how environmental concerns have developed historically, and to get an overview of the environmental regime in terms of how environmental issues are currently addressed, governed, implemented, and enforced.

Environmental concerns date back to the Middle Ages, although limited to a national context. National laws on the environment addressing specific pollutants or to protect particular forests or water resources were enacted in Belgium, France, the Netherlands, and the United Kingdom.[3] International environmental law emerged about a century ago. The first global convention entered into force was the 1902 Convention for the Protection of Birds Useful to Agriculture,[4] designed to protect designated species of wildlife.[5]

One could argue that the international nature of environmental law arose out of the need to protect living organisms, whose movements were inevitably trans-boundary. The characteristics of the early international environmental treaties were often regional rather than global.[6] The notion of environmental protection changed fundamentally at the end of the 1960s when, for the first time, a satellite provided a global view of the planet and environmental concerns began to be addressed in a more global manner.

[3] See A. Kiss, *International Environmental Law*, Transnational Publishers, New York, 2000, 3.

[4] Convention for Protection of birds Useful to Agriculture, 19 March 1902, 102 *BFSP* 969, 191 CTS 91.

[5] See A. Kiss, *International Environmental Law*, Transnational Publishers, New York, 2000, 55.

[6] For instance, the 1909 agreement respecting boundary waters was between the United States and Canada and the 1933 London convention relative to preservation of Fauna and Flora in their natural States applied only to largely colonial Africa. See A. Kiss, *International Environmental Law*, Transnational Publishers, New York, 2000, 56.

The evolution of international environmental law was driven by the recognition that ecological interdependence is not merely a domestic matter but has international implications that need to be addressed by international laws and regulations.[7] This body of law developed based on the notion that, although the principle of sovereignty[8] is respected in the environmental field, states are not absolved from their responsibility to maintain the environment in a good state. It was not sufficient to rely only on domestic measures to prevent environmental harm; rather, an approach by the international community as a whole would be required. Indeed, by the 1970s, the application of the principle of state sovereignty had become less frequent with the growing need for the international community to cooperate to protect the environment.[9] The United Nations Conference on the Human Environment held in Stockholm in 1972 became the starting point for the development of international environmental law as a separate branch of the body of international law.[10] The balance between the states' sovereign rights over natural resources[11] and their obligation to protect the environment has become increasingly tilted toward the obligation, particularly in terms of increased multilateral environmental legislation over the last 30 years.[12]

3.2. The Institutional Framework

Currently, there is no single governing authority responsible for the existing wide range of environmental protection activities. Indeed, such activities are executed by a number of states and numerous organizations, including regional entities and United Nations organizations, each involving different efforts and actions.[13] Most notably, the United Nations Environmental Programme

---

[7] See P. Sands, *Principles of International Environmental Law*, Cambridge University Press, Cambridge, 2003, 3.

[8] A State has exclusive jurisdiction in its territory to adopt laws and enforce them, administer the territory and judge disputes.

[9] See P. Sands, *Principles of International Environmental Law*, Cambridge University Press, Cambridge, 2003, 236.

[10] See P. Malanczuk, *Akehurst's Modern Introduction to International Law*, Routledge, 1997, 241.

[11] General Assembly Resolution. 1803 (XVII), 17 U.N. GAOR Supp. (No.17) at 15, U.N. Doc. A/5217 (1962).

[12] See K. Kaline & K. Raustiala, International Environmental Agreements and Remote Sensing Technologies, Background paper prepared for the workshop on Remote Sensing and Environmental Treaties: Building more Effective Linkages, December 4–5, New York, 2000, 5.

[13] See generally A. Kiss, *International Environmental Law*, Transnational Publishers, New York, 2000, 3, 81–129.

(UNEP) has served as a catalyst in the development of many international environmental treaties and non-binding instruments.[14]

Many other UN specialized agencies with different areas of expertise engage in activities related to the protection of the environment. For instance, the United Nations Education, Science, and Culture Organization (UNESCO) has developed treaties related to the protection of world heritage sites, biosphere reserves, and wetlands, and has played a major role in enhancing the protection of these sites around the globe. The World Meteorological Organization (WMO) has facilitated cooperation in meteorological observation and rapid exchange of such information, particularly related to climate change.[15] The International Maritime Organization (IMO), established to address marine pollution, has introduced a series of measures designated to prevent accidents.[16]

The institutional framework surrounding environmental protection is complex and scattered, with many structural and functional parallels amongst international organizations that are each engaged in similar and overlapping activities.[17] With increased numbers of organizations and programmes being conducted, a more centralized approach – such as coordination of all environmental-related activities in one jurisdiction – is desirable to reduce the duplication of efforts amongst the diverse ongoing activities.

Environmental protection is implemented at two levels: internationally and domestically. Implementation takes place internationally through the establishment of organizations such as secretariats, as well as through holding meetings of parties to a MEA (multilateral environmental agreement) such as Conferences of Parties. The various secretariats, commissions, and committees exist primarily as a result of the birth of conventions.[18] Domestically, states implement environmental protection at three levels: first, by adopting national implementation measures; second, by ensuring compliance with such national measures; and third, by fulfilling international obligations such as reporting.[19] In other

---

[14] See P. Sands, *Principles of International Environmental Law*, Cambridge University Press, Cambridge, 2003, 83.

[15] See P. Sands, *Principles of International Environmental Law*, Cambridge University Press, Cambridge, 2003, 98.

[16] See A. Kiss, *International Environmental Law*, Transnational Publishers, New York, 2000, 98.

[17] See A. Kiss, *International Environmental Law*, Transnational Publishers, New York, 2000, 82. See also E. Louka, *International Environmental Law: fairness, effectiveness, and world order*, Cambridge University Press, New York, 2006, 15.

[18] Article 8 of the World Heritage Convention establishes an intergovernmental Committee for the Protection of the Cultural and Natural Heritage of Outstanding Universal Value.

[19] See P. Sands, *Principles of International Environmental Law*, Cambridge University Press, Cambridge, 2003, 174.

words, effective implementation hinges on actions taken at the domestic level and how each state adopts measures, implements them effectively, and enforces them within its national framework. Although international organizations and other non-governmental actors can advise or apply pressure on states towards implementation, much is up to the state that actually takes concrete measures.

Enforcement requires action to take measures to ensure that international legal obligations are met or to obtain decisions from a court or other appropriate bodies in the case of violation of these obligations.[20] Domestically, environmental obligations can be enforced by the national court in case of a violation of the national rules regarding implementation. Internationally, enforcement may occur under the leadership of one or more states that have a major role in enforcing rules, or by an international organization and/or non-state actors, although both have a limited role compared to individual states.[21]

Enforcement has taken place from affected neighbouring states in case of serious transboundary harm;[22] however, there are many examples of states' reluctance to enforce their obligation towards the environment – such as the Soviet Union in the incident at Chernobyl in 1986.[23] International organizations play an important role in the development of environmental law,[24] but their enforcement function is limited.[25] This is illustrated, for instance, in the proposals from certain governments that suggested UNESCO should be competent to enter into sovereign states to monitor their compliance with the World Heritage Convention;[26] the proposals did not gain sufficient support and were eventually withdrawn.[27] With respect to non-state actors, their role remains largely at the national level, although they have been involved in the process of

---

[20] See P. Sands, *Principles of International Environmental Law*, Cambridge University Press, Cambridge, 2003, 182.

[21] See P. Sands, *Principles of International Environmental Law*, Cambridge University Press, Cambridge, 2003, 182.

[22] For instance, the Trail Smelter Arbitration (United States v. Canada) 3 RIAA (1941) 1905. American farmers suffered damage of sulphur dioxide as a result of smelter operation in Canada. See A. Kiss, *International Environmental Law*, Transnational Publishers, New York, 2000, 274–277.

[23] See P. Sands, *Principles of International Environmental Law*, Cambridge University Press, Cambridge, 2003, 191.

[24] For instance, UNESCO adopted the World Heritage Convention and IMO adopted International Convention for the Prevention of Pollution from Ships.

[25] See P. Sands, *Principles of International Environmental Law*, Cambridge University Press, Cambridge, 2003, 191.

[26] Convention Concerning the Protection of World Cultural and Natural Heritage 1972, 1037 *UNTS* 151. [Hereafter: World Heritage Convention]

[27] See J.C. Kunich, World Heritage in danger in the hotspots, 78-2 *Indiana Law Journal* (2003), 651.

international enforcement.[28] One can argue that environmental protection is primarily state-based in nature: implementation is generally left to the goodwill and capacity of each state, and enforcement takes place either within the state or by affected neighbouring states.

Implementation and enforcement are also affected by the situation of states, regardless of their willingness to meet their environmental obligations. Indeed, states vary from one to another[29] in terms of capabilities and priorities, and states go through changes in terms of political, economic, and social stability. For example, in May of 1995, the Russian Federation, Belarus, Bulgaria, Poland, and Ukraine expressed concerns over their inability to be in full compliance with their obligations under the 1987 Montreal Protocol for the year 1996, because of their severe economic circumstances.[30] It is difficult to expect a high degree of commitment and spontaneous implementation from all states at all times without enforcement; therefore, implementation and enforcement efforts also need to come from outside the states to make the environmental regime more effective. Ideally, the regime should function to influence the behaviour of states to the benefit of the environment and highlight the commitments of states so that any environmental problems addressed in particular are improved.

3.3. Legal Norms

Dailler and Pellet contend that international environmental law is composed of the international legal rules necessary for the protection of the global ecosystem.[31] What constitutes 'international environmental law' is not established as such, but is a reference to a broad body of law across the different disciplines dealing with the environment. This clearly indicates the interdisciplinary nature of international environmental law and its dependence on various social, economic, scientific, and political factors.

The sources of environmental law are international agreements, both multilateral and bilateral; other international acts;[32] international customary law;

---

[28] See P. Sands, *Principles of International Environmental Law*, Cambridge University Press, Cambridge, 2003, 195.

[29] Some are democratic, autocratic, having strong environmental ministers. See K. Kaline & K. Raustiala, International Environmental Agreements and Remote Sensing Technologies, Background paper prepared for the workshop on Remote Sensing and Environmental Treaties: Building more Effective Linkages, December 4–5, New York, 2000, 13.

[30] See A. Kiss, *International Environmental Law*, Transnational Publishers, New York, 2000, 595.

[31] P. Dailler & A. Pellet, *Droit International Public*, 7th edition, LGDJ, Paris, 2002, 1271.

[32] Other international acts include those adopted by international organizations and by States in the form of non-binding declarations or action plans.

general principles of international law; and other subsidiary sources.[33] Various sources of law endorse rules surrounding environmental protection differently. Some establish general codes of conduct whilst others give rise to specific obligations.

The most fundamental principles of environmental law are endorsed in gentlemen's agreements such as the 1972 Stockholm Declaration on the Human Environment[34] and the 1992 Rio Declaration on the Environment and Development.[35] These declarations spell out central concepts and legal principles regarding environmental protection. They include the principle of sustainable development;[36] the obligation not to cause transboundary harm;[37] the obligation to take preventive and precautionary measures;[38] international cooperation in protecting the environment;[39] the right to environmental information;[40] and compensation to be made by the polluter for environmental damage.[41]

International customary law derives from a few cases including the Trail Smelter Case,[42] which established the obligation not to cause transboundary harm; the importance of an assessment of the consequence of an activity causing significant risks;[43] and the cost of environmental harm is to be borne by the polluter.[44]

Other subsidiary sources include case law related to the environment, such as the case Gabcikovo-Nagimaros concerning the dispute between Hungary and Slovakia that established the obligation to cooperate in good faith and the need to reconcile the economic development with protection of the environment.[45]

---

[33] Decisions of courts and tribunals and the writing of jurists. See P. Sands, *Principles of International Environmental Law*, Cambridge University Press, Cambridge, 2003, 153.

[34] Stockholm Declaration of the United Nations on the Human Environment, UN Document A/Conf/48/14(1972).[Hereafter: Stockholm Declaration]

[35] Rio Declaration on Environment and Development, 31 *ILM* 874(1992). [Hereafter: Rio Declaration]

[36] Principle 1 and 4 of the Rio Declaration.

[37] Principle 21 of the Stockholm Declaration, Principle 2 of Rio Declaration.

[38] Principle 15 of the Rio Declaration.

[39] Principle 24 of the Stockholm Declaration.

[40] Principle 1 of the Stockholm Declaration

[41] Principle 16 of the Rio Declaration.

[42] Trail Smelter Arbitration (United States v. Canada) 16 April 1938 3 RIAA 190781941.

[43] See Commentaries on the Draft Articles on prevention of Transboundary harm from hazardous activities, ILC 2001.

[44] In the Trail Smelter case, the Canadian government was obliged to pay $350,000 to settle damages claims before January 1, 1932. From the date to October 1937, the tribunal awarded $78,000 for damage to cleared and uncleared land, See A. Kiss, *International Environmental Law*, Transnational Publishers, New York, 2000, 274–275.

[45] *Gabcikovo-Nagimaros Project* case (Hungary/Slovakia), ICJ Reports, 1997.

The ways principles are affirmed in case law are more specific than in the declarations.

The environmental agreements consist of: 1) a number of multilateral environmental legislations that document agreement amongst more than two states, and that address different areas of the environment including land, atmosphere, and ocean; and 2) bilateral agreements that create specific obligations between two states to cooperate, such as the agreement between the United States and Canada on the water quality of the Great Lakes. The duty to notify is incorporated into several bilateral agreements concerning environmental disasters along shared boarders, and in numerous agreements on nuclear accidents.[46]

Overall, declarations by environmental conferences, as well as case law, create norms for states with respect to the environment, whereas sets of environmental legislation create more specific commitments. Bilateral agreements and multilateral agreements create specific obligations, which induce better performance of parties and thus facilitate better protection of the environment. The contribution of bilateral agreements to overall environmental protection appears to be limited compared to that of multilateral environmental agreements, because they are designed for agreement between two states on specific environmental matters in a relatively narrow context. On the other hand, multilateral environmental agreements are binding upon multiple state parties, address diverse issues, and have greater potential for influencing the behaviours of states to the benefit of the environment, thereby leading to enhanced levels of protection.

3.3.1. *General Principles of Environmental Law*

The content of principles endorsed by gentlemen's agreements constitutes certain general principles of environmental law, including sustainable development, the obligation not to cause transboundary harm, common but differentiated responsibility, the prevention and precautionary principle, the right to environmental information, and the 'polluter pays' principle.

The concept of sustainable development lies in the coexistence of environmental protection and economic development. The term 'sustainable development' first emerged in the Report of the World Commission on Environment and Development, which defined it as "development that meets the needs of the present without compromising the ability of future generations to meet their own needs."[47] Economic development is inevitably accompanied by a certain

---

[46] See A. Kiss, *International Environmental Law*, Transnational Publishers, New York, 2000, 285.

[47] See A. Kiss, *International Environmental Law*, Transnational Publishers, New York, 2000, 248. See also P. Sands, *Principles of International Environmental Law*, Cambridge University Press, Cambridge, 2003, 252.

degree of adverse effect upon the environment.[48] It is thus critical to balance the development of the nation whilst minimizing the adverse effect on the environment.

Whilst the international community respects states' sovereignty, particularly with regard to the management of natural resources within sovereign territories, states have the responsibility not to cause environmental harm outside their territory. Commonly, treaty obligations concerning environmental protection limit states parties' sovereignty as they must be executed on territories of state parties, for instance, to protect species of wild fauna and flora, and to prevent air pollution.[49]

Common but differentiated responsibility is a twofold concept of recognition for common responsibility of states and states' varying degrees of contribution to the creation of a particular environmental problem, as well as their ability to prevent, reduce, and control such a threat.[50] This concept places more responsibility to protect the environment upon developed countries. Although responsibility is common to all states, higher standards of conduct are explicitly set for developed states on the grounds that they have often contributed most to causing problems[51] and are better equipped to deal with the consequences.

The prevention principle is intended to prevent damage to the environment or to reduce, limit, or control activities that may have an adverse impact upon the environment. This principle is based on the fact that it is frequently impossible to recover from certain environmental harms such as the extinction of a species of fauna or flora, erosion, and the dumping of long-life pollutants into the sea.[52]

The primary obligation that derives from the principle of prevention is the prior assessment of those activities that are potentially harmful to the environment. Prior assessment prevents environmental harm by identifying precautionary measures, even if the potential hazard to the environment is not proven. In order to prevent harm beforehand, it is necessary to assess the impact of the activities. The environmental impact assessment is a concrete measure of the prevention and precautionary principle and is endorsed in different legislations. The principle is also implemented by operational programmes.

---

[48] Urbanization reduces the nature and use of fossil carbon, as combustion fuels are inevitable in various economic sectors from transport, domestic heating, industrial production and electricity generation.

[49] See A. Kiss, *International Environmental Law*, Transnational Publishers, New York, 2000, 259.

[50] See P. Sands, *Principles of International Environmental Law*, Cambridge University Press, Cambridge, 2003, 286.

[51] See P.W. Birnie & A.E. Boyle, International Law and Environment, Oxford University Press, New York, 2002, 101.

[52] See A. Kiss, *International Environmental Law*, Transnational Publishers, New York, 2000, 263.

The principle of cooperation is the foundation of general international law and it is critical to environmental protection. International cooperation is required to conserve the environment in its totality, both within states' territorial jurisdiction and for areas outside state territory.[53] International cooperation in environmental protection is put forward through the creation of organizations, concrete programmes, and numerous agreements.

A right to information, in the narrower sense, can mean the freedom to seek information; more broadly, it can mean the right to access information as well as to receive it.[54] The duty to provide and the right to obtain access to information on the environment – whether for the public at large or for specific categories of persons – is a recent but firmly entrenched development in international law.[55] The obligation to provide information means that states may be held responsible if they fail to provide such information. The duty to provide information regarding the environment is not applied to all cases, but limited to certain hazardous activities such as operations that are nuclear-related, as well as to cases of shared interests such as shared air space and shared rivers.

The 'polluter pays' principle is the most specific amongst these legal principles and is a concrete measure to ensure environmental protection. This principle is essentially an economic policy for allocating the costs of pollution or environmental damage borne by public authorities.[56] The principle creates incentives for the states or other entities to prevent environmental harm. The most concrete implementation of the 'polluter pays' principle is compensation made for environmental damage following environmental disputes. Past cases clearly demonstrate that polluters have been held responsible and made to compensate for the environmental harm they had caused.

Whilst multilateral environmental agreements (MEAs) concretely implement the general principles of environmental law, the effectiveness of such agreements varies depending upon their specific content. The strength and power of these agreements as legal instruments hinges significantly on the presence of specific obligations toward better protection and the inclusion of sanctions in case of violation of such obligations, as well as concrete procedures and mechanism for meeting such obligations. The characteristics of MEAs and the provisions of different sets of legislation with respect to commitments and sanctions are discussed in more detail below.

---

[53] See A. Kiss, *International Environmental Law*, Transnational Publishers, New York, 2000, 259.

[54] See A. Kiss, *International Environmental Law*, Transnational Publishers, New York, 2000, 147.

[55] See P. Sands, *Principles of International Environmental Law*, Cambridge University Press, Cambridge, 2003, 852.

[56] See P.W. Birnie & A.E. Boyle, International Law and Environment, Oxford University Press, New York, 2002, 88.

3.3.2. *The Examination of Multilateral Environmental Agreements*

To date, 242 MEAs exist.[57] Although each one is different, there are certain shared characteristics and tendencies. This section provides an overall picture of MEAs, analyses their general trends, and continues with more specific analysis of the characteristics of a MEA, followed by examples of MEAs with different levels of effectiveness as a legal instrument.

In order to analyze the MEA as a legal instrument, one needs first to understand the structure of environmental legislation. Some instruments have a twofold structure for addressing a particular environmental problem, calling for action first and setting up concrete commitments later. In contrast, other conventions establish a very general set of commitments first, which then create a framework for the negotiation of more specialized agreements known as protocols.[58] Examples of environmental agreements which have the twofold structure described above include the 1992 Framework Convention on Climate Change and the Kyoto Protocol, which will be examined in depth later in this Chapter.

The overall tendency of MEAs – amongst numerous conventions – is that they tend to impose concrete obligations on member states through certain environmental legislation, particularly which addresses the shared environment or that deals with hazardous activities such as nuclear-related operations or operations with high risk of transboundary harm.

Hazardous operations and operations with high risk of transboundary harm often give rise to concrete obligations such as a duty to notify. For instance, the duty to notify and provide information is stated in Article 2 of the 1986 Vienna Convention on Early Notification in the case of Nuclear Accident or Radiological Emergency.[59] A duty to inform is also included in Article 11 of the Bonne

---

[57] http://www.unep.org/Law/Law_instruments/. Last accessed on 25 April 2010.

[58] See K. Kaline & K. Raustiala, International Environmental Agreements and Remote Sensing Technologies, Background paper prepared for the workshop on Remote Sensing and Environmental Treaties: Building more Effective Linkages, December 4–5, New York, 2000, 8.

[59] 1986 Vienna Convention on Early Notification in the case of Nuclear Accident or Radiological Emergency, IAEA INRIRC 335, 25 IM 1370(1986). Article 2 stipulates that "In the event of an accident specified in article 1, the State Party referred to in that article shall:
(a) forthwith notify, directly or through the International Atomic Energy Agency, those States which are or may be physically affected as specified in article 1 and the Agency of the nuclear accident, its nature, the time of its occurrence and its exact location where appropriate; and
(b) promptly provide the States referred to in sub-paragraph (a) directly or through the Agency, and the Agency with such available information relevant to minimizing the radiological consequences in those States, as specified in Article 5.

Convention for the Protection of the Rhine against Chemical Pollution.[60] A duty to exchange information and a duty to notify are stipulated in Articles 13 and 14, respectively, of the 1992 Convention on the Protection and Use of Transboundary Water Courses and International Lakes,[61] Finally, Article 1 of the Convention for the Control of Maritime Pollution from Vessels gives rise to an obligation not to discharge harmful substances into the sea; this Convention will be discussed in depth later in this Chapter.

Many MEAs are outside the categories of operations with high risk of transboundary harm and hazardous operations; thus the obligations they impose are not sufficiently effective for the better protection of the environment. Particularly, obligations endorsed in the legislation dealing with a state's natural resource within its sovereign territory, such as protection of fauna and flora, appear to be weak.

The Ramsar Convention on Wetlands,[62] for instance, merely obliges states to list wetlands of importance within their territory and to use them wisely, but it does not commit states to regulate or protect wetlands in any specific manner.[63] Such an obligation helps to clarify the current situation with respect to the problem addressed, but does not take steps toward solving that problem. Likewise, the Biodiversity Convention[64] endorses general obligations that are vague,[65] requiring parties to identify components of biological diversity important for conservation, monitor the components, and evaluate the underlying causes of the diversity loss.[66] Again, a state achieves a more specific understanding

[60] Bonn Convention for the Protection of the Rhine against Chemical Pollution, 3 December 1976, 16 *ILM* 242 (1977). When a government party to this Convention notes a sudden and sizeable increase in Annex I and II substances in the waters of the Rhine or has knowledge of an accident that may result in seriously endangering the quality of those waters, it will inform without delay the International Commission and the Contracting Parties likely to be affected according to a procedure to be established by the International Commission.

[61] Convention on the Protection and Use of Transboundary Water Courses and International Lakes, Mar. 17, 1992, 31 *ILM* 1312. "The Riparian States shall without delay inform each about any critical situation that may have transboundary impacts".

[62] Ramsar Convention on Wetlands of International Importance especially as Waterfowl Habitat, 2 February 1971, 996 *UNTS* 245. [Hereafter: Ramsar Convention]

[63] Article 2-1 of the Ramsar Convention states: "Each Contracting Party shall designate suitable wetlands within its territory for inclusion in a List of Wetlands of International Importance".

[64] Convention on Biological Diversity, 5 June 1992, 31 *ILM* 822 (1992). [Hereafter: Biodiversity Convention].

[65] For instance, Article 7 of the Biodiversity Convention states: "Each contracting party shall as far as possible and as appropriate (...) identify components of biological diversity important for its conservation and sustainable use".

[66] Article 7 of the Biodiversity Convention.

of the existing problems through compliance with this Convention, but has no obligation to actually enhance the conservation of biosphere reserves. Compliance with these types of conventions is relatively easy to achieve; however, the compliance does not actually bring much improvement to the state of the environment.

MEAs can be roughly categorized into three types in terms of strengths as a legal instrument: ones that lack concrete obligations to realize the objective of the convention; ones with concrete obligations created without the specific mechanisms for implementation and enforcement; and ones with concrete obligations accompanied by specific mechanisms of implementation and enforcement.

For a detailed analysis, this section focuses on three specific instruments addressing different environmental issues from each category above: the World Heritage Convention from the first category; the Convention for the Control of Maritime Pollution from Vessels from the second category; and the Kyoto Protocol introduced by the Framework Convention on Climate Change from the third category. The analysis is devoted to an examination of the legislation as a legal instrument and the identification of components that can be consolidated to attain the objective of the convention.

3.3.2.1. *The World Heritage Convention*

In 1972, the work suggested by several member states of the United Nations Educational, Scientific, and Cultural Organization (UNESCO) was consolidated and entered into force through the adoption by UNESCO of the Convention Concerning the Protection of the World Cultural and Natural Heritage.[67] This Convention aims to safeguard sites which are considered to have outstanding universal values; these sites are then inscribed on the World Heritage List. World heritage means both cultural heritage, such as monuments, and natural heritage sites, such as national parks and tropical rain forests. As of March 2010, there are 890 sites on the World Heritage List comprising 689 cultural, 176 natural, and 25 mixed sites in 148 states.[68]

The World Heritage Convention recognises the duty of States Parties to ensure the identification, protection, conservation, presentation, and transmission of cultural and natural heritage sites to future generations.[69] It calls States Parties to take comprehensive measures in preserving the world heritage sites.[70]

---

[67] In general see C. Redgwell, The World Heritage Convention and Other Conventions Relating to the Protection of the Natural Heritage, *The World Heritage Convention: a commentary*, Oxford University Press, Oxford, 2008, 377–396.

[68] See http://whc.unesco.org/en/list. Last accessed on 20 April 2010.

[69] Article 4 of the World Heritage Convention.

[70] Article 5 of the World Heritage Convention states: "Each country shall endeavour to take the appropriate legal, scientific, technical, administrative and financial

The overall goal of the Convention is to establish a system of international cooperation and assistance designed to support States Parties to the Convention in efforts to identify and conserve the sites.[71]

The Convention is inadequate to a certain extent. It does not contain concrete provisions as to how a system of international cooperation and assistance should be established and/or what sort of measures should be taken by States Parties in efforts to identify and conserve sites. The reality is that each state party has the responsibility to identify and conserve its sites. For instance, the application for a site to be inscribed on the World Heritage List must come from the country where the site is located. No site may be placed on the list without the consent of the nation concerned. This underlines that the sovereignty of each state is respected to such an extent that identification and conservation hinges on the goodwill of the individual state party. Rather than creating a concrete mechanism for the preservation of sites by the international community, the World Heritage Convention serves as a general framework and leaves the rest to each state for taking specific measures for conservation.

Since identification and preservation of sites is left to individual nations, the extent of preservation varies greatly. Some countries have better measures and/or efforts than others. Amongst all, developing countries often do not have adequate measures and/or financial resources to identify sites properly and to ensure the preservation of inscribed sites. Indeed, the fact that developing countries tend to have fewer world heritage sites is reflected in impartial geographical distribution of world heritage sites: they are more apt to be concentrated in developed nations, particularly in Europe.[72] For example, the nation with highest number of inscribed sites is Italy.[73] The implications of this fact are that although a number of sites with recognized outstanding universal value vary by states, inscription is largely affected by the abilities and capacities of states to nominate and identify such sites.

The World Heritage Convention does not have provisions for sanctions. It is silent as to consequences if the parties do not respect their obligation under the Convention. Indeed, for certain sites, the enforcement of the World Heritage Convention has not been adequate.[74]

---

measures necessary for the identification, protection, conservation, presentation and rehabilitation of this heritage."

[71] Article 7 of the World Heritage Convention.

[72] Amongst nations with high number of sites are Italy, Spain, Germany, France, and the United Kingdom.

[73] Italy has 45 inscribed sites.

[74] Armed conflict and war, earthquakes and other natural disasters, pollution, poaching, uncontrolled urbanization, and unchecked tourist development pose major problems to World Heritage sites.

The World Heritage Committee has recognized this fact by inscribing 30 World Heritage sites on the List of World Heritage in Danger, "a list of the property appearing in the World Heritage List for conservation for which major operations are necessary and for which assistance has been requested."[75] This means that at least 30 World Heritage sites have been identified as losing the main values for which they were inscribed; there may be more unreported sites in danger. With respect to sites in danger, the Committee or UNESCO can list them without the consent of the host state, but cannot take action without a request coming from that state. If state consent is forthcoming, then the Committee can take immediate action to address the situation, and this has led to many successful restorations and international safeguarding campaigns.[76] Amongst others, successful restoration cases include the Giza Pyramids in Egypt,[77] and Angkor Wat in Cambodia.[78]

Both site identification and ensuring preservation could be improved if a functional mechanism for implementation and enforcement of the World Heritage Convention is established and strengthened. Effective implementation and enforcement of the World Heritage Convention can be achieved, for instance, by enhancing the role of UNESCO and/or the World Heritage Committee[79] as well as through participation by entities in operational programmes. In this way, although the World Heritage Convention itself is not a very strong legal instrument, it is consolidated by projects particularly associated with site restoration programmes and international campaigns.

---

[75] Article 11. 4 of the World Heritage Convention.

[76] Over the years, 26 international safeguarding campaigns were organized, costing altogether close to US$1 billion.

[77] These pyramids were threatened in 1995 by a highway project near Cairo which would have seriously damaged the value of this archaeological site. Negotiations with the Egyptian Government resulted in a number of alternative solutions that replaced the disputed project.

[78] One of the most important archaeological sites in South-East Asia, Angkor Archaeological Park contains the magnificent remains of the different capitals of the Khmer Empire, from the 9th to the 15th century. In 1993, UNESCO embarked upon an ambitious plan to safeguard and develop the historical site carried out by the Division of Cultural Heritage in close cooperation with the World Heritage Centre. Illicit excavation, pillaging of archaeological sites, and land mines were the main problems. The World Heritage Committee, having noted that these threats to the site no longer existed and that the numerous conservation and restoration activities coordinated by UNESCO were successful, removed the site from the List of World Heritage in Danger in 2004.

[79] The Committee defines criteria for inclusion to the World Heritage List, coordinates and encourages the studies and research needed for making the list (Art. 11.7), receives and studies requests for international assistance (Art. 13), and decides on action to be taken on sites for which international assistance is requested.

Implementation of such projects can well be strengthened by the use of satellite remote sensing.

3.3.2.2. *The International Convention for the Prevention of Pollution from Ships*
The 1973 International Convention for the Prevention of Pollution from Ships[80] (MARPOL) is designed for the prevention of pollution caused by the discharge of oil and other harmful substances from ships.[81] The growth in maritime transport and an increasing concern for the marine environment led to the creation of the International Maritime Organization (IMO), the first international body devoted exclusively to maritime safety and environmental matters.[82] The IMO developed the 1973 MARPOL and the subsequent 1978 Protocol to MARPOL. The objective of the Convention is to completely eliminate intentional pollution and minimize accidental discharge. The following discussion illustrates how efforts toward reduction are addressed under MARPOL.

The scope of the Convention is quite broad, dealing with all forms of marine pollution from ships except the disposal of land-generated waste into the sea by dumping, and that associated with sea-bed mineral resources operations.[83] MARPOL imposes obligations to prevent ships from polluting the marine environment by discharging harmful substances,[84] and provides for sanction in the case of violation.[85]

The obligation imposed under the Convention is the most precise and comprehensive of any international environmental agreement.[86] MARPOL established that: 1) specific obligation is imposed on coastal states to exercise enforcement over the vessels that pollute the seas in their territory; and 2) evidence is necessary to initiate proceedings.

Under MARPOL, implementation is specific because it delegates surveillance power to the state parties: port and coastal states have the responsibility to

---

[80] International Convention for the Prevention of Pollution from Ships, November 1973, 1313 *UNTS* 3. [Hereafter: MARPOL]

[81] See further O. Özacayit, MARPOL and its Amendments, 10–5 *The Journal of International Maritime Law* (2004), 444–453, R. Becker, 73/78 : an overview in international environmmental enforcement, 10-2 *The Georgetown International Environmental Law Review*, 625–642.

[82] MARPOL consists of a principal convention, three protocols and five annexes, each accompanied by nine appendices. Since 1959, the IMO has promoted the adoption of over thirty conventions and protocols, and adopted more than 700 codes and recommendations concerning maritime safety, the prevention of pollution, and other issues of environment protection.

[83] See Article 3 of MARPOL.

[84] See Article 1.1 of MARPOL.

[85] See Article 4 of MARPOL.

[86] See P. Sands, *Principles of International Environmental Law*, Cambridge University Press, Cambridge, 2003, 440.

detect violations and the right to inspect vessels in their ports and offshore terminals.[87]

It is coastal states that exercise enforcement over the vessels that pollute the seas in their territory,[88] and not the IMO. The Convention gives the IMO authority to receive information from governments on proceedings taken regarding any violation of the Convention by ships, and reports on the polluting incidents and the results of investigations;[89] but implementation and enforcement is the role of the coastal states.

Article 4 of MARPOL provides that evidence is required to initiate proceedings against any violation of the Convention;[90] however, it is silent as to detail. It does not specify what constitutes admissible evidence or define accepted methods of collecting evidence, or list the requirements for admittance of evidence into the proceedings. Conventionally, a method of proof to identify the polluter is to take a sample of the oil spill and match it with the oil in the fuel tanks of the alleged offending vessel.[91] The farther the location of an incident from the coast, the more difficult it is to detect a violation and inspect the ship using conventional means.

MARPOL contributed to a dramatic reduction in oil pollution. It is estimated that operational discharges of oil dropped 85% between 1973 and 1990.[92] One can conclude that MARPOL is an effective Convention, and can be strengthened if concrete mechanisms for implementation and enforcement are established in a systematic way. Such mechanisms would include enhancing the role of the IMO in implementation and enforcement, as well as strengthening the process of implementation and enforcement through making the most of available tools for regular surveillance and methods for obtaining evidence. In this way, violation can be detected more easily and

---

[87] See Article 6 of MARPOL.

[88] Article 4.2 of MARPOL states that "Any violation of the requirement of the present convention within the jurisdiction of any party to the Convention shall be prohibited and sanction shall be established therefore under the law of that party. Whenever any violation occurs, the Party shall either: a) cause proceedings to be taken in accordance with its law; or b) furnish to the Administration of the ship such information as evidence as may be in its possession that a violation has occurred".

[89] See A. Kiss, *International Environmental Law*, Transnational Publishers, New York, 2000, 38.

[90] Article 4.1–4.3 of MARPOL. If the Administration is informed of such a violation and is satisfied that sufficient evidence is available to enable proceedings to be brought in respect of the alleged violation, it shall cause such proceedings to be taken as soon as possible.

[91] See P. De Selding, Courts Limit Use of Satellite Imagery for Prosecuting Suspected Polluters, *Space News*, 2 March 2005.

[92] See S. Hampton, P.R. Kelly. & R. Carter, Tank Vessel Operations, Seabirds, and Chronic Oil Pollution in California, 31 *Marine Ornithology* (2003), 30.

proceedings are facilitated that will lead to a greater reduction of overall discharges.

3.3.2.3. *The Framework Convention on Climate Change*

The Framework Convention on Climate Change (FCCC)[93] addresses global warming.[94] The objective of the FCCC is to stabilize the concentrations of all greenhouse gases (GHG) in the atmosphere at a level that would prevent dangerous anthropogenic interference with the climate system.[95] Such interference can be caused by land use change such as agricultural practice and deforestation, as well as human energy use.[96] The biggest anthropogenic cause lies in human induced activities: use of fossil carbons as combustion fuels in transport, domestic heating, industrial production, and electricity generation.[97] The principle concept of the FCCC lies in the leadership and expected efforts of developed nations, which contribute extensively to greenhouse gas emission, in preventing the adverse effect to climate systems.

The Convention was intended to achieve "stabilization of greenhouse gas concentrations in the atmosphere at a level that would prevent dangerous anthropogenic interference with the climate system."[98] The nature of the Convention remained general; although it recognized the necessity of returning to earlier levels of anthropogenic emissions of carbon dioxide and other greenhouse gases, it took a voluntary approach without legally binding obligations,[99] and set no specific schedules and targets for limiting such emissions.[100] The Convention created a general framework, but not a functioning regime to establish the commitments of state parties.

[93] United Nations Framework Convention on Climate Change, *ILM* 31(1992) 849. [Hereafter: FCCC]

[94] See further R.R. Churchill & D.A.C. Freestone, *International law and global climate change*, Graham & Trotman, London, 1991. See also F. Yamin and J. Depledge, *International Climate Change regime: a guide to rules, institutions and procedures*, Cambridge University Press, Cambridge, 2004.

[95] See A. Kiss, *International Environmental Law*, Transnational Publishers, New York, 2000, 514.

[96] See K. Sagawa, *Global Warming and Responsibility of America*, Fujiwara Press, Tokyo, 2002, 27. The IPCC notes that "approximately three-quarters of the anthropogenic emissions of carbon dioxide into the atmosphere during the past 20 years was due to the burning of fossil fuels, and the rest was largely due to land-use change, particularly deforestation".

[97] See K. Sagawa, *Global Warming and Responsibility of America*, Fujiwara Press, Tokyo, 2002, 33.

[98] Article 2 of the FCCC.

[99] See K. Sagawa, *Global Warming and Responsibility of America*, Fujiwara Press, Tokyo, 2002, 33.

[100] See A. Kiss, *International Environmental Law*, Transnational Publishers, New York, 2000, 514.

The Kyoto Protocol, which came into force in 2005, strengthens parties' obligations under the Framework Convention on Climate Change by imposing legally binding commitments upon developed nations to reduce six greenhouse gases during 2008–12, using the emission level of the reference year of 1990.[101] The obligation is only placed on developed nations listed in Annex A, and the emission reduction target imposed varies by each party.[102] The Kyoto Protocol obliges parties to control the changing amounts of greenhouse gases emissions and sequestration at a national level and commits them to a reduction by providing a mechanism for offsetting greenhouse gases emissions. The Protocol imposes an obligation to report on national land use change and forestry activities,[103] and provides concrete implementation mechanisms such as the Joint Implementation,[104] the Clean Development Mechanism,[105] and Emission Trading[106] to facilitate parties in meeting the target as well as giving opportunities for non-Parties to participate.

Establishing the baseline and reporting the changes in the level of emissions measured compared to that of 1990 are the basis of the commitment by the parties.[107] Changes are measured based on the fact that carbon is stored globally in different living systems; for example, there are systems that store more carbon than they emit which are known as carbon sinks.[108] The total emission is calculated by adding or reducing sinks: forests have the highest sink potential and hence, afforestation and reforestation are counted toward the emission targets, whereas deforestation is counted against the emission targets.[109] The Kyoto Protocol created specific measures to control emissions at a national level.

Innovative international measures for implementation were established through the Joint Implementation, Clean Development Mechanism, and

---

[101] See further D. Freestone, & C. Streck *Legal Aspects of Implementing the Kyoto Protocol mechanisms: making Kyoto work*, Oxford University Press, Oxford, 2005.

[102] For instance, Japan is placed at 6 % reduction, the EU at 8% and Canada at 6 %.

[103] See Article 3.3 of the Kyoto Protocol.

[104] See Article 4 of the Kyoto Protocol.

[105] See Article 12 of the Kyoto Protocol.

[106] See Article 17 of the Kyoto Protocol.

[107] Article 2-2 of the Kyoto Protocol states: "The net changes in greenhouse gas emissions by sources and removals by sinks resulting from direct human-induced land-use change and forestry activities, limited to afforestation, reforestation and deforestation since 1990, measured as variables changes in carbon stocks in each commitment period, shall be used to meet the commitments under this Article of each Party included in Annex I".

[108] See S. Oberthur & H.E. Ott, *The Kyoto Protocol International Climate Policy for the 21st Century*, Springer, Berlin, 1999, 131.

[109] Article 3.3 of the Kyoto Protocol provides that not only changes in emissions from sources but also certain removals from sinks shall be used to meet the targets of Annex B countries.

Emission Trading, known as the 'Kyoto Mechanism'.[110] The Joint Implementation is undertaken by the developed nations to facilitate meeting the target through conducting projects aimed at the reduction of emissions together. The actual reduction achieved within such a project will be counted toward the emission target.[111] In contrast, the Clean Development Mechanism is for developed nations to assist developing nations to invest in a project aimed at reduction of emissions or increase of sinks; the actual reduction achieved in such projects will be counted toward the emission target of the state parties.[112] The Emission Trading Mechanism is the trading of greenhouse gas emission credits between those countries that do not meet their obligations and those that still have capacities in their levels of emission.[113] Under this mechanism, the nations with difficulties in meeting the target can do so more easily through purchasing the emission credits from those with surplus capacity, thus creating incentives for both.

The Kyoto Protocol establishes a functional regime with concrete obligations and implementation mechanisms, including regular meetings of Conferences of Parties as well as a number of actual projects. Nevertheless, complying with the reporting obligation as well as meeting the target is a challenge. Establishing a comprehensive and reliable national inventory and keeping it updated to make the best estimates of land use change and forestry activities are considered to be essential toward meeting targets. Furthermore, rigorous measures need to be taken domestically to reduce greenhouse gas emissions, involving the active participation by governmental organizations, private entities, and individuals. Remote sensing from space can contribute in aspects of establishing national inventory and implementing projects.

To review, there are three reasons why the current environmental regime suffers from inadequacies. First, environmental protection is primarily at a state level and commitment is left to each state's capacity, which is affected by a number of factors: some states do not have capacity and/or appropriate technology to meet their obligation under the environmental agreements, whilst other states may be reluctant to respect their environmental obligations. Furthermore, the role of international organizations and non-governmental

---

[110] See S. Oberthur & H.E. Ott, *The Kyoto Protocol International Climate Policy for the 21st Century*, Springer, Berlin, 1999, 96.

[111] See Article 6 of the Kyoto Protocol.

[112] Article 12 states that (a) Parties not included in Annex I will benefit from project activities resulting in certified emission limitation and reduction commitments under Article 3. (b) Parties included in Annex I may use certified emission reductions accruing from such project activities to contribute to compliance with part of their quantified emission limitation and reduction commitments under Article 3.

[113] See S. Oberthur & H.E. Ott, *The Kyoto Protocol International Climate Policy for the 21st Century*, Springer, Berlin, 1999, 187.

actors in implementation and enforcement is relatively limited. Enhanced levels of participation from these actors have a high potential for improving the state of the environment.

Second, all MEAs are not functioning to the full extent needed to contribute to better protection. The majority of these agreements are not effective, as they have neither concrete obligations and commitments toward environmental protection nor implementation and enforcement mechanisms. Many environmental treaties address the problems and call for cooperation in understanding and efforts to improve the state of affairs, yet fail to have the powerful effect on parties to change their behaviours that are adverse to the environment. Consequently, compliance with such environmental treaties will not contribute much to improve the state of the environment. On the other hand, a MEA that is a powerful legal instrument is also not necessarily successful in terms of achieving its objective because of inadequate implementation and enforcement and/or the difficulties of meeting obligations – all of which are daunting tasks.

By looking at three specific MEAs, one can conclude that whilst weaknesses as a legal instrument may limit the actual environmental protection to a certain extent, even the MEAs that are powerful legal instruments do not necessarily guarantee better protection. Indeed, positive results may be achieved with a relatively weak legal instrument, insofar as it is complemented by concrete measures for implementation. Hence, maximizing the effect of existing MEAs with effective operations and concrete measures for protection is the key to success.

In summary, even if a MEA is a strong legal instrument, its mere presence does not necessarily lead to the enhanced performance of parties; rather, it needs to be accompanied by effective implementation and enforcement. Currently, despite the imminent need for a vigorous protection of the environment, the regime is inadequate to meet that need: the problems continue to be addressed, but have not been solved sufficiently. Clearly this must change. What is necessary is not just to maintain the current state of the environment, but rather to introduce drastic measures to enhance levels of protection. The current regime needs reinforcement of some kind and satellite remote sensing is ideally suited for this task. The next section examines how remote sensing can consolidate the current regime and contribute to enhanced levels of environmental protection.

3.4. How Satellite Remote Sensing can Support Environmental Protection

Remote sensing by orbiting satellites can make up for the inadequacies of the current regime by strengthening implementation and enforcement. In this section, the benefit of the use of remote sensing for the protection of the environment is examined: first, why and how remote sensing can assist environmental protection; and second, the implementation of environmental legislation through the use of remote sensing.

Considering the nature of the inadequacies of environmental protection, satellite remote sensing can bring extensive benefits. It can consolidate the institutional shortcomings of the environmental regime, such as the limited roles played by relevant non-state actors. Furthermore, it can maximize the effect of MEAs by providing a vital tool for implementation and enforcement.

It is important to understand some background of satellite remote sensing, which makes it a powerful tool to support environmental protection and to complement the inadequacy of the current regime. The reasons that make satellite remote sensing effective for environmental applications can be analysed in terms of both technical and legal aspects.

Satellite remote sensing is suited for environmental protection from several technical standpoints. First, it enables us to make environmental maps and assessments of different regions including the atmosphere, land, and oceans.[114] Major applications for monitoring land include the assessment of land and forest cover change, which can identify urbanization, desertification, and deforestation.[115] Satellite remote sensing is useful for the study of coastal features, including mapping of water lines, coral reefs, and wetlands.[116] Satellite monitoring of the oceans includes studies of climate, surface biology, ice conditions, and pollution.[117] Different types of sensors are used depending upon the applications. For instance, infrared instruments can monitor ocean currents and temperature anomalies, and the data can be useful not only for weather predictions but also to study plankton and fish populations or for revealing the nature of El Nino. Synthetic aperture radar (SAR) can monitor ice distributions and ice movement, and can be used to warn shipping companies of icebergs, as well as to help climate scientists in studies of environmental change.[118]

---

[114] For a detail of different environmental applications, see Nieuwenhuis, G..J.A., R.A.Vaughan, and M. Molenaar, *Operational Remote Sensing for Sustainable Development*, Balkema, Rotterdam, 1999. See also *Droit télédétection et environnement: actes du colloque international: le droit face aux techniques de télédétections par satellite au service de l'environnement*, S. Courteix (Eds.), Strasbourg, 2-3-4 juin 1993, Sides, Antony, 1994.

[115] See e.g., F. Blasco, Déforestation et désertification, *Droit télédétection et environnement: actes du colloque international : le droit face aux techniques de télédétections par satellite au service de l'environnement*, S. Courteix (Eds.), Strasbourg, 2-3-4 juin 1993, Sides, Antony, 1994, 61–75.

[116] See U.R. Rao, Space for Sustainable Development, *Recent Trends in International Space Law and Policy*, Lancers Books, New Delhi, 1997, 46.

[117] See D.H. Staelin & J. Kerekes, Remote Sensing Capabilities, *Heaven and Earth: Civilian Uses of New Earth Space*, Martinus Nijhoff Publishers, Hague, 1997, 174.

[118] See D.H. Staelin & J. Kerekes, Remote Sensing Capabilities, *Heaven and Earth: Civilian Uses of New Earth Space*, Martinus Nijhoff Publishers, Hague, 1997, 174.

Second, remote sensing can provide information on otherwise inaccessible areas such as islands, deep forests, and mountainous areas at high attitudes. Areas that need to be observed for environmental monitoring are sometimes so inaccessible, remote, and/or vast that observation by other means such as helicopter and airplanes is impractical. Indeed, satellite remote sensing has been used to detect illegal fishing in the French territory of the Kerguelen Islands on the Southern Indian Ocean – one of the most remote places on the globe.[119] Furthermore, satellites can cover the entire globe, thus allowing derived information to be compared between different geographic areas from different time periods.[120] Whilst it permits the rapid collection of data even on a near real time basis, remote sensing also offers access to the archive of vast data sets to enable the assessment of the timing and frequency of a particular operation that adversely affects the environment, such as an illegal landfill.[121] Not only it is well-suited for monitoring specific areas of interest, remote sensing also permits the assessment and analysis of changes in local, regional, and global environments over longer periods of time through the comparison of data from the past with that of the present.

Through its comprehensive approach, remote sensing would permit us to gain an improved understanding of the environment and would also enable predictions to be made for the future about such phenomena as global warming. Finally, as mentioned in the introduction, remotely sensed data from satellites can be easily integrated with other sources of information – such as statistical information, population density, geographic information systems, and aerial photography – to provide useful knowledge in the form of detailed environmental assessments ready for immediate use by decision makers.[122] Satellite data can be combined with GPS and GIS to make a more accurate mapping of the state of the environment in terms of location, as well as to assess the interacting factors such as population growth that have significant impact on the environment. Remote sensing supplements the existing information sources, which are often limited, to generate more comprehensive and reliable information, which will then permit more effective assessment and analysis.

As already examined in Chapter 2, remote sensing is a highly effective tool for the observation of the Earth's environment across national boundaries, as it permits monitoring of the entire globe without infringing states' sovereignty.

---

119 "Envisat Radar Protecting Patagonian Toothfish From Pirates", *Space News*, November 18, 2005.

120 See J. Aschbacher, Monitoring environmental treaties using earth observation. Available on the Internet at http://www.vertic.org/assets/VY02_Aschbacher.pdf. Last accessed on 15 April 2010.

121 See R. Purdy, Satellites: A New Era for Environmental Compliance? 3-5 *Journal for European Environmental and Planning Law* (2006), 406.

122 See further H. George, Remote sensing of Earth resources: emerging opportunities for developing countries, 14 *Space Policy* (1998), 27–37.

Indeed, satellites can sense over any location without the prior consent of the sensed state, whether or not that state is in agreement with such observation. This is a clear contrast to aerial monitoring of the territories of other states, which is not permissible without each state's consent. As has been discussed, the concept of sovereignty by states is one of the major obstacles for ensuring protection of the environment within sovereign territories as well as in areas beyond national jurisdiction. Indeed, regarding the international marine oil pollution regime, the initial attempt for regulating oil spills on the high seas essentially failed, as the spills were unobservable with the conventional technologies; hence, rules were a weak deterrent.[123] Remote sensing from space, with the capabilities to reveal both environmental conditions and behaviour of states or entities regardless of their location, acts as a powerful deterrent to states causing transboundary environmental harm, as well as harm within the sovereign territory.[124]

The use of remote sensing for environmental protection is supported specifically by the legal regime of remote sensing.[125] The Remote Sensing Principles provision related to environmental protection, Principle X, states that: "remote sensing shall promote the protection of the Earth's natural environment. To this end, states participating in remote sensing activities that have identified information in their possession that is capable of averting any phenomenon harmful to the Earth's environment shall disclose such information to states concerned."[126] This provision is general, and does not concretely establish data availability or terms and conditions for access to and use of data, which would assist the beneficiaries on such occasions. If specific data policies benefiting environmental application were to be established at an international level, remote sensing data could be utilized much more extensively toward better environmental protection.

Access to environmental information, supported by the 1990 EC Directive on Access to Environmental Information,[127] amongst others, can be effectively

---

[123] See K. Kaline & K. Raustiala, International Environmental Agreements and Remote Sensing Technologies, Background paper prepared for the workshop on Remote Sensing and Environmental Treaties: Building more Effective Linkages, December 4–5, New York, 2000, 15.

[124] See K. Kaline & K. Raustiala, International Environmental Agreements and Remote Sensing Technologies, Background paper prepared for the workshop on Remote Sensing and Environmental Treaties: Building more Effective Linkages, December 4–5, New York, 2000, 31.

[125] See further G. Catalano Sgrosso, Sharing of Remote Sensing Data Concerning Environmental Protection for Public Benefit, *Proceedings of the Thirty-Ninth Colloquium on the Law of the Outer Space,* IISL, AIAA, Washington, 1995, 94.

[126] Principle X of the UN Remote Sensing Principles.

[127] Directive 2003/4/EC of the European Parliament and of the Council of 28 January 2003 on public access to environmental information and repealing Council Directive 90/313/EEC. [Hereafter: Environmental Directive].

implemented by remote sensing from space. Article 7 of the Directive states that: "Member states shall take the necessary measures to ensure that public authorities organize the environmental information, which is relevant to their functions and which is held by or for them, with a view to its active and systematic dissemination to the public, in particular by means of computer telecommunication and/or electronic technology, where available."[128] Satellite data easily accessible via Internet is a good example of the possibilities open to the public for access to environmental information. For example, a vast amount of Landsat data in archives is accessible to the public and made available at only marginal cost. This data can be used for environmental applications, such as land cover change and assessments of deforestation. Increasing international cooperation and programmes for environmental observation using satellites are steps forward for the enhanced access to environmental information. For instance, the intergovernmental Group on Earth Observation (GEO) coordinates different environmental monitoring platforms, including the space-borne ones, to implement the Global Observation Earth Observation System of Systems (GEOSS),[129] which is designed to gain better understanding of environmental factors that affect humans.

3.4.1. *How Remote Sensing can Strengthen Environmental Protection: General Observations*

This section provides general observations as to how remote sensing can bring greater robustness to environmental protection in terms of mitigating the institutional shortcomings as well as in assisting with different phases of environmental protection.

3.4.1.1. *A Tool to Meet Environmental Obligations*

Remote sensing can enhance the performance of states under their agreements by providing concrete tools for use in meeting environmental obligations. Some states lack capacities to implement and comply with MEAs, despite their willingness to do so, and lack of capacity is often a major source of non-compliance.[130] Remote sensing helps those states with limited capacities to implement an MEA because it is an effective observation and assessment tool.

---

[128] Article 7 of the Environmental Directive.

[129] Presentation Brochure on Group on Earth Observation, 2004, 2. See further, R. Ohlemacher, The Earth Observation Summit: hearing a new era for international cooperation, 19 *Space Policy* (2003), 277–281.

[130] See K. Kaline & K. Raustiala, International Environmental Agreements and Remote Sensing Technologies, Background paper prepared for the workshop on Remote Sensing and Environmental Treaties: Building more Effective Linkages, December 4–5, New York, 2000, 29.

Data acquired on non-compliance can help outside experts and officials from international bodies to assist the state having a problem with compliance in evaluating the causes of non-compliance and developing possible remedies.[131] Indeed, improved conservation of endangered world heritage sites has been implemented through the use of satellite data. Furthermore, remote sensing can consolidate the content of reporting under the Kyoto Protocol by providing a significant information source. As mentioned previously, remote sensing can be of help with – and may be sometimes the only practicable means of achieving – the daunting tasks of making and updating an inventory of land use, land use change, and forestry activity, while complementing current limitations in reporting (as seen in the earlier example of Italy). These issues are explored more fully in the next section.

3.4.1.2. *Enhanced Participation by International Organizations and Non-State Actors*

Remote sensing could enhance both the participation and operation of international organizations and non-state actors; it could also enhance the centralized functions of MEAs such as the committees and secretariats. Remote sensing can enhance the role of international organizations in implementing concrete programmes conducted at their initiatives. For instance, the Food and Agriculture Organization (FAO) has used remote sensing data to create AFRICOVER, a 1:250,000 map of Africa in support of combating desertification.[132] In addition, remote sensing facilitates participation by private entities in environmental protection by assisting private actors in voluntarily achieving compliance with domestic-level regulations to strengthen the implementation of international commitments.[133] In recent years, there has been growing interest amongst private entities including media, car manufacturers, and steel companies in operating remote sensing satellites for purposes related to the environment. Obtaining up-to-date information on the state of the environment is essential

[131] See K. Kaline & K. Raustiala, International Environmental Agreements and Remote Sensing Technologies, Background paper prepared for the workshop on Remote Sensing and Environmental Treaties: Building more Effective Linkages, December 4–5, New York, 2000, 29.

[132] See K. Kaline & K. Raustiala, International Environmental Agreements and Remote Sensing Technologies, Background paper prepared for the workshop on Remote Sensing and Environmental Treaties: Building more Effective Linkages, December 4–5, New York, 2000, 27.

[133] See K. Kaline & K. Raustiala, International Environmental Agreements and Remote Sensing Technologies, Background paper prepared for the workshop on Remote Sensing and Environmental Treaties: Building more Effective Linkages, December 4–5, New York, 2000, 30. For instance, major oil multinational Texaco is developing remote sensing capability in assisting the pursuit of environmentally related policies and measures.

for the effective implementation and enforcement of environmental measures. International organizations and other non-state actors can take the initiative for better protection by being both users and operators of remote sensing satellites. Remote sensing can also help the executing bodies through supplying information to assist them in the process of implementation, reviewing the implementation, and evaluating the cause of implementation failures.[134] For instance, use of remote sensing assisted the Committee with the nomination process under the World Heritage Convention.

3.4.1.3. *Consolidation of Different Modes of Environmental Protection*

Remote sensing can strengthen the different modes of environmental protection such as negotiation, identifying problems, and enforcement; it can be used as a basis for negotiation; and it can be used for implementation and enforcement. The evidence of tropical forest destruction identified through satellites in the 1990s gave an impetus to Mexico and Guatemala to create a 4-million acre biosphere reserve along their shared border.[135] Remote sensing also can serve as a tool to help understand the environmental problem more specifically. For example, NOAA AVHRR and Landsat TM data were used to map water and plant distributions in two Ramsar sites – the Khingan Nature Reserve and the Ganukan Landscape Refuge in Far East Russia – and to classify the different parts of these sites into three categories: the wettest part with sedges and reeds; the modestly wet part with sedges and grasses; and the dry part with various kinds of grasses.[136] Precise understanding of environmental conditions and the associated problems helps decision-makers to make more appropriate choices.

[134] See K. Kaline & K. Raustiala, International Environmental Agreements and Remote Sensing Technologies, Background paper prepared for the workshop on Remote Sensing and Environmental Treaties: Building more Effective Linkages, December 4–5, New York, 2000, 29.

[135] See K. Kaline & K. Raustiala, International Environmental Agreements and Remote Sensing Technologies, Background paper prepared for the workshop on Remote Sensing and Environmental Treaties: Building more Effective Linkages, December 4–5, New York, 2000, 28. A Landsat image clearly identified the contrast between the Mexican side that was largely deforested and the Guatemalan side that was largely undamaged, and this catalyzed the first meetings in decades between Mexico and Guatemala to discuss border land management. See also A. De Sherbinin, and C. Giri, Remote Sensing in Support of Multilateral Environmental Agreements: What Have we Learned from Pilot Applications?, Open Meeting of Human Dimensions of Global Environmental Change Research Community, Rio de Janeiro, 6–8 October 2001, 5.

[136] See K. Kaline & K. Raustiala, International Environmental Agreements and Remote Sensing Technologies, Background paper prepared for the workshop on Remote Sensing and Environmental Treaties: Building more Effective Linkages, December 4–5, New York, 2000, 8.

Remote sensing can provide a measure for review of MEA implementation.[137] Some environmental agreements require national reports on implementation. Remote sensing facilitates the integration of data at a national level with those at both regional and global levels.[138] For instance, assessments of forests in different countries in Europe can be integrated to produce results for all of Europe and assist in its compliance with the Kyoto Protocol in meeting the reduction target of emission of green house gas imposed on Europe as a whole. Remote sensing can facilitate the review of contribution levels from each country and their impact on overall greenhouse gas reduction.

Remote sensing can be used to detect and identify non-compliance, as well as reveal the source and extent of environmental damage to facilitate effective response and enforcement. Following the Chernobyl accident in 1987, satellite data played a pivotal role in revealing the extent of the disaster and prompting the former Soviet Union to respond. The western press had no information regarding the accident until Landsat 5 imaged the accident site followed by the SPOT 1 satellite a few days later, giving the press the solid information base to reveal the accident.[139] It is believed that the satellite images prompted the Soviet Union to respond and disclose the critical information concerning the disaster to the rest of the world.[140] Satellite imagery could be used to serve as a deterrent for criminal behaviour and assist in establishing more efficient regulatory compliance.[141] A major environmental enforcement tool on marine pollution is the satellite detection and mapping of oil slicks.[142] For example, a satellite clearly captured the oil slick released into the Persian Gulf from Kuwait in 1991.[143]

---

[137] See K. Kaline & K. Raustiala, International Environmental Agreements and Remote Sensing Technologies, Background paper prepared for the workshop on Remote Sensing and Environmental Treaties: Building more Effective Linkages, December 4–5, New York, 2000, 29.

[138] See K. Kaline & K. Raustiala, International Environmental Agreements and Remote Sensing Technologies, Background paper prepared for the workshop on Remote Sensing and Environmental Treaties: Building more Effective Linkages, December 4–5, New York, 2000, 29.

[139] See R.J. Aamoth, Use of Remote Sensing Imagery by the News Media, *Heaven and Earth: Civilian Uses of Near-Earth Space*, Martinus Nijhoff Publishers, Hague, 1997, 140.

[140] See P.D. Zimmerman, Superpower Relations, and Public Diplomacy, *Commercial Observation Satellites and International Security*, the Macmillan Press, Hampshire, 1990, 39.

[141] See R. Purdy, Satellites: A New Era for Environmental Compliance?,3-5 *Journal for European Environmental and Planning Law* (2006), 406.

[142] See U.R. Rao, Space for Sustainable Development, *Recent Trends in International Space Law and Policy,* Lancers Books, New Delhi, 1997, 47.

[143] See D.H. Staelin & J. Kerekes, Remote Sensing Capabilities, *Heaven and Earth: Civilian Uses of New Earth Space*, Martinus Nijhoff Publishers, Hague, 1997, 174.

Remote sensing data may be used as evidence in the courtroom to enforce compensation for environmental damage;[144] it can identify the polluter – such as the vessel suspected of an oil spill – so that the polluter can be required to make compensation. Data can also be used as a basis to resolve disputes over transboundary harm.

3.4.2. *How Remote Sensing can Specifically Strengthen the Effects of Multilateral Environmental Agreements*

A considerable number of MEAs benefit from observations by remote sensing, including the Biodiversity Convention, the Ramsar Convention,[145] the UN Desertification Convention[146] and, most of all, the three MEAs that were examined in detail in the previous section. This section provides a further examination of how these three agreements benefit from remote sensing, which balances their inadequacies – as identified above – and strengthens both implementation and enforcement.

3.4.2.1. *The World Heritage Convention*

Satellite remote sensing is useful for the implementation of the World Heritage Convention.[147] The use of remote sensing is suited for the preservation of world heritage sites, as they belong to the host state and the mandate to protect the sites is given to such state; hence, the approach from the international community as a whole in ensuring protection has been difficult. UNESCO has developed an operational network of satellites to monitor and study world heritage sites using remote sensing and has produced successful results.

Since 2001, UNESCO and ESA undertook an initiative to demonstrate the application of Earth observation and other space technologies (e.g., navigation and positioning communication) in support of implementing the World Heritage Convention, and to establish a framework of cooperation open to

---

[144] See S. Courteix, Toward the legal Recognition of a new method of proof for the defence of the Environment, *Proceedings of the Thirty-Seventh Colloquium on the Law of Outer Space,* IISL, AIAA, 1994, 226.

[145] Globwetland, which uses satellite images to help states to better assess the condition and monitor the wetland.

[146] There is a project called DESERTWATCH, ESA's Data User Programme, to develop an EO-based information system to support national and regional authorities in reporting to the Convention. AFRICOVER, is created to support the combat of desertification.

[147] This section is an updated and edited version of A. Ito, Legal Aspects of Implementing World Heritage Convention Using Remote Sensing Data, *Proceedings of the Forty-Fifth Colloquium on the Law of Outer Space*, IISL, AIAA, Washington, 2005, 70–79.

space agencies, research institutions, non-governmental organizations (NGOs), and the private sector.[148] Since then, the initiative has been operating successfully and a number of space agencies have joined. Current partners include: the Canadian Space Agency, Catholic University of Louvain of Belgium, the Chinese Academy of Science, the French Space Agency – CNES, the Argentinean Space Agency – CONAE, the German Space Agency – DLR, Eurisy, GEOSPACE, the International Space University – ISU, and the Japanese Aeronautical Exploration Agency – JAXA. Partners to be signed include the Jet Propulsion Laboratory (JPL), the National Aeronautics and Space Administration (NASA), the National Remote Sensing Centre of India (NRSC), GeoEye, Spot Image, the Turkish Academy of Sciences, the University Corporation for Atmospheric Research of the United States, and the University of Ghent. Some governments recognized as partners have provided funding for projects and local conservation authorities, whose sites are selected for the projects under the 'Open Initiative'. This is a good illustration of putting together space agencies and non-state actors such as universities and private institutions for participation in environmental protection.

The aim of the initiative is to strengthen the capacity to monitor the status of conservation of World Heritage sites using satellite images in those countries requesting to participate. Indeed, it has helped those states who need assistance in complying with the World Heritage Convention, and has created local capacity in developing countries that are less familiar with the application of space technologies to natural and cultural heritage sites. Such regions include the World Heritage sites of the Virunga National Park, the Kahuzi Beiga National Park in Democratic Republic of Congo, and the Bwindi Impenetrable Forest National Park in Uganda. Due to influences from conflict within the region, influx of refugees, and illegal clearing of forest, the habitats of gorillas have been put in danger.[149] The terrain is mountainous, including some of Africa's highest peaks reaching over 5,000 m in altitude; consequently the area has been inaccessible and poorly mapped. As a result of the completion of the initial phase of the project, accurate maps of the region derived from satellite images have been provided for the first time to local end-users. During the project, satellite images taken over the sites between 1990 and 2003 were compared to get a view of the state of conservation in the region and assess the adverse impacts on the environment.[150] The products included digital elevation models, vegetation and land use change maps, as well as 1:200,000 and 1:50,000 cartography to help

[148] Open Initiative on the use of space technologies to support the World Heritage Convention.

[149] "Satellite maps will ease plight of endangered mountain gorillas", ESA NEWS, 8 April 2005.

[150] Brochure on "UNESCO – using Space Technologies for the conservation of Natural and Cultural Heritage" 2005.

the conservation authorities and non-governmental organizations working in and around the parks.[151]

By combining the GIS layers provided by the project with Global Positioning System (GPS) data collected in the field, gaps in existing anti-poacher patrol coverage were identified and readdressed, and the migration of gorilla groups could be followed across national boundaries.[152] The mapping could not have been done without the use of Earth observation from satellites, which provided an efficient, cost-effective, and frequently updated means of surveying the most inaccessible regions of interest.[153] Less than 700 Mountain Gorillas remain alive and the project has significantly contributed to preserving their habitat.[154]

The UNESCO List of World Heritage in Danger is designed to call the attention of the international community to the conditions that threaten the sites and to encourage corrective action. Amongst the 890 sites inscribed as of March 2010, 31 sites are considered to be in danger. The list includes several parks and nature preserves.[155] It has to be noted here that the Virunga and Kahuzi-Biega national parks selected under the 'Monitoring of Gorillas' Habitats' project, are on the List of World Heritage in Danger,[156] which is a list of sites where priority assistance is given due to their aggravated conditions.

Remote sensing also helped nations to identify new sites for nomination. Indeed, satellite remote sensing has been useful to study atolls at some of the richest coral reefs in the Pacific, which are considered to be potential World Heritage sites. For example, UNESCO participated in a project to study the Ailinginae Atoll, which is situated north of the Republic of the Marshall Islands. The map of the coral reef was made based upon high-resolution satellite images;

---

[151] Brochure on "UNESCO – using Space Technologies for the conservation of Natural and Cultural Heritage" 2005.

[152] See Foret Rwanda: des satellites au secours des gorilles *GEO*, no. 319, September 2005,102–105.

[153] *Data User element DUE_DUP Directory 2002–2004*, ESA, Paris, 2004, 32.

[154] See Foret Rwanda: des satellites au secours des gorilles *GEO*, no. 319, September 2005,102–105.

[155] Natural sites on the World Heritage in Danger list as of September 2008 include: 1. Manovo-Gounda St. Floris National Park in the Central African Republic, 2. Comoé National Park, 3. the Mount Nimba Nature Reserve in the Ivory Coast, 4. the Virunga, 5. Garamba, 6. Kahuzi-Biega, 7. Salonga National Parks and 8 Okapi Wildlife Reserve all in Democratic Republic of Congo, 9 Galapagos Island in Ecuador, 10. Simien National Park in Ethiopia, 11 Mount Nimba Strict Nature Reserve in Guinea, 12. Manas Wildlife Sanctuary in India, 13. Air and Ténéré Natural Reserves in Niger, and 14. Niokolo-Koba National Park in Senegal.

[156] Inscribing a site on the List of World Heritage in Danger allows the World heritage Committee to allocate immediate assistance from the Wold Heritage Fund to the endangered property.

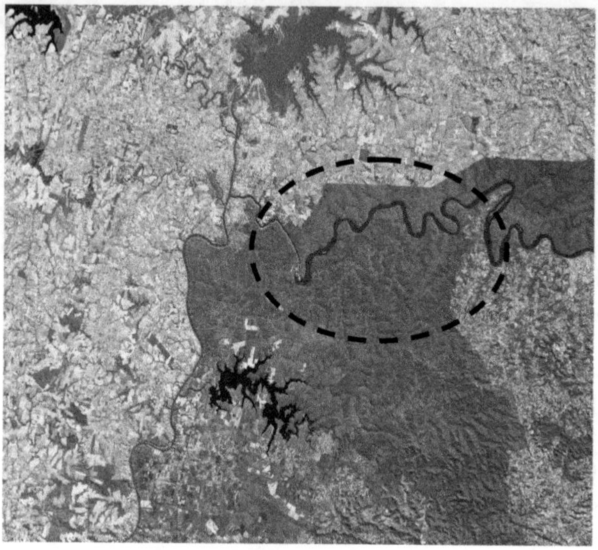

Figure 3.1: UK-DMC2 image of Iguaçu, 13 June 2010.

this map will now be used as a reliable source for the World Heritage Committee to determine whether the site meets the criteria to be added to the World Heritage List[157].

Satellite images can be used to illustrate successful implementation of the Convention as well as to clarify a problem. Figure 3.1. shows a satellite image taken by UK-DMC2 illustrating the state of the environment surrounding the World Heritage site of Iguaçu National Park encompassed by both Brazil and Argentina.

Within the black oval, the area above the river is the Brazilian side, and area below the river is the Argentinean side. It can be seen that the area of preserved forest on Argentina's side is far greater than that of the Brazil, where there is extensive agricultural cultivation in the areas surrounding Iguaçu.

Currently, a joint project in collaboration with CONAE is carried out to assist Argentina and Brazil in their conservation efforts. Satellite images are being used to survey the surrounding areas of Iguaçu and Iguaçu National Parks

[157] D. Siciliano, and K. Joyce, Mapping for Conservation: Using QUICKBIRD satellite imagery to assess marine and terrestrial resources of Ailinginae Atoll, Republic of the Marshall Islands, a proposed World Heritage Site, Communication to *30th International Symposium Remote Sensing of the Environment (ISRSE),* Nov 10–14, 2003, Honolulu, Hawaii.

and the analysis of satellite imagery provides the local conservation authorities with a basis for examining the feasibility of enlarging the protected areas.[158]

Projects under the Open Initiative demonstrated that satellite remote sensing has successfully assisted implementation of the World Heritage Convention complementing the inadequacies of the Convention as a legal instrument. The successful outcome of the initiative has also invoked the potential use of space technologies to enforce the World Heritage Convention, such as monitoring the status of conservation from outside. From the institutional point of view, it has helped governments that have difficulties in conserving the inscribed sites or that do not have capacities to nominate potential sites.

The programme has facilitated the participation of UNESCO as well as local conservation authorities, space agencies, private entities, and universities in implementing concrete projects. This operational mechanism based upon remote sensing has brought together entities at different levels to share their capacity and knowledge for better protection.

3.4.2.2. *International Convention for the Prevention of Pollution from Ships*

MARPOL benefits from remote sensing because remote sensing assists the detection of discharges at sea, assesses their extent, and provides a method of obtaining the evidence required under the Convention. Indeed, the IMO recognizes the utility of satellites images to detect oil spills.[159]

Satellite remote sensing, particularly Synthetic Aperture Radar (SAR), is suited for pollution detection because oil slicks change the surface roughness on bodies of water under windy conditions, and this shows up as changes in backscatter on radar instruments.[160] Remote sensing is useful in several scenarios of oil spill control such as large area surveillance, site-specific monitoring, and

---

[158] Information can be found on the Internet at http://www.unesco.org/science/remotesensing.
Other ongoing projects under the Open Initiative include: Satellite Imagery to Observe and Safeguard Archaeology Southern Iraq Uruk-Warka; Inventory and Preservation of Frozen Tombs in Altai Mountains across Russia, Kazakhstan, Mongolia and China; Decision Support System for Archaeological Sites in Guatemala, Using Remote Sensing to Provide an Effective Management Tool for Machu Picchu; and UNESCO & ESA Using Space Technologies to Assist Mesoamerica with the Biological Corridor.

[159] Resolution IMO A 152 of 17 November 1983 provides the authority for the use of satellites images as evidence in enforcement actions by listing them alongside accepted methods of obtaining evidence such as photography, sea water sampling, and the examination of ship log books.

[160] See A. De Sherbinin & C. Giri, Remote Sensing in Support of Multilateral Environmental Agreements: What Have we Learned from Pilot Applications?, Open Meeting of Human Dimensions of Global Environmental Change Research Community, Rio de Janeiro, 6–8 October 2001, 12.

tactical assistance in emergencies.[161] Particularly it is well-suited to target those areas where oil spills are suspected, based upon information from ocean surveillance.[162]

Satellite data have been used in a number of occasions to identify the polluter, assess the extent of oil spills, and monitor behaviours of ships. Satellite imagery played a pivotal role in identifying the vessel responsible for the oil spill in the straight of Malacca, and was used by the authority to initiate proceedings in the 1996 Song San Case.[163] ERS satellite imagery, together with laboratory analysis, helped the Marine and Port Authority of Singapore to identify the source of the pollution.[164]

Damage assessments have also been facilitated through use of remote sensing data. The Disaster Charter has been activated by the Civil Protection Unit of European Commission in the case of oil spill monitoring on 17 January, 2003 for the oil spill off the Coast of Galicia in Spain, and on 10 October 2002 for the oil spill in the Gulf of Aden.[165]

Satellite images are increasingly used for verification and enforcement purposes. The use of the remote sensing data supported cleaning operations related to an oil spill and the ensuing insurance claims in Indonesia. In the oil spill incident[166] by the Natuna Sea tanker that ran aground next to Batam Island in early October 2000, remote sensing data from the SPOT and ERS satellites guided the cleaning operations as well as substantiated insurance claims by the affected countries – Indonesia, Malaysia and Singapore – for refund of the cleaning costs incurred and the loss of revenues from affected fishing and aquaculture activities.[167] Insurance claims require proof to justify the claims, and

---

[161] See A. De Sherbinin & C. Giri, Remote Sensing in Support of Multilateral Environmental Agreements: What Have we Learned from Pilot Applications?, Open Meeting of Human Dimensions of Global Environmental Change Research Community, Rio de Janeiro, 6–8 October 2001, 12.

[162] See A. De Sherbinin & C. Giri, Remote Sensing in Support of Multilateral Environmental Agreements: What Have we Learned from Pilot Applications?, Open Meeting of Human Dimensions of Global Environmental Change Research Community, Rio de Janeiro, 6–8 October 2001, 12.

[163] See C. Amerasinghe, *Evidence in International Litigation*, Martinus Nijhoff Publishers, Leiden, 2005, 394–395.

[164] See E. Nabet, Legal Aspects of the Use and Applications of Remote Sensing in South East Asia, 5 *Singapore Journal of International and Comparative Law* (2001), 161.

[165] See http://www.disasterscharter.org/. Last accessed on 18 April, 2010.

[166] Over 7,000 tons of crude oil was discharged into the sea. See E. Nabet, Legal Aspects of the Use and Applications of Remote Sensing in South East Asia, 5 *Singapore Journal of International and Comparative Law* (2001) 162.

[167] See E. Nabet, Legal Aspects of the Use and Applications of Remote Sensing in South East Asia, 5 *Singapore Journal of International and Comparative Law* (2001), 162.

satellite images can play a vital role. There are growing needs for use of satellite images for such enforcement purposes.

Limitations exist, however, as to the use of satellite imagery as an enforcement tool for oil spill detection. The radar images generally give an unacceptable number of false positives, so the technology is only applicable as a surveillance tool in conjunction with Infrared Red and UV sensors, used for reconnaissance and confirmation of potential oil slicks.[168] In the Son Sang Case, the ERS satellite data was not used as sole evidence, but was combined with other data. The reliability of data is an important and ongoing issue.

One can conclude that satellite remote sensing can strengthen the weak legal regime by providing a viable tool to implement and enforce MARPOL more effectively. It has served as a vital tool for monitoring the extent of damage associated with oil spills and is increasingly used as a method of acquiring evidence for detection of oil slicks. Yet, satellite remote sensing still needs improvements as a tool for verification; specifically, for the data to be accepted as sole and reliable evidence to be used in a court. Guidelines need to be established within the legal framework of remote sensing for uses of imagery for verification purpose, such as audit trail to keep track of processes for data generation, processing, and analysis.[169]

3.4.2.3. *The Kyoto Protocol*

The Kyoto Protocol provides a concrete mechanism to offset the emission of green house gases by increasing carbon sinks – most notably, forests. In order to implement the protocol, both land use changes induced by human activities and forestry activities need to be monitored, and reliable estimates of the resulting changes in the carbon sinks need to be made.[170] In accordance with the obligations imposed on the developed nations[171] to reduce the greenhouse gas emission by at least 5% in the period 2008–2012 compared to a reference year of 1990, it is necessary to establish national inventories of the base year. First, land and forest cover must be recorded and the forest biomass must be expressed in carbon stocks for the base year; then, the demonstration of the emission

---

[168] See A. De Sherbinin & C. Giri, Remote Sensing in Support of Multilateral Environmental Agreements: What Have we Learned from Pilot Applications?, Open Meeting of Human Dimensions of Global Environmental Change Research Community, Rio de Janeiro, 6–8 October 2001, 12.

[169] The issue of an audit trail will be examined in Chapter 5.

[170] See *Data User element DUE_DUP Directory 2002–2004*, ESA, Paris, 2004, 25.

[171] Australia, Austria, Belgium, Bulgaria, Canada, Czech Republic, Denmark, Estonia, Finland, France, Germany, Greece, Hungary, Iceland, Ireland, Italy, Japan, Latvia, Lichtensetein, Luxembourg, Monaco, Netherlands, New Zealand, Norway, Poland, Portugal, Romania, Russian Federation, Slovakia, Spain, Sweden, Switzerland, UK and USA.

reduction in the particular commitment period compared to the base line is needed.[172] To meet the reporting obligation, a nation-wide monitoring of land use change and forestry activities is required.[173] Consequently, a large amount of reliable data needs to be created and national inventories updated. Remote sensing can play a vital role in establishing and updating the inventory of carbon stocks.

Remote sensing has enabled useful studies related to creating inventories. For instance, ESA's Kyoto Inventory service – a three-year project designed to generate information products to support reporting obligation under the Kyoto Protocol – has demonstrated that satellite images improved the available information sources by effectively mapping forest and land use change associated with carbon sinks or sources.[174] Data from satellites including ERS, Landsat, Proba, and SPOT were utilised to produce forest maps and land use change maps covering 1990, 1997, and 2002, across 200,234 square kilometres of European territory.[175] The studies were based on active participation of private entities[176] and results helped the ministries as well as national environmental agencies.

The case studies generally found the remote sensing data to be a helpful information source: the Swiss case study showed that overall product accuracy ranged from 90% to 95% and similar results were obtained in Italy, with an accuracy of 89.9 % to 94.4%.[177] On the other hand, studies highlighted that accuracy of data and analysis still need improvement. For instance, in the Swiss case there was confusion between the open forest and grassland in a small part of the area covered;[178] also, the Spanish case had occasional misclassification of open forest with a 5–15% tree crown cover classified as shrubland.[179]

Remote sensing can help developing nations to measure the state of forest carbon for potential benefits to emission trading. Developing countries con-

---

[172] See J. Aschbacher, Monitoring Environmental Treaties using Earth Observation, Verification Year Book, 179.

[173] See *Data User element DUE_DUP Directory 2002–2004*, ESA, Paris, 2004, 25.

[174] See *Data User element DUE_DUP Directory 2002–2004*, ESA, Paris, 2004, 25.

[175] "Satellites support Kyoto Protocol through forest mapping service," *ESA News*, 11 November 2005.

[176] The case study took place in Italy, Spain, the Netherlands, and Switzerland. In the programme, the Italian company Intecs, a private entity, took a lead in implementation of the programme through an international consortium of governmental and private partners.

[177] Used to be located at http://kyoto-inv.pisa.intecs.it/. Present location unknown.

[178] Used to be located at http://kyoto-inv.pisa.intecs.it/. Present location unknown.

[179] Used to be located at http://kyoto-inv.pisa.intecs.it/. Present location unknown.

sider the possibilities for entering the carbon market by protecting their tropical forests or reforesting large areas.[180] Remote sensing facilitates the developing nations' participation in the Kyoto mechanism.

The above examples illustrate how remote sensing can supplement the current efforts by state parties and support their challenging reporting obligations by providing an effective mean to create and update a national inventory. It encourages developing countries as well as private entities to become vigorously engaged in projects based on Clean Development Mechanism, Joint Implementation, and Emissions Trading under the Kyoto Protocol. Remote sensing technology offers a far more substantial tool for the assessment of changes in afforestation, deforestation, and reforestation if accessibility of the data and accuracy of derived information are improved.

3.4.3. *Use of Satellite Imagery as Evidence*

Satellite images have been used for enforcement purposes, as a tool for verification and, more specifically, as evidence.[181] The images are useful to gain understanding of changing phenomena, effectively illustrating a particular environmental pollution in progress and identifying the source of that pollution.

3.4.3.1. *Cases Involving the Satellite Images as Evidence*

Satellite images have been used for verification purposes and to support the settlement of environment-related disputes. Satellite images have been presented in the international court as well as in the national courts of several states.

Starting with an international case, satellite images were introduced in *Mali v. Burkina Faso*,[182] where the Court considered that maps could not constitute a binding document or a territorial title by themselves, unless the parities concerned had expressed their acceptance.[183]

---

[180] See A. De Sherbinin & C. Giri, Remote Sensing in Support of Multilateral Environmental Agreements: What Have we Learned from Pilot Applications?, Open Meeting of Human Dimensions of Global Environmental Change Research Community, Rio de Janeiro, 6–8 October 2001, 12.

[181] For a use of satellite image for verification purpose, see further J.K. Hettling, *Satellite for verification and enforcement of public international law*, Heymann, Köln, 2008.

[182] *Mali v. Burkina Faso*, 1986 ICJ 554. ICJ Report.

[183] Report on the legal Aspects of the privatisation of space activities: Remote Sensing and National Space Legislation, Berlin Conference 2004, International Law Bar Association, 12.

Satellite images were introduced in a dispute regarding the boundary between Namibia and Botswana[184] to determine where exactly the agreed border – the main channel of the river – was located.[185] LANDSAT images taken in 1975 and aerial photography taken between 1925 and 1985, along with other evidence, were used by experts to define the location of the channel.[186] The ICJ used the satellite data only to a minimal degree to actually determine the boundary.[187]

Satellite images were also introduced in a case concerning maritime delimitation and a question of territorial boundaries, *Qatar v. Bahrain*,[188] on 16 March, 2001. The dispute concerned the existence of a natural channel that was later blocked by the construction of a refinery.[189] SPOT data were used at the oral hearing for Bahrain, and Landsat data were used to image the shallow water areas as a measure of accuracy of the bathymetric charts.[190]

Finally, in the *Nigeria v. Cameroon*[191] case, the entire land boundary – from Lake Chad in the north to the Bakassi Peninsula in the south – was disputed by the two states, both claiming the village of Tipsan to be in their territory. Nigeria used a satellite image to show ICJ the location of Tipsan to be in the Nigerian territory; however, Cameroon interpreted Nigeria's satellite imagery differently to claim that Tipsan was in Cameroon's territory.[192] Because the same image was interpreted differently by the two parties, it generated confusion for the Court. This case highlights the significance of establishing a common standard for the accuracy of the interpretation of satellite imagery.

---

[184] *Namibia v. Botswana*, 13 December 1999 ICJ Report.

[185] The ICJ heard the arguments to determine the specific location of the river's main channel, defined by parameters including depth and width of the channel, the rate of flow of the river, bed profile, and navigability. Namibia claimed the channel was one of the southern channels whilst Botswana claimed it was the northern one.

[186] See K.J. Markowitz, Legal Challenge and Market Rewards to the use and Acceptance of Remote Sensing and Digital Information as Evidence, 12 *Duke Environmental Law and Policy Forum* (September 2002), 255. The Court determined in 1999 that the boundary between Botswana and Namibia follows the line of deepest soundings in the Northern Channel of the Chobe river around Kashikili / Sedudu Island. Kashikili / Sedudu island forms a part of territory of the republic of Botswana and shall enjoy equal treatment in the two channels around Kashikili / Sedudu Island.

[187] See NPA Group, *Final Report BNSC Sectors Studies Programme Applications of Earth Observation to Legal Sector*, BNSC, London, 2001, 74.

[188] *Qatar v. Bahrain*, 2001 ICJ Report.

[189] See NPA Group, *Final Report BNSC Sectors Studies Programme Applications of Earth Observation to Legal Sector*, BNSC, London, 2001, 38.

[190] NPA Group, *Final Report BNSC Sectors Studies Programme Applications of Earth Observation to Legal Sector*, BNSC, London, 2001, 38.

[191] *Nigeria v. Cameroon*, 10 October 2002 ICJ Report.

[192] See NPA Group, *Final Report BNSC Sectors Studies Programme Applications of Earth Observation to Legal Sector*, BNSC, London, 2001, 73.

Moving on to national cases, satellite images have been used in national courts such as in the United States, India, and Singapore. The United States has the most extensive experience with using satellite data as evidence in court. There are a considerable number of cases in which satellite data were submitted as evidence. In *United States v. Reserve Mining*,[193] satellite images were used to show the dispersion of taconite tailings by a mining company.[194] It was held that discharge of tailings into waters violated the Clean Water Act and generated a threat to public health.[195]

In *Gasser v. United States*,[196] the Court rejected the evidence of an expert witness that flooding had increased in a particular area, whilst accepting satellite imagery on the same point as evidence.[197] Often satellite images were used in combination with aerial photography: as in the case of *Nickel Plate Mining Co. v. Office of Surface Mining Reclamation and Enforcement;*,[198] or used as an illustrative tool/ supporting evidence in *Oklahoma Water Resource Board v. Texas County Irrigation and Water Resource Association.*[199]

In the United States, use of satellite images for verification purposes has facilitated settlement of claims. Gabrynowicz argues that the satellite image serves as such persuasive evidence that some attorneys agree to settle out of court rather than proceeding with the claim.[200] Indeed, in a case concerning Florida wetlands, presentation to the court of aerial photographs from 1971 to 1979,

---

[193] *United States v. Reserve Mining*, 380 F. Supp. 11 (D. Minn. 1974). In this case, the United States and the State of Minnesota brought an action against a taconite mining and processing company to prevent the company from continuing the discharge of taconite tailings into the waters of Lake Superior.

[194] See Crowsey Incorporated, *Legal Assistant Guide to Legal Applications for Geospatial Information*, June 10, 2002, 6.

[195] See K.J. Markowitz, Legal Challenge and Market Rewards to the use and Acceptance of Remote Sensing and Digital Information as Evidence, 12 *Duke Environmental Law and Policy Forum* (September 2002), 256.

[196] *Gasser v. United States*, 14 Cl. Ct. 476(1988). Plaintiff, owner of property interests in Baja California, Mexico, alleged that the construction and operation of Hoover Dam and Glen Canyon Dam in the United States resulted in a substantially reduced flow of water in the Colorado River, causing the development of a sediment blockage below their properties. They alleged that the river has either directly overflowed their land or caused elevated groundwater conditions, resulting in the taking of a permanent flowage easement.

[197] See R. Abeyratne, Satellite Imagery as Evidence at Pre Trial and Trial hearings, *German Journal of Air and Space Law*, 2003, 232.

[198] *Nickel Plate Mining Co. v. Office of Surface Mining Reclamation and Enforcement*, NX5-122-R (Department Interior 1987).

[199] *Oklahoma Water Resource Board v. Texas County Irrigation and Water Resource Association*, 711 P.2d 38 (OK 1984).

[200] See F. Frederick, Arizona uses IKONOS Imagery in Dispute with Developer, *Space News*, September 5, 2005.

combined with LANDSAT imagery for site orientation and proximity analysis, led to an out-of-court settlement by the defendant.[201]

On the other hand, there were some cases where satellite images were considered to be inconclusive, as in *Chevron USA, Inc. v. US EPA*,[202] or cases submitted to the court but not relied upon, as demonstrated in the case of *Wetsel-Oviatt Lumber Co. v. United States*.[203]

In India, remote sensing data has been used in a court to establish encroachment of forest land. In one case, a particular surface water pool which had gone dry was encroached upon and utilised for cultivation, and remote sensing data was used to provide proof to the district administration that there had indeed been a water pool in the past at that site.[204]

As mentioned previously, satellite images were used as evidence to identify the polluter responsible for oil spills. The Marine and Port Authority of Singapore collected samples of the oil slick and compared them with samples taken from the ship after its arrival at the port.[205] Consequently, the Singaporean tanker Song San was identified as the polluter and legal proceedings in Singapore were initiated based on the satellite data together with a laboratory analysis.[206] Subsequently, the ship owner was fined SGD 1.2 million as an estimated clean up costs incurred by the Republic of Singapore.[207]

The cases that we have looked at demonstrate that satellite images have been used in different scenarios to support the contentions of disputing parties. Yet these cases have also highlighted the limitations of satellite images as evidence,

---

[201] See American Society for Photogrammetry and Remote Sensing, *Earth Observation Systems legal considerations for 90's*, American Bar Association, Chicago, 1990, 38–39.

Aerial photograph were used to depict representative vegetation and disturbance areas to direct the on-site field investigations and to document the existence of these areas.

[202] *Chevron USA, Inc. v. US EPA*, 658 F. 2d 271 (5th Cir.,1981).

[203] *Wetsel-Oviatt Lumber Co. v. United States*, 40 Fed. Cl. 557(1998). This case involved ground water elimination, in which the Oklahoma Supreme Court reversed the lower court's decision to grant an oil company a permit to withdraw groundwater for an oil recovery process. Aerial photograph were used to depict representative vegetation and disturbance areas to direct the on-site field investigations and to document the existence of these areas.

[204] See B.I. Deekshatlu, Overview of Legal Aspect of Remote Sensing, *Recent Trends in International Space Law and Policy*, Lancers Books, New Delhi, 1997, 233.

[205] See P.B. De Selding, Courts Limit Use of Satellite Imagery for Prosecuting Suspected Polluters, *Space News*, 2 March 2005.

[206] See N.J. Brehon, Le Satellite au Service de l'Environnement, *Air & Cosmos* (May 25 2001), 52.

[207] See E. Nabet, Legal Aspects of the Use and Applications of Remote Sensing in South East Asia, 5 *Singapore Journal of International and Comparative Law* (2001), 161.

since they were occasionally not relied upon or did not clarify the disputed point. In order to overcome the current limitations, it is necessary to identify the gap between what is required as evidence and the characteristics of satellite images. To do this, it is useful to look at how remote sensing data would fit in the framework of evidence.

3.4.3.2. *The Examination of Remote Sensing Data as Evidence*

Use of satellite images as evidence is still in its infancy.[208] The previous section illustrated that satellite imagery often has been used either as a source of information on which the court based its opinion, or accepted by the court as complementarily to other forms of evidence upon which to decide the case.[209] It is worthwhile to explore how evidence is established in an international and national context and how satellite images would fit into such frameworks. This section first provides the definition of evidence and types of admissible evidence. Then, the section analyses the value of satellite images amongst other forms of evidence, and the requirements for admission of evidence.

Munir contends that "evidence is anything, which demonstrates, clarifies or shows the truth of a fact."[210] The aim of this section is to consider how satellite images can be used to demonstrate or clarify the truth; for this, it is useful to investigate national evidence law with respect to satellite images, using examples from US evidence law and German evidence law.

Satellite images can serve as reference materials to the court in making decisions directly or indirectly when admitted either as demonstrative evidence stipulated under Rule 1006 of the Federal Rules of Evidence[211]or as scientific evidence under Rule 702–703.[212]

Rule 1006 of the Federal Rules of Evidence states that "the contents or voluminous writings, recordings, or photographs which cannot conveniently be examined in court may be presented in the form of a chart, summary or calculation."[213] Hence, information obtained through satellite remote sensing, after being transformed into recognizable images, can serve directly as evidence to demonstrate a particular fact.

---

[208] See R. Purdy, Satellites: A New Era for Environmental Compliance? 3-5 *Journal for European Environmental and Planning Law* (2006), 406.

[209] Examples include *Oklahoma Water Resource Board v. Texas County Irrigation and Water Resource Association, Song Son Case, ICJ delimitation case* and *Nickel Plate Mining Co. v. Office of Surface Mining reclamation and Enforcement.*

[210] See A.B. Munir, Electronic Commerce: Legal and Policy Frameworks, *Cyber Law Policies and Challenges*, Butterworth Asia, Kuala Lumpur, 1999, 208.

[211] Rule 1006 of the Federal Rules of Evidence.

[212] Rules 702 and 703 of the Federal Rules of Evidence.

[213] Rule 1006 of the Federal Rule of Evidence.

On the other hand, scientific evidence is knowledge upon which an expert has based an opinion, as stated in Rule 702 of the Federal Rules of Evidence.[214] If accepted as scientific evidence, satellite imagery could be used to demonstrate a fact through contributing to another form of evidence – expert opinion.

In Germany, admissible evidence includes but is not limited to, witness, expert witness, written opinions of witnesses and expert witness, visual evidence and local inspections, public records and files, public documents, and other enquiries of any kind.[215] It can be construed that the satellite image falls under the definition of admissible evidence since photographs are categorized as visual evidence; it would appear that the satellite imagery physically presented in a court proceedings is regarded as visual evidence rather than private records,[216] as it demonstrates a fact through visual representation.

In the examples above on a German case, under the existing framework surrounding evidence, satellite images are not referred to specifically, but can nevertheless be categorized as admissible evidence just like a photograph. The issue to consider now is whether the form of evidence determines the degrees of reliance (e.g., one form of evidence carries more weight with the Court than another form).

The weight placed upon different types of evidence – such as witness testimony, or photographs – varies by jurisdiction and depends upon the circumstances of each case. More weight is given to a recording in the United States courts than in the French court procedures.[217] In some US cases witness testimony was more heavily relied upon than photographs and satellite images, but vice versa in other cases. Photographs were introduced in the case *DIC of Delaware, Inc.*, et al. *v. Tehran Redevelopment Corporation* et al.,[218] but were given little evidentiary weight by the Tribunal. Furthermore, in *Nickel Plate Mining Co. v. Office of Surface Mining Reclamation and Enforcement (OSMRE)*,[219] maps made of aerial photographs and satellite images presented as evidence by

[214] Rule 702 of the Federal Rules of Evidence states that "if scientific, technical, or other specialized knowledge will assist the trier of fact to understand the evidence or to determine a fact in issue, a witness qualified as an expert knowledge, skill, experience, training or education, may testify thereto in the form of an opinion or otherwise."

[215] See C. Arzt, Use of Satellite Imagery in Legal Proceedings, 24-4/5 *Air & Space Law* (1999), 197.

[216] See C. Arzt, Use of Satellite Imagery in Legal Proceedings, 24-4/5 *Air & Space Law* (1999), 197.

[217] See S. Courteix, Toward the legal Recognition of a new method of proof for the defence of the Environment, *Proceedings of the Thirty-Seventh Colloquium on the Law of Outer Space*, IISL, AIAA, 1994, 228.

[218] *DIC of Delaware, Inc.*, et al. *v. Tehran Redevelopment Corporation* et al, 8 Iran-US CTR 144 (1985). [Hereafter: DIC of Delaware case]. See C. Ameransinghe, *Evidence in International Litigation*, Martinus Nijhoff Publishers, Leiden, 2005, 394–395.

[219] See *Nickel Plate Mining Co. v. Office of Surface Mining Reclamation and Enforcement*, NX5-122-R (Department Interior 1987).

OSMRE were admitted by the Court; however, the Court gave greater weight to Nickel Plate's witnesses, who had actually seen the area themselves. On the other hand, as mentioned previously, in *Gasser v. USA*, the Court rejected the evidence of an expert witness that flooding had increased in a particular area, whilst accepted satellite imagery on the same point as evidence to the contrary.[220]

There appears to be no absolute hierarchy of different types of evidence in terms of admissibility. The value of evidence lies in its ability to persuade the court.[221] Such persuasion hinges on both relevance and reliability. Indeed, in the *Daubert v. Merrel Dow Pharm* case,[222] the Trial Court emphasized that scientific evidence will be admitted with caution and it needs to be ensured that the evidence submitted is not only relevant but also reliable.[223]

In the United States, relevant evidence is defined as "evidence having any tendency to make the existence of any fact that is of consequence to the determination of the action more probable or less probable than it would be without the evidence."[224] Evidence that does not address the disputed point would not be taken into consideration. As data collected at a time irrelevant to the disputed point is useless, the date(s) of data collection is a critical element for the admissibility of evidence.

Evidence submitted without sufficient information or background on the evidence itself does not qualify as 'relevant evidence'. Indeed, in the *DIC of Delaware case* the admissibility of photographs was questioned because no evidence was available to the tribunal as to: who took the photographs; the qualifications of the person who provided the commentary to each picture; when the photography was performed (the date and place that the pictures were taken); or who was responsible for the alleged defects shown.[225] Clearly, availability of information on the evidence is critical.

With regard to reliability of evidence, Rule 95 of the Rules of Procedure and Evidence of the International Criminal Tribunal for the Former Yugoslavia (ICTY) states that "no evidence shall be admissible if obtained by methods

---

[220] See American Society for Photogrammetry and Remote Sensing, *Earth Observation Systems legal considerations for 90's.*, American Bar Association, Chicago, 1990, 232.

[221] See C. Arzt, Use of Satellite Imagery in Legal Proceedings, 24-4/5 *Air & Space Law* (1999), 208.

[222] *Daubert v. Merrell Dow Pharmaceuticals*, 509 U.S. 579 (1993).

[223] See *Daubert v. Merrell Dow Pharmaceuticals*. This obligation is stated under Rule 104(a) of Federal Rule of Evidence stipulating "Preliminary questions concerning the qualification of a person to be a witness, the existence of a privilege, or the admissibility of evidence shall be determined by the court, subject to the provisions of subdivision (b). In making its determination it is not bound by the rules of evidence except those with respect to privileges."

[224] Rule 401 of the Federal Rules of Evidence.

[225] See C. Amerasinghe, *Evidence in International Litigation*, Martinus Nijhoff Publishers, Leiden, 2005, 394–395.

which cast substantial doubt on its reliability or if its admission is antithetical to, and would seriously damage, the integrity of proceedings."[226] Although the rule is applicable to criminal proceedings only, it can be a reference to the requirements for admittance of evidence in other proceedings. The key element affirmed here is the reliability of both the method of collecting evidence and the content.

Applying the above criteria to satellite images, they may be inadmissible if the sensor is defective, or if the images are unclear or interpreted incorrectly. Unclear images would only confuse the court instead of clarifying a fact for the court. Likewise, evidence that is misinterpreted by the processing specialist and/or data expert is likely to mislead the court. Indeed, the court appears to seek whether objective measures were taken to verify the accuracy of data prior to its admittance into evidence. In the aforementioned *West – Oviatt Lumber v. United States*,[227] although a satellite image was introduced as evidence, the court found faults with the assessments, including the failure of the US Forest Service to supply ground-truth information associated with the image;[228] therefore, the court did not rely on the submitted satellite image. It can be inferred from this case that admissibility will be more likely if data is ground verified and, if possible, other measures are also taken to verify accuracy.

It is important to ensure that the evidence in question shows the truth of a fact through a process that establishes the genuineness of the information. In fact, authentication is regarded essential for admissibility of evidence.[229] The aforementioned MARPOL has a relevant section that refers to authentication as a requirement. Annex 2 of Resolution IMO A 152 of 17 November 1983 provides that all satellite observation, photographs, and documentation must be supported by a signed certificate guaranteeing their authenticity.[230]

In the United States, litigators who offer satellite imagery into evidence at trial have to qualify an expert to introduce and explain such imagery, authenticate it, and prove the contents of data carried in the images – whether the imagery is accepted as scientific evidence or illustrative evidence.[231]

---

[226] UN Doc IT/32/Rev.3, 30 January 1995, 57.

[227] See B.I. Deekshatlu, Overview of legal aspect of remote sensing, *Recent Trends in International Space Law and Policy*, Lancers Books, New Delhi, 1997.

[228] See K.J. Markowitz, Legal Challenge and Market Rewards to the Use and Acceptance of Remote Sensing and Digital Information as Evidence, 12 *Duke Environmental Law and Policy Forum* (Sept. 2002), 257.

[229] See A. Gahtan, *Electronic Evidence*, Carswell Publishing, Toronto, 1999, 157.

[230] See NPA Group, *Final Report BNSC Sectors Studies Programme Applications of Earth Observation to Legal Sector*, BNSC, London, 2001, 64.

[231] See American Society for Photogrammetry and Remote Sensing, *Earth Observation Systems legal considerations for 90's*, American Bar Association, Chicago, 1990, 234. See US Federal Rules of Evidence, 901(a).

It should be emphasized that expert opinion can be tricky. Expert opinion is generally recognized to support the authenticity and accuracy of data; however, it may raise issues as to the expert's objectivity. A satellite image accompanied by expert testimony does not necessary convince the court as to its authenticity. Opinions are subjective by nature and can even be manipulative. Situations may arise where expert opinion on the interpretation of data on one side may be challenged by experts on the opposite side.[232] This happened in reality in the ICJ case when, in the dispute resolution between Cameroon and Nigeria, the parties presented conflicting interpretations of the satellite image; because of the different opinions on the satellite image, presentation of the images did not clarify the point in dispute and was thus ineffective as evidence.[233] US evidence law takes a cautious stance as to such evidence which is admissible as to one party whilst not admissible as to the other.[234] Admissibility of satellite images will be improved if the court can confirm their authenticity and accuracy objectively.

From the foregoing analysis, it appears that the law of evidence, as well as case law, has not yet developed to a sufficient extent to provide uniform rules for admissibility of satellite imagery, but has highlighted that it is primarily a question of authenticity and accuracy of evidence, which is a fundamental element that a court would seek for any other evidence submitted. The value of satellite data as evidence will be recognized more widely and established globally if its reliability, authenticity, and accuracy are ensured in a systematic and objective manner. For this, harmonized rules at the international level are desirable.

3.5. The Need for Improvement to the Legal Regime of Remote Sensing

As has been discussed, the widespread use of satellite data for environmental applications is not yet realized because of the uncertain data reliability, as well as barriers to their accessibility as a solid information source. These specific problems need to be further explored and analyzed to consider what kind of improvements are required to the current legal regime.

---

232 See B.I. Deekshatlu, Overview of legal aspect of remote sensing, *Recent Trends in International Space Law and Policy,* Lancers Books, New Delhi, 1997, 233.

233 See Reports on the legal aspects of the privatisation of space activities: Remote Sensing and National Space Legislation, Berlin Conference 2004, International Bar Association, 12–13.

234 Rule 105 of the Federal Rules of Evidence states: "When evidence which is admissible as to one party or for one purpose but not admissible as to another party or for another purpose is admitted, the court, upon request, shall restrict the evidence to its proper scope and instruct the jury accordingly."

3.5.1. *The Problem of Reliability of Satellite Data: Technical Observations*

The reliability of satellite data has, in fact, been a major obstacle for its use for enforcement purposes – including use as evidence in court. As previously mentioned, the studies related to the Kyoto Protocol and MARPOL indicated barriers to establishing the reliability of data in support of these agreements. An examination of the problems with reliability requires an understanding of the nature of satellite data, as well as the processes involved before its transformation into useful information.

3.5.1.1. *The Nature of Satellite Images*

The technical nature of digital imagery easily exposes satellite data to intentional and unintentional errors. Digital satellite imagery differs from traditional analogue photography as there is no original negative; instead digital imaging relies on the visual representation of numerical data.[235] In that regard, satellite images may involve certain elements of subjectivity as data is transmitted to Earth in a digital format, which will then requires human operations to produce the recognizable satellite images.[236] If the image enhancement and interpretation is conducted manually, assumptions by the image processors and interpreters are inevitable. Consequently, products resulting from manual processing may potentially involve unintentional human errors. Furthermore, satellite data, just like other forms of electronic evidence, is more susceptible than paper-based evidence to data manipulation.[237] Satellite imagery – just as photographs and computer generated visual evidence – can be manipulated,[238] especially with current technology, where one can process or enhance the data in varying degrees to bring out the feature of interest and, specifically, to change the pixels' spectral values in a satellite image.[239] This results in scepticism over the use of such data as evidence in court. Such shortcomings of digital images can be overcome if measures are taken to ensure that evidence is free from data manipulation.

3.5.1.2. *How the Reliability of Satellite Data should be Established*

The fundamental measure for ensuring the reliability of digital imagery is to verify the processes of how data is collected, processed, and interpreted by experts; however, there are no unified mechanisms or guidelines for reference.

---

[235] See R. Purdy, Satellites: A New Era for Environmental Compliance? 3-5 *Journal for European Environmental and Planning Law* (2006), 411.

[236] See Crowsey Incorporated, *Legal Assistant Guide to Legal Applications for geospatial information*, June 10, 2002, 24.

[237] See A. Gahtan, *Electronic Evidence*, Carswell Publishing, Toronto, 1999, 73.

[238] See S.H. Hodge, Satellite data and environmental law: technology ripe for litigation application, 14-2 *Pace Environmental Law Review* (1997), 710–713.

[239] See Crowsey Incorporated, *Legal Assistant Guide to Legal Applications for geospatial information*, June 10, 2002, 24.

Presently, data generators, processing wholesalers, and information extractors are all conducting data handling activities differently. Varying methods of processing are adopted and different approaches are taken as to the interpretation and analysis of images. No criteria are established internationally as to the different ways of processing and interpretation and there is no consideration of the implications that these differences may have on the resulting products. Consequently, it is difficult for the users, expert witness, and third parties – such as the court – to provide objective judgment on the authenticity of presented evidence and accuracy of its content.

In considering how mechanisms and guidelines for data authentication and accuracy validation should be established, it is useful to refer to the *DIC of Delaware case*. This case illustrated that the unknown data history and lack of information about the data hampered its admissibility. Comprehensive metadata and fidelity in data history are major factors for establishing reliability. Mechanisms and guidelines that incorporate such elements should be established in an international context. In this way, different types of products from multiple sources can be compared more effectively and assessed objectively to determine whether the data and/or derived products in question are suitable as evidence for admittance to a court.

3.5.2. *The Need for Improvement to the Legal Framework in Association with Data Policy*

Another area for improvement concerns inadequate data policies on remote sensing data for the purpose of environmental protection. A user-friendly data environment is essential for any community concerned with environmental protection to benefit from the full extent of its available resources. A coordinated EO programme for environmental applications such as Global Monitoring for the Environment and Security (GMES), for instance, is a good subject for consideration for improved data policies. The GMES space-based network will involve the use of existing satellites, both national and international, as well as the development of new infrastructure, including new satellites to be built collectively by ESA and the European Commission.[240] GMES looks at applications tied to specific environmental treaties such as the Kyoto Protocol, as well as a number of other applications from marine services to atmospheric monitoring.[241] The problem is that numerous programmes within the framework of GMES exist, but it is difficult for users to understand the scope and mechanism for a given program. Most of all, users have trouble actually accessing and using

[240] See T. Hitchens, International Satellite Innovation and Cooperation, *Military Satellites 2006*, April 18 2006, 5.

[241] For an overview of GMES and details of different services of applications see http://www.gmes.info/. Last accessed on 10 June, 2010.

the data that are made available under each programme. Such a situation is not limited to GMES, but reflects a trend that can be seen in any programme designed for environmental protection. Key problems lie in high-threshold access to data, and data sharing beyond a single programme is discouraged by the data supplier.

3.5.2.1. *Data Accessibility*

The time and resources required to search for existing data poses a major obstacle for access to the data, but perhaps, the foremost hurdle is the cost factor. Often, a large amount of data is required for the monitoring and assessment of the state of the environment and meeting such data needs is costly. In the project to assist the Ramsar Convention, a significant cost was incurred to map all 1,400 Ramsar wetland sites around the globe.[242] The significant cost for data is, on one hand, burdensome to users; on the other hand, it is inevitable from the perspective of data providers. The high cost of establishing and maintaining a space remote sensing system has led to the common practice of satellite operators distributing data on a commercial basis.[243]

Environmental agencies have no special access to data for environmental applications, although in reality, data has been made available for such applications on favourable conditions. For example, the developers of the VEGETATION sensor have offered to provide the data to the scientific community free-of-charge to assist in the Millennium Ecosystem Assessment.[244] In fact, uniformity has not been achieved for the cost of access to data for the purpose of environmental protection. The reality is that data is made available at a favourable cost only to certain users under specific arrangements or programmes through good will on the part of the providers. (e.g., data access by UNESCO under the Open Initiative) The fact remains that data availability and accessibility established under the UN Remote Sensing Principles, as discussed in Chapter 2, does not confer privileges upon any users or favour specific uses. Broader communities would be able to benefit from lower threshold access to data if a data policy that allows preferential treatment for environmental applications were to be formerly adopted.

---

[242] See R. Harris, and R. Browning, *Data Policy Assessment for GMES Final Reports*, EVK2-CT-2002-80012-DPAG., University of London, London, 2004, 59.

[243] See P.R.M. Serra & M.N. Barbosa, The point of view of an user country:A) Brazil, *Droit télédétection et environnement: actes du colloque international: le droit face aux techniques de télédétections par satellite au service de l'environnement*, S. Courteix (Eds.), Strasbourg, 2-3-4 juin 1993, Sides, Antony, 1994, 209.

[244] See K. Kaline & K. Raustiala, International Environmental Agreements and Remote Sensing Technologies, Background paper prepared for the workshop on Remote Sensing and Environmental Treaties: Building more Effective Linkages, December 4–5, New York, 2000, 31.

3.5.2.2. *Data Sharing*

The second problem concerns the issues of data sharing. Currently, data made available for a specific project appear to stay within the project. Such data are for designated users under the project and are not available to other relevant communities. Indeed, the same data is needed for different projects; nevertheless, restrictive policies prevent the data from being shared effectively. For instance, many researchers in GMES projects, such as BIOPRESS, have faced enormous difficulties obtaining data from the Natura 2000, an ecological network which is mapping 12,000 to 15,000 biodiversity sites in Europe, because the dataset is not released for reasons of confidentiality.[245] The same datasets may be purchased individually by different entities – an unnecessary duplication of effort.

Data use is not rationalized at present, but there is considerable room for rationalization, since it is likely that the data requirements for environmental applications overlap. For example, there are some overlaps between the sites registered under the World Heritage Convention, the Biodiversity Convention, and the Ramsar Convention. The Laguna Blanca of Argentina, the Dalai Lake of China, and Lac Saint–Pierre of Canada are all registered both as Biosphere reserves[246] and as wetlands. Other overlaps also exist in the world heritage sites and biosphere reserves: *inter alia*, the central Amazon conservation complex of Brazil, the Dja Faunal Reserve of Cameroon, the Waterton Glacier International Peace Park in Canada, and the Virunga National Park in the Democratic Republic of Congo are all inscribed in both categories. These overlaps demonstrate the potential for rationalization of data use. The codification of data policies to allow effective data sharing would benefit on-going and future environmental applications that involve demanding data needs. Such a rationalization in environmental assessment will reduce the costs and hurdles for access to the data, contribute to more widespread use of data, and is likely to enhance the overall levels of achievement for environmental protection. In these ways, both issues of limited data accessibility and data sharing give rise to the need for an overall data policy that facilitates environmental applications and uses.

3.6. Conclusion

Today, environmental degradation is one of the most pressing concerns for human kind to undertake. The rate of environmental degradation has exceeded

---

245 See R. Harris, and R. Browning, *Data Policy Assessment for GMES Final Reports*, EVK2-CT-2002-80012-DPAG., University of London, London, 2004, 92.

246 A list of biosphere reserves is available at www.unesco.org/mab/BRs/BRlist.shtml.

our countermeasures; thus, the need to take drastic environmental measures is urgent. It is critical that the regime surrounding environmental protection function to its fullest extent in achieving this challenging task. The current environmental regime does not necessarily have to change, but efforts to maximize the effect of existing MEAs and to call for increased participation of various actors in consolidating domestic measures are essential.

Satellite remote sensing can contribute extensively to make up for inadequacies in the existing regime, and, for certain MEAs, remote sensing serves as a powerful tool for implementation. Regardless of the nature of a legal instrument – whether or not it endorses concrete obligations to influence the behaviour detrimental to the environment, or whether or not it addresses urgent environmental matters affecting mankind or those that are less urgent – remote sensing can make MEAs more effective by serving as a basis for implementation of projects and by adding a significant source of information. It can facilitate the participation of international organizations and other non-state actors by opening up the possibilities to take measures through more outside/beyond state framework under the initiatives of international organizations and other non-state actors.

The full benefits of this powerful tool have not yet been realized. Major obstacles exist as to the use of data for enforcement purposes, including evidentiary hindrances, despite the growing recognition for such applications. The nature of satellite images makes them susceptible to errors and manipulation which, coupled with the lack of an international framework addressing the digital evidence, has limited their use for verification. It is necessary to ensure the authenticity and accuracy of data through the use of systematic and objective procedures.

Current limitations with regard to data availability and accessibility hamper meeting the demanding needs of users for routine monitoring of the changing state of the Earth's environment and for planning accordingly for the appropriate measures to better protect it. Data access and use can be rationalized and made more user friendly, particularly considering the right of the public to access environmental information that is increasingly recognized.

The role of the legal framework of remote sensing is crucial in promoting data use, achieving effective data sharing, and enhancing data utility for environmental applications. In order to maximize satellite remote sensing as a tool to achieve a higher level of environmental protection, it is critical to establish clear data policies that ensure easier access, as well as to ensure the integrity of the data. Attention to these matters would facilitate and proliferate the use of data, generate greater confidence in the use of EO data for critical applications, and ultimately contribute extensively to meeting the challenging tasks of protecting the environment for the benefit of mankind.

Chapter Four

Disaster Management

4.1. Introduction

The world is becoming increasingly vulnerable to both natural and man-made disasters. In the decade from 1997 to 2006, the number of reported disasters increased from 4,241 to 6,806, and the reported deaths doubled from 600,000 to 1.2 million compared to the previous decade.[1] On the other hand, mankind's capabilities to cope with disasters have been enhanced by the advance of technology – including satellite remote sensing. Remote sensing satellites have been used not only for monitoring the disasters themselves but also in the aftermath of the disasters: for example, it has proven particularly useful in the case of the 2004 Indian Ocean Tsunami.

Although it is not possible to prevent natural disasters, their impacts can be mitigated. Space technology, and specifically remote sensing from space, can contribute extensively to disaster management. Its capability to provide both global coverage and the ability to make repeated observations of specific areas of interest makes satellite remote sensing well suited for monitoring the progress and effect of a diverse range of disasters.

Space technology has been extensively used in various stages of disaster management: risk assessment, mitigation, response, and reconstruction. At each of these stages, space technology contributes significantly – particularly through the use of remote sensing data. Satellite images captured before and after the

[1] See *World Disaster Report* 2007, International Federation of Red Cross and Red Crescent Societies, 174.

disaster are compared to identify the scale of the disaster and to assess the extent and nature of the damage. Aid is then directed on the basis of the information derived from satellite images.

The current legal regime of remote sensing remains inadequate, particularly with regard to the liability issues identified in Chapter 2, which are not addressed by the general disaster management legal framework; hence, critical issues are raised in international disaster relief as to the uncertain liability regime affecting the interests of both donors and recipients of the aid. Chapter 4 explores these issues, clarifying the framework of disaster management, demonstrating the need for an improved liability regime, and offering an appropriate model for disaster response.

4.2. Framework of Disaster Management

The framework for disaster management is much more un-coordinated and lacking in structure than that for environment monitoring, as illustrated here through examination of the background on disaster management, the current practice regarding disaster management, the current legal framework. First of all, some background on disaster management should be given, including the definition of disaster management, the scope and different phases of disaster management, and the actors involved.

4.2.1. *Definition*

A disaster is "an event concentrated in time and space, in which a society or a community undergoes severe danger and incurs such losses to its members that the social structure is disrupted and the fulfilment of all or some of the essential functions of the society is prevented."[2] Disasters can be categorized into two types, natural disasters and man-made disasters. According to a traditional view, natural disasters are those resulting from the effects of natural phenomenon and they may be further divided according to meteorological, topological, or tectonic causes.[3] Examples of natural disasters include earthquakes, floods, avalanches, typhoons, tornados, hurricanes, cyclones, drought, erosion, volcanic eruptions, and epidemics.

Man-made disasters that occur as a result of human activity, whether accidentally or from negligence, can be further categorized into those known as

---

[2] B. Jakovljevic, International Disaster Relief Law, 34 *Israel Yearbook on Human Rights* (2004), 251.

[3] See P. Macalister-Smith, *International Humanitarian Assistance Disaster Relief Actions in International Law and Organization*, Martinus Nijhoff Publishers, Geneva, 1985, 2.

technological disasters and those resulting from deliberate actions known as armed conflict situations.[4] The former includes oil spills, fires, train accidents, and nuclear accidents, whilst the latter includes civil disturbances, terrorism, internal strife, riots, and persecutions based on the violation of certain basic human rights.[5] It should be noted that there are also disasters that fall between the two categories, such as landslides, which occur largely from natural causes, but are partly due to large-scale deforestation.

The previous chapter on environment touched upon some aspects of man-made environmental disasters, such as nuclear operations; in this Chapter, the focus will generally be on natural disasters. Disaster management is a comprehensive term, which refers to all activities concerned with prevention, preparedness, and response.[6] Each phase can be further categorized into various operations.

4.2.2. *Different Phases of Disaster Management*

Different phases of disaster management involve various operations to help the population mitigate the impact of disasters, respond appropriately to them, recover from the aftermath, and return to a normal state. Each phase contributes to saving lives; these different phases of disaster management are often labelled as prevention, preparedness, response, and recovery and reconstruction.

4.2.2.1. *Prevention*

We cannot generally prevent natural disasters, but the impact can often be mitigated by taking appropriate measures beforehand. Prevention includes the evaluation of hazards, the analysis of vulnerability and risk, and the establishment of legislation and regulatory measures.[7] A concrete example of the evaluation of hazards is the creation of a hazard map, which is used to identify vulnerable areas and to support the effective evacuation of the threatened population, when appropriate. For instance, the volcanic eruption that occurred in Mount

---

[4] See P. Macalister-Smith, *International Humanitarian Assistance Disaster Relief Actions in International Law and Organization*, Martinus Nijhoff Publishers, Geneva, 1985, 2.

[5] See B. Jakovljevic, International Disaster Relief Law, 34 *Israel Yearbook on Human Rights* (2004), 251.

[6] See Office of the United Nations Disaster Relief Coordinator (UNDRO), *Space Application and Dissemination of Disaster-Related Data*, Expert Meeting Geneva, 14–17 June 1983, 13.

[7] See Office of the United Nations Disaster Relief Coordinator (UNDRO), *Space Application and Dissemination of Disaster-Related Data*, Expert Meeting Geneva, 14–17 June 1983, 14.

Usu in Japan in 2000 resulted in no loss of life, thanks to the hazard maps that were created and distributed to residents beforehand, leading to a smooth evacuation operation.[8]

4.2.2.2. *Preparedness*

Disaster preparedness actions are intended to minimize loss of life and damage to infrastructure, as well as to organize and facilitate timely and effective rescue, relief, and rehabilitation in the area where the disaster occurs.[9] Preparedness is concerned with forecasting, warning, organization for, and management of disaster situations – including preparation of operational plans.[10] The creation of effective emergency plans is critical. Following an earthquake in 2001, El Salvador's response was gravely limited because there was no emergency plan.[11]

Forecasting is the attempt to predict the occurrence of disasters beforehand, and may be conducted for earthquake, tsunami, severe weather conditions (such as tropical storms), and volcanic eruptions. National weather services, as well as regional meteorological centres, provide storm forecasts.

Warning is defined as a short timescale forecast when a disaster is known to be imminent. Technological developments have increased the availability, reliability, and accuracy of short-term disaster warnings, particularly in cases of tropical storms, high rainfall, floods, and volcanic eruptions.[12] Warning makes it possible to mitigate the impact of certain types of disasters; indeed, it saves people's lives by facilitating suitable displacement of potentially affected population prior to the disaster. For instance, following the eruption of Mount Pinatubo in June 1991, the Philippine Institute of Volcanology and Seismology, together with the USGS, issued an early warning to the communities at risk; as a result, 35,000 people were safely evacuated before the eruption.[13]

---

[8] See Increasing Natural Disasters, *Newton*, Newton Press, Tokyo, 2004, 38.

[9] See Office of the United Nations Disaster Relief Coordinator (UNDRO), *Space Application and Dissemination of Disaster-Related Data*, Expert Meeting Geneva, 14–17 June 1983, 15.

[10] See Office of the United Nations Disaster Relief Coordinator (UNDRO), *Space Application and Dissemination of Disaster-Related Data*, Expert Meeting Geneva, 14–17 June 1983, 15.

[11] See C. Villacis, Latin American Cases Hurricane Mitch, Flash Floods and Landslides in Venezuela, El Salvador Earthquakes, *Report: Recovery and reconstruction reports of the worldwide disasters and their comparative study,* Asian Disaster Reduction Center, Kobe, 2005, 21.

[12] See Office of the United Nations Disaster Relief Coordinator (UNDRO), *Space Application and Dissemination of Disaster-Related Data*, Expert Meeting Geneva, 14–17 June 1983, 16.

[13] See De Guzman, E.M., Eruption of Mount Pinatubo in the Philippines, *Report: Recovery and reconstruction reports of the worldwide disasters and their comparative study*, Asian Disaster Reduction Center, Kobe, 2005, 10.

Tsunami warnings can also mitigate the impact, as potentially affected populations can move away from the sea shore following such a warning. For example, NOAA operates two tsunami warning centres – the Alaska Tsunami Warning Center, which covers Alaska, British Columbia, Washington, Oregon; and the California and the Pacific Tsunami Warning Center, which provides warnings to almost every country around the Pacific Rim and to most of the Pacific Island states.

The damage to homes and infrastructure threatened by or resulting from major disasters in many cases necessitates the temporary relocation of at least part of the affected population[14] to evacuation centres in order to mitigate loss of life – particularly in case of volcanic eruption, earthquake, floods, fire, and landslides. Evacuation plans need to be made based on accurate information clearly indicating the dangerous and safe areas and the extent and severity of the impending disaster, as there are inherent dangers with large-scale evacuations and these must not outweigh the risk from the disaster itself. Of course, the choice of the evacuation destinations must be suitably selected, so as not to pose a risk greater than the original disaster. Large-scale disasters have required considerable numbers of people to be evacuated; following the flood that hit 24 provinces in China in 2007, 3 million persons were evacuated.[15] Precise instructions for the evacuation and prompt relocation of the affected population to appropriate areas are critical if the evacuation is to be beneficial.

4.2.2.3. *Response*

Disaster response is defined as the actions taken following a disaster, and entails the assessment of the impact of disasters, search and rescue operations, and the provision of appropriate relief provided in accordance with such an assessment. This section addresses different types of operations involved in disaster response.

The first step should consist of an assessment of the situation: both as to the extent of damage as well as to relief requirements.[16] The extent of damage is often unknown immediately following a disaster, particularly in the case of catastrophic disasters such as the Indian Ocean Tsunami in 2004, the cyclone that hit Myanmar, and the earthquake that struck China in May

[14] See Office of the United Nations Disaster Relief Coordinator (UNDRO), *Space Application and Dissemination of Disaster-Related Data*, Expert Meeting Geneva, 14–17 June 1983, 17.

[15] Widespread flooding caused over 500 deaths and did enormous damage to dams, roads, buildings, and crops. The cities of Fuyang and Huainan in Anhui province, were particularly hard hit.

[16] See Office of the United Nations Disaster Relief Coordinator (UNDRO), *Space Application and Dissemination of Disaster-Related Data*, Expert Meeting Geneva, 14–17 June 1983, 17.

of 2008. By conducting such assessments, the governments of afflicted states or other organizations gain understanding of the precise condition and extent of damage in order to give appropriate instructions for aid. For an accurate assessment, timely access to relevant information is critical. Indeed, following floods and landslides in Venezuela in December 1999, the lack of information impeded the initial evaluation of damage and delayed subsequent response activities.[17]

Many disasters – especially floods, earthquakes, and landslides – require subsequent search and rescue operations[18] which are critical in reducing the losses amongst the affected populations that are often stranded and/or injured. Such operations are conducted by aid workers with special equipment together with medical teams, recovery dogs, and ambulance units, as well as conducted by military forces – usually the national army of the afflicted states[19] and by troops sent from foreign countries. Search and rescue operations can also be carried out by local residents themselves, as when local residents rescued 98% of the 50,000 people who were pulled from beneath the rubble following the 1999 Marmara Earthquake in Turkey.[20]

Relief is the assistance and aid provided to people in need, implying actions taken to restore the affected population back to the level that existed prior to the disaster.[21] Relief can be both financial and material. Financial aid comes in the form of grants and loans from foreign governments and international organizations. Material aid usually takes the form of access to goods[22] and services[23] by the victims, and is an essential part of any relief operation.

---

[17] See C. Villacis, Latin American Cases Hurricane Mitch, Flash Floods and Landslides in Venezuela, El Salvador Earthquakes, *Report: Recovery and reconstruction reports of the worldwide disasters and their comparative study,* Asian Disaster Reduction Center, Kobe, 2005, 21.

[18] See Office of the United Nations Disaster Relief Coordinator (UNDRO), *Space Application and Dissemination of Disaster-Related Data*, Expert Meeting Geneva, 14–17 June 1983, 17.

[19] For example, 50,000 Turkish army troops began search and rescue operations three days after the Marmara earthquake. A.K. Sinha & S. Srivastava, Survey Report on the Reconstruction following the Earthquake in Marmara, *Report: Recovery and reconstruction reports of the worldwide disasters and their comparative study*, Asian Disaster Reduction Center, Kobe, 2005, 5.

[20] A.K. Sinha & S. Srivastava, Survey Report on the Reconstruction following the Earthquake in Marmara, *Report: Recovery and reconstruction reports of the worldwide disasters and their comparative study*, Asian Disaster Reduction Center, Kobe, 2005, 5.

[21] See B. Jakovljevic, International Disaster Relief Law, 34 *Israel Yearbook on Human Rights* (2004), 252.

[22] Goods include food, water, medicine, temporary shelters, clothing, blankets, and sanitation.

[23] Services include medical services such as vaccination, tracing service, communication, and transportation.

4.2.2.4. *Recovery and Reconstruction*

Recovery and reconstruction is the phase during which afflicted countries recover from the devastation in the aftermath of disasters and attempt to get back to a normal state. Examples of recovery and reconstruction include rebuilding infrastructure such as railways, water supply systems, communication and power; building of roads, bridges, homes, hospitals, schools, river and irrigation systems; restoration of economic activity; and restoration of damaged ecosystems.[24] Recovery and reconstruction often take months or even years, depending upon the scale and extent of the disaster.

Having identified the different phases of disaster management, the nature of each phase should be examined. There is more international assistance provided in the post-disaster phase than in the pre-disaster phase. Generally, the prevention and preparedness phase, such as risk assessment, is conducted at a national level; thus, taking the necessary measures is left to each individual government. Disparities exist across the variety of states, however, because of their individual capabilities and capacities for disaster prevention and preparation. Often less-developed nations do not have such capabilities. Some regions such as Asia are more frequently struck by natural disasters than other regions.[25] Although there are emerging initiatives to render international assistance in prevention and preparation for disasters to disaster-prone states,[26] it is during the response, recovery, and reconstruction phases that states can gain more international assistance.[27] In fact, international cooperation and implementation of disaster management programmes is most active in such phases.

4.2.3. *Actors*

It is primarily the responsibility of the afflicted state to respond to disasters and render assistance to satisfy the needs of its affected population.[28]

---

[24] See A.K. Sinha & S. Srivastava, Survey Report on the Reconstruction following the Earthquake in Marmara, *Report: Recovery and reconstruction reports of the worldwide disasters and their comparative study*, Asian Disaster Reduction Center, Kobe, 2005, 8.

[25] Asia is particularly vulnerable to disasters as it accounts for approximately 90% of the world crisis victims and more than 50% of the total losses suffered around the globe. See *Asian Natural Disasters Data Book*, Asian Disaster Reduction Centre, Kobe, 2002, 15.

[26] For example, the initiative by UNESCO to establish Tsunami Early Warning Systems in the Indian Ocean rim following the Indian Ocean Tsunami.

[27] There are many examples of assistance provided by international organizations in the response phases. For example, UNDP conducted evaluation and FAO coordinated provision of food following floods and landslides in Venezuela in December 1999.

[28] Guiding Principle 4 of the General Assembly Resolution 46/182; UN Doc. A/RES/46/182(1991), which led to the creation of the Department of Humanitarian

The government of the stricken country assesses the impact of potential disasters, plans relief operations, and gives directions for providing aid, evacuation, and rescue operations. National and local authorities of the afflicted state play an important role in emergency relief operations. Foreign governments provide material aid as well as financial aid such as loans and grants or a combination of both. For example, following the earthquake in El Salvador in 2001, the government of Spain sent material aid, including relief supplies, four doctors, and five search and rescue experts. Spain also offered of financial aid, amounting to a loan of $36 million for relief efforts.[29]

There are a number of international organizations involved in disaster management. Many specialized United Nations agencies take part in relief for natural disasters including, but not limited to, the United Nations High Commissioner for Refugees (UNHCR), the Office for the Coordination of Humanitarian Affairs (UNOCHA), the United Nations Children's Fund (UNICEF), the United National Development Programme (UNDP), the United Nations Environmental Programme (UNEP), the United Nations Education Science and Cultural Organization (UNESCO), the World Food Programme (WFP), the World Health Organization (WHO), and the International Telecommunication Union (ITU). These UN bodies provide humanitarian assistance as well as financial aid.

Intergovernmental organizations, such as the European Union (EU), the North Atlantic Treaty Organization (NATO), and the Association of South East Asian Nations (ASEAN) are also actively involved in relief efforts. Amongst these organizations, it should be mentioned that the European Union has been playing an active role in humanitarian assistance through the European Community Humanitarian Office (ECHO), which has been involved extensively in providing assistance not only to European nations but to countries outside Europe.[30]

NGOs play an active role in the work of international assistance. Unrestricted by political factors, the NGOs are often the first to arrive at the

Affairs states that "Each State has the responsibility first and foremost to take care of the victims of natural disasters and other emergencies occurring on its territory. Hence, the affected State has the primary role in the initiation, organization, coordination, and implementation of humanitarian assistance within its territory".

[29] See C. Villacis, Latin American Cases Hurricane Mitch, Flash Floods and Landslides in Venezuela, El Salvador Earthquakes, *[Report] Recovery and reconstruction reports of the worldwide disasters and their comparative study,* Asian Disaster Reduction Center, Kobe, 2005, 20.

[30] For example, ECHO provided approximately US$ 7.3 million in response to the earthquake in El Salvador in 2001. See A.K. Sinha & S. Srivastava, Survey Report on the Reconstruction following the Earthquake in Marmara, *[Report] Recovery and reconstruction reports of the worldwide disasters and their comparative study*, Asian Disaster Reduction Center, Kobe, 2005, 5.

disaster site.[31] The International Federation of the Red Cross and the Red Crescent Societies[32] are amongst the world's largest providers of humanitarian aid to the victims of natural disasters, both through the local work of over 100 national Red Cross and Red Crescent Societies themselves, and through the Federation's international support for that work.[33] In addition, many other voluntary organizations are engaged in relief operations,[34] such as Médecins Sans Frontières,[35] which is dedicated to providing medical assistance by sending medical teams, equipment, and supplies following natural and man-made disasters.[36] Last, but not least, the active role played by volunteers should be mentioned. During disasters, a large number of people with no pre-planned role typically arrive at the scene to offer assistance.[37]

Operations are not always effectively coordinated. The various actors involved in relief action, such as states and various organizations responding to disasters, often operate independently from one another with little interaction amongst themselves. Coordination is necessary to avoid confusion, overlapping efforts, and waste, and to establish efficient contacts between the donors and beneficiaries.[38] It has been noted that, particularly in the medium term reconstruction projects, international agencies were often working at cross-purposes.[39] The report on Hurricane Mitch that struck Central America in October 1998 points out that the substantial foreign assistance that came in the aftermath of the disaster was deployed in many cases largely to implement redundant and

---

[31] R. Evans, The Humanitarian Challenge: A Foreign Policy Perspective, 6-2 *African Security Review* (1997), 2.

[32] Created in 1919, the International Red Cross consists of the International Committee of the Red Cross and the League of Red Cross Societies; the International Red Cross consists of the International Committee of the Red Cross and the League of Red Cross Societies as well as over 100 internationally recognized National Red Cross and Red Crescent Societies.

[33] See P. Walker, Victims of natural disaster and the right to humanitarian assistance: a practitioner's view, 325 *International Review of the Red Cross* (1998), 611–617.

[34] See Office of the United Nations Disaster Relief Coordinator (UNDRO), *Space Application and Dissemination of Disaster-Related Data*, Expert Meeting Geneva, 14–17 June 1983, 19.

[35] Médecins Sans Frontières was founded by a small group of French doctors in 1971. See http://www.msf.org/.

[36] For example, following February the 1998 Afghanistan earthquake, it provided immediate medical assistance and emergency surgery.

[37] J.R. Wible, Volunteer's Liability Issues, Medical Reserve Corps National Leadership and Training Conference, April 17–20, 2007, 1.

[38] B. Jakovljevic, International Disaster Relief Law, 34 *Israel Yearbook on Human Rights* (2004), 269.

[39] See A.K. Sinha & S. Srivastava, Survey Report on the Reconstruction following the Earthquake in Marmara, *[Report] Recovery and reconstruction reports of the worldwide disasters and their comparative study*, Asian Disaster Reduction Center, Kobe, 2005, 19.

uncoordinated projects.[40] Such uncoordinated operations are largely ineffective and may even have harmful consequences. Furthermore, coordination was sometimes avoided deliberately. For instance, following the 2005 Indian Ocean Tsunami, the United States received criticism for its decision to set up its own coalition of governments to coordinate relief in the region, rather than working under the umbrella of UN efforts.[41] It is, therefore, very important that operations are coordinated so as to enhance efficiency and effectiveness.

4.2.4. *The Current Practice Regarding Disaster Management*

As noted above, various actors are involved in disaster response with different response times and often with little coordination. In this section, the trend of current practice will be analysed. Current practice is characterized by case-by-case responses and voluntary implementation efforts, but it is also encumbered with political and operational problems.

4.2.4.1. *The Case-by-case Response*

Reponses to disasters are essentially on a case-by-case basis; some countries respond better than others[42] and the disparity reflects their levels of preparedness and individual capacities. There were no search and rescue teams in El Salvador following the earthquake in 2001; the official rescue operation began only when international teams arrived from Mexico, Colombia, Venezuela, and Spain.[43] Even within the same nation, there may be significant variations in disaster response, in part because of different local government structures and the presence or absence of operational disaster field offices. For example, following the Northridge earthquake that happened in California in 1995, the US Federal Emergency Management Agency (FEMA) responded quickly – partly

[40] See A.K. Sinha & S. Srivastava, Survey Report on the Reconstruction following the Earthquake in Marmara, *[Report] Recovery and reconstruction reports of the worldwide disasters and their comparative study*, Asian Disaster Reduction Center, Kobe, 2005, 19.

[41] N. Lepp, Disaster Relief Politics Complicate South Asia Effort, *The Dominion*, January 3 2005, 1.

[42] For instance, following Indian Ocean Tsunami, the Indonesian president declared a national disaster, but unlike Sri Lanka, did not immediately call for international aid. See *World Disaster Report 2005*, International Federation of Red Cross and Red Crescent Societies, 82.

[43] See C. Villacis, Latin American Cases Hurricane Mitch, Flash Floods and Landslides in Venezuela, El Salvador Earthquakes, *[Report] Recovery and reconstruction reports of the worldwide disasters and their comparative study*, Asian Disaster Reduction Center, Kobe, 2005, 20.

because it had a local Disaster Field Office *in situ*,[44] in contrast to the slow response to Hurricane Katrina in 2005, where no local field office was in place. Even in an instance of response to the same disaster, there have been disparities amongst the responses across the affected areas. For example, the World Disaster Report 2005 indicates that following the Indian Ocean Tsunami, Ache, Indonesia experienced tremendous breakdowns in cooperation amongst various international actors providing assistance, leading to disparities where some areas were over-served and others were ignored.[45] The foregoing examples illustrate how the conduct of governments and organizations – particularly those of NGOs – in responding to disasters are not well regulated; thus, relief operations are often unpredictable and ineffective. These circumstances give rise to administrative problems, such as registrations of NGOs,[46] questions of taxes placed on relief supplies, entry visas for relief personnel,[47] and the liability for damage faced by relief personnel.

4.2.4.2. *Political Impediments Against Effective Provision of Aid*

As emphasized earlier, the fundamental principle for disaster response is that afflicted states decide whether to request and accept international assistance in case of a disaster.[48] This allows for a high degree of discretion by governments in disaster response, regardless of the imminent needs. The response can be affected by self-interest of governments who may be generous with political allies, but grant little or no aid to those countries with whom they have bad diplomatic relations, even when the objective needs of the suffering population is higher.[49] Some affected governments have also been reluctant to allow prompt

[44] See T. Yamashita, *Intelligence Strategy for Risk Corresponding Society*, Nikkei BP Publisher, Tokyo, 126.

[45] Information black hole in Aceh, *World Disaster Report 2005*, International Federation of Red Cross and Red Crescent Societies, 2005, 81.

[46] In Thailand, it has been reported that the process of registration involves lengthy paperwork. Following the Indian Ocean Tsunami, only a small portion of the NGOs involved in the tsunami relief and reconstruction have registered with Thai authorities.

[47] Personnel of non-registered NGOs, granted no work permit, had to enter the country on short term tourist visas, requiring them to leave and reenter the country every month.

[48] UNGA Resolution 46/182; UN Doc. A/RES/46/182(1991). Strengthening of the coordination of humanitarian emergency assistance of the United Nations, Annex I. 3 states that "humanitarian assistance should be provided with the consent of the affected country and in principle on the basis of an appeal by the affected country."

[49] See R. Evans, The Humanitarian Challenge: A Foreign Policy Perspective, 6-2 *African Security Review* (1997), 2.

external humanitarian assistance.[50] Following the February 1998 earthquake in Afghanistan, victims in Ghanj, the most affected village, had not received any international aid for six days – due partly to Afghanistan's purist Islamic Taliban militia that controlled approaches to the region.[51]

There are occasions when, strictly for political reasons, affected states have actually refused the aid offered from foreign governments. Following Hurricane Katrina, the US State Department rejected such an offer from Cuba on the grounds that the United States did not have full diplomatic relations with Cuba.[52] Likewise, following the Indian Ocean Tsunami, India refused to accept any foreign aid and denied the access of NGOs to the Andaman Islands, partly because of the presence of a key military base on Car Nicobar Island.[53] Internal relief operations have also been hampered by the political situation within a country. For instance, following the 2001 earthquake in El Salvador, many of the most affected areas such as Santa Tecla received little aid because of their municipalities' political affiliation with the opposition party.[54] Furthermore, following the cyclone that struck Myanmar in the beginning of May 2008, the military government of Myanmar refused international aid for weeks, resulting in unnecessary loss of life. In these ways, politics can have a disproportionately negative impact on disaster response and relief operations.

4.2.4.3. *The Practical Implementation on a Voluntary Basis*

The practical implementation of disaster response is achieved on a voluntary basis, or through agreements, rather than governed by any humanitarian law that obliges assistance. In reality, relief is a voluntary donation in principle, as no one is bound to offer it.[55] The aid provided is largely in the form of large and small donations or contributions from public and private donors. Examples of

---

[50] An example of delay in request for international assistance is a severe earthquake in 1990 that hit the Gilan province in Iran, killing more than 50,000 people and destroying entire villages. Cf. R.J. Hardcastle & A.T.L. Chua, 325-1 *International Review of the Red Cross* (1998), 589.

[51] "Politics of the Afghan earthquake", *The Economist*, 14 March 1998.

[52] S. Zunes, Arrogance in the Face of Disaster – The Bush Administration's Refusal of Cuba's Offer for Medical Assistance Following Katrina, *Foreign Policy in Focus,* October 19 2005, 1.

[53] See N. Lepp, Disaster Relief Politics Complicate South Asia Effort, *The Dominion,* January 3 2005, 1.

[54] See C. Villacis, Latin American Cases Hurricane Mitch, Flash Floods and Landslides in Venezuela, El Salvador Earthquakes, *[Report] Recovery and reconstruction reports of the worldwide disasters and their comparative study* Asian Disaster Reduction Center, Kobe, 2005, 21.

[55] See B. Jakovljevic, International Disaster Relief Law, 34 *Israel Yearbook on Human Rights* (2004), 264.

donations following disasters are not limited to financial aid but include medical supplies, tents, shovels, and pick axes,[56] and even a mobile hospital.[57] Furthermore, the response by volunteers from the general public has been substantial. In 1989, sixty percent of the population of San Francisco and seventy percent of the population in Santa Cruz, California responded to the Loma Prieta Earthquake.[58] In the aftermath of the 911 attack in 2001, over 40,000 unsolicited volunteers arrived in New York.[59]

Disaster relief, which should form a major portion of the rapid response system, is seen as a charitable exercise by both the donor and recipients.[60] Agreements are largely specific, often limiting the response to receive and provide assistance of a certain type. The agreement between the United Nations and the governments of Sweden and Peru provided that a technical Cadre Unit of the Swedish Stand-by Force would assist for up to six months in reconstruction of devastated areas in case of disasters in Peru.[61] As is generally the case with bilateral agreements, the agreement specifying provision of assistance can, in a sense, be characterized as voluntary, since it is concluded based on the respective governments' will to create binding obligations. Bilateral aid is at the discretion of the granting government, which may choose to render assistance, or not, according to their national circumstances – such as domestic pressures, public opinion, and budgetary constraints.[62] In contrast, aid of a humanitarian nature provided during armed conflicts is given occasionally without consent

---

[56] See C. Villacis, Latin American Cases Hurricane Mitch, Flash Floods and Landslides in Venezuela, El Salvador Earthquakes, *[Report] Recovery and reconstruction reports of the worldwide disasters and their comparative study,* Asian Disaster Reduction Center, Kobe, 2005, 21.

[57] See C. Villacis, Latin American Cases Hurricane Mitch, Flash Floods and Landslides in Venezuela, El Salvador Earthquakes, *[Report] Recovery and reconstruction reports of the worldwide disasters and their comparative study,* Asian Disaster Reduction Center, Kobe, 2005, 22.

[58] See J.R. Wible, Volunteer's Liability Issues, Medical Reserve Corps National Leadership and Training Conference, April 17–20, 2007, 1.

[59] L.S. Fernandez, Strategies for Managing Volunteers during Incident response: A Systems Approach. Homeland Security Affairs, 2-3 *Online Journal of the Naval Postgraduate School Center for homeland Defense and Security* (2006). Cf. J. Wible, J. Volunteer's Liability Issues Medical Reserve Corps National Leadership and Training Conference April 17-20, 2007, 1.

[60] See R. Evans, The Humanitarian Challenge: A Foreign Policy Perspective, 6-2 *African Security Review* (1997), 1.

[61] Following an earthquake in 1970, the service was made available in accordance with this agreement. See P. Macalister-Smith, *International Humanitarian Assistance Disaster Relief Actions in International Law and Organization*, Martinus Nijhoff Publishers, Geneva, 1985, 2, 151.

[62] See R. Evans, The Humanitarian Challenge: A Foreign Policy Perspective, 6-2 *African Security Review* (1997), 2.

from the affected state, and is a form of intervention – an issue that will be explored more fully later in this Chapter.

4.3. Legal Framework

There is no definite, broadly accepted source of international law that spells out legal standards, procedures, rights, and obligations pertaining to disaster response and assistance.[63] As stated previously, the organization and provision of international humanitarian assistance in disaster situations is usually undertaken between states on a bilateral basis or through the established international agencies holding the appropriate responsibilities.[64] The rules associated with disaster management exist in a rather fragmented and scattered arrangement in the form of bilateral and multilateral agreements; instruments such as United Nations General Assembly Resolutions;[65] guidelines; rules of organizations; recommendations; and declarations of principles. A number of instruments are often established stimulated by the need following disasters, or on an ad hoc basis.

4.3.1. *The Historical Development*

It is worthwhile to look at the historical development of the rules regarding disaster management. By looking at the historical background, one can study the stance and behaviour of governments toward disaster response, and make observations on the process that has evolved to form the present legal framework. Various efforts have been made to codify disaster relief rules, specifically with regard to the obligation to render assistance in the case of disaster;

---

[63] D.P. Fideler, Indian Ocean Tsunami and International Law, *ASIL Insight* (January 2005), 2.

[64] See P. Macalister-Smith, *International Humanitarian Assistance Disaster Relief Actions in International Law and Organization*, Martinus Nijhoff Publishers, Geneva, 1985, 151.

[65] Relevant Resolutions include 1) UNGA Resolution 2816 Assistance in cases of natural disaster and other disaster situations, 2) UNGA Resolution 36/225;UN Doc. A/RES/36/225(1981) Strengthening the capacity of the United Nations system to respond to natural disasters and other disaster situations, 3) UNGA Resolution 43/131; UN Doc. A/RES/43/131(1988) Humanitarian Assistance to victims of natural disasters and similar emergency situations, 4) 1990 UNGA Resolution 45/100;UN Doc. A/RES/45/100(1990) Humanitarian Assistance to victims of natural disasters and similar emergency situations, and 5) UNGA Resolution 57/150: UN Doc. A/RES/57/150(2002) Strengthening effectiveness and coordination of international urban search and rescue operations.

however, these attempts have not been successful in creating an overall legal framework addressing such rights and obligations. Attempts to formulate such a regime on a multilateral basis have either failed, or have been codified but then withdrawn because they lacked sufficient support. The following discussion tracks the chronological development of the legal framework.

In 1982, the United Nations Institute for Training and Research (UNITAR) attempted to bridge the gap of existing humanitarian law, which covered armed conflicts but not natural disasters, and considered how such a regime should regulate disaster relief.[66] After considering different approaches, specifically the conclusion of multilateral, regional, and bilateral agreements, UNITAR concluded that, given the political and legal circumstances, bilateral agreements were the most practical and feasible solution at that time.[67] In the Model Rules for Disaster Relief Operations,[68] Rule 1 of the Model Rules of Bilateral Agreement between Assisting State and Receiving State specifies that "The Receiving State shall designate a national relief authority to coordinate the relief activities, documentation and procedures required."[69] Therefore, the coordination was primarily in the hands of afflicted states. The resulting proposals, which had no legal effect themselves, did not challenge the existing principles and practice surrounding disaster management – most notably, that relief will be provided at the request or with the consent of the recipient state.[70]

States showed reluctance to codify a concrete disaster response law applicable to all states for any disaster. Attempts to establish concrete rules regarding disaster management have, in effect, failed. Amongst other proposals, the 1984 UNDRO Draft Convention on Expediting the Delivery of Emergency Assistance of the Disaster Relief,[71] was withdrawn.[72] This Draft Convention was

[66] See P. Macalister-Smith, *International Humanitarian Assistance Disaster Relief Actions in International Law and Organization*, Martinus Nijhoff Publishers, Geneva, 1985, 155.

[67] See Y. Beigbeder, *The role and status of international humanitarian volunteers and organizations*, Martinus Nijhoff Publishers, Dordrecht, 1991, 375.

[68] M. Baradei et al., Model Rules for Disaster Operations, 8 Policy and Efficacy Studies, UNITAR, 1982. They are designed to regulate the emergency phase of the disaster when time is recognized to be the critical factor.

[69] Rule 1 of the Model Rules of Bilateral Agreement between Assisting State and Receiving State.

[70] See Y. Beigbeder, *The role and status of international humanitarian volunteers and organizations*, Martinus Nijhoff Publishers, Dordrecht, 1991, 375–377.

[71] U.N. Doc. A/39/267/Add.2-E/1984/96/Add.2, 18 June, 1984. It established that the receiving State shall have within its territory responsibility for facilitating the coordination of operations to meet the situation created by the disaster. [Hereafter: UNDRO Draft Convention].

[72] See Y. Beigbeder, *The role and status of international humanitarian volunteers and organizations*, Martinus Nijhoff Publishers, Dordrecht, 1991, 379.

characterised by concrete provisions; in particular, it had a specific provision on liability. It provided that the receiving state would bear all risks and claims, and it recognised the liability of the receiving state for the loss caused to the assisting state during its assistance, stating that "Receiving State shall compensate the assisting State or organization for the death of, or temporary or permanent injury to, its personnel, as well as for loss." Furthermore, whilst the UNDRO Draft Convention provides that the assisting state is to bear all risks and claims, it recognizes the liability of the assisting state in causing intentional damage as follows: "recourse actions can be brought against the personnel of an Assisting State or organization only in respect of damage or injury which they have caused intentionally or recklessly with the knowledge that damage may result."[73]

The UN General Assembly Resolution of 8 December 1988 on "Humanitarian assistance to victims of natural disasters and similar emergency situations"[74] – originally proposed by France to clearly establish the right to receive and the duty to provide humanitarian assistance – met strong opposition from numerous nations at the drafting stage and the final version that was adopted deviated substantially from the initial French document.[75] Amongst its other provisions, the document "reaffirms also the sovereignty of affected States and their primary role in the initiation, organization, co-ordination and implementation of humanitarian assistance within their respective territories"[76] and "appeals, therefore, to all States to give their support to these organizations working to provide humanitarian assistance, where needed, to the victims of natural disasters and similar emergency situations."[77] In effect, the Resolution inferred that the right to receive humanitarian assistance is primarily based on the policy of the afflicted state and merely emphasized the recommendation to all states to provide relief to an afflicted state.

Rarely has a legal instrument on disaster response been codified. There was an attempt or movement toward codification, but this did not materialize in the end. For instance, on 8 March 1989, the UNDRO working group reaffirmed that a new international Convention or Declaration on disaster relief was neither necessary nor desirable.[78] Even when certain legal instruments have been codified, sufficient signatories have not been received for such instruments and consequently, they do not have much effect. For instance, in the United States, an attempt to create a comprehensive regional treaty on disaster assistance led to adoption of the Inter-American Convention to Facilitate Disaster Assistance.[79]

[73] Article 22 (4) of the UNDRO Draft Convention.

[74] UN Doc A/RES/43/131(1988).

[75] See Y. Beigbeder, *The role and status of international humanitarian volunteers and organizations*, Martinus Nijhoff Publishers, Dordrecht, 1991, 380–383.

[76] Principle 2, UN General Assembly Resolution of 8 December 1988.

[77] Principle 5, UN General Assembly Resolution of 8 December 1988.

[78] See Y. Beigbeder, *The role and status of international humanitarian volunteers and organizations*, Martinus Nijhoff Publishers, Dordrecht, 1991, 383.

[79] Inter American Convention to Facilitate Disaster Assistance, 7 June 1991.

This Convention requires states to designate a National Coordinating Authority,[80] gives the responsibility for the coordination of assistance to the assisting state,[81] and includes specific provisions on entry and movement of supplies;[82] however, there are currently only three parties to the treaty.[83]

One can conclude here that states have been reluctant to enter into a binding regime that imposes obligations and involves responsibility in rendering assistance whenever and wherever disaster hits. It can well be seen that neither the receiving nor providing side wishes to codify and formalize either the right to receive relief or the obligation to provide relief for all potential cases; rather, they wish to retain discretion in responding to each case. It appears that governments will be more willing to respond to concrete needs in particular cases rather than to accept legal formulations framed from a context of generalities or assumptions.[84] The history of unsuccessful formulations and codifications of general rules surrounding disaster management, and/or reluctance to adopt unified law, confirms the tendency of states to adopt a position that favours state sovereignty and discretion regarding providing and receiving aid, which inevitably results in a case-by-case response.

4.3.2. *The Analysis of Existing Rules*

Unified rules on disaster management do not exist, and the rights and obligations involved are uncertain. Therefore, it is necessary to examine further the existing rules associated with providing and receiving aid.

Certain rules that may be applicable to disasters can be found at the level of general international law. The obligation to provide assistance is established implicitly through the principles of good neighbourliness and humanity. 'Principles of good neighbourliness' means 'the willingness to cooperate with neighbouring states to inform a potentially affected state of relevant data and a willingness to take appropriate steps to address the legitimate concerns of those potentially impacted by the activities in another state'.[85]

---

80 Article III of the Inter American Convention to Facilitate Disaster Assistance.

81 Article IV of the Inter American Convention to Facilitate Disaster Assistance.

82 Article V of the Inter American Convention to Facilitate Disaster Assistance.

83 Background paper on International Disaster Response Laws, Rules and Principles for the FEMA Legal Round table on Foreign Assistance in the United States, Washington DC, November 28, 2005, 3.

The three parties are Panama, Peru, and Uruguay.

84 See P. Macalister-Smith, *International Humanitarian Assistance Disaster Relief Actions in International Law and Organization*, Martinus Nijhoff Publishers, Geneva, 1985, 167.

85 Transboundary Assessment overarching principles, Council Resolution no. 95-7, 13 October, 1995, 2.

Other rules specifically address disaster situations, which can be categorized as either international or national. The general tendency is that binding multilateral agreements, such as treaties, are limited to dealing with a small number of specified matters such as customs and telecommunications, or have limited scope – applying only within a region, or binding only certain parties. On the other hand, soft law – a collective of rules that are neither strictly binding nor lacking in legal significance to help other states in peril (including cases of natural disasters) – presently address disaster management issues in a more comprehensive way, including overall standards and accountability; however, both the applicability and enforcement of these instruments are questionable.

4.3.2.1. *International Instruments*

The international instruments are, *inter alia*, international treaties and conventions, and UN General Assembly Resolutions and principles endorsed by the relevant organizations. Amongst these instruments, some are in the category of binding law, whilst others fall into the 'soft law' category.

There are not many legal instruments at multilateral levels. Amongst those are multilateral treaties addressing specific matters, such as the 1973 Kyoto Convention on the Simplification and Harmonization of Customs Procedure,[86] and the Tampere Convention on the Provision of Telecommunication Resources for Disaster Mitigation and Relief[87] that facilitate providing prompt telecommunication assistance to mitigate the impact of a disaster, as well as regional agreements such as the ASEAN Agreement on Disaster Management and Emergency Response.[88]

The Tampere Convention – unanimously adopted on 18 June 1998 by the 75 countries attending the Intergovernmental Conference on Emergency Telecommunications (ICET-98) in Tampere, Finland – provides the procedures for telecommunication assistance, recognizing the right of a state to direct, control, and coordinate assistance within its territory provided under the Convention.[89] The Tampere Convention is the first treaty of its kind to provide privileges and immunities to the personnel of NGOs, and it exempts relief agencies from taxation and duties.[90]

---

[86] Kyoto Convention on the Simplification and Harmonization of Customs Procedure, *TIAS* 6633.

[87] Tampere Convention on the Provision of Telecommunication Resources for Disaster Mitigation and Relief, 1439 *UNTS* 275. [Hereafter: Tampere Convention]

[88] ASEAN Agreement on Disaster Management and Emergency Response, Vientiane, 26 July 2005.

[89] "Tampere Convention: saving lives through emergency telecommunications", *UN Press Release*, IHA/983 L/I4385/SAG/309, 7 January, 2005.

[90] "Tampere Convention: saving lives through emergency telecommunications", *UN Press Release*, IHA/983 L/I4385/SAG/309, 7 January, 2005.

The Tampere Convention does not impose an obligation to provide assistance. Indeed, the provision is that "Each State Party to which a request for telecommunication assistance is directed, shall promptly determine and notify the requesting State Party whether it will render the assistance requested."[91] Thus, under the Convention, no parties will be held responsible under the law for refusing to provide resources or for unsuccessful attempts to render such resources. No right or duty to receive and render assistance is clearly established nor enforced under the Convention. In fact, the effect of the Convention on disaster relief is uncertain: none of its provisions were specifically invoked in the relief operation following the Indian Ocean Tsunami, despite Sri Lanka's ratification of the Tampere Convention.[92]

The ASEAN regional multilateral treaty on disaster management, signed by the Foreign Ministers of ASEAN in July 2005, strengthens prevention, monitoring, and loss mitigation measures for response to disasters in the region. It is a comprehensive treaty, which addresses various matters – such as the entry and operations of relief personnel, goods, and equipment – yet its applicability is limited to ASEAN Member Countries. The treaty will enter into force upon ratification by all ASEAN Member Countries. Thus far, only Thailand has ratified, whilst other ASEAN Member Countries are still in the process of ratifying the agreement. If adopted, the treaty would create an innovative mechanism within the ASEAN secretariat to facilitate communications and coordination between international actors and affected governments.[93] It is important to note here that even whilst the agreement is not in force, ASEAN Member Countries have begun to implement many of the provisions.[94] Thus, it can be argued that it is the willingness of government for cooperation that actually facilities coordination and assistance rather than the legal effect of the treaty forcing the parties to do so.

There are several General Assembly Resolutions addressing the issues of disaster response, which deal with general rules such as respect for sovereignty in making decisions to request and accept international assistance, and the mandate of a stricken country to respond to disasters. UN General Assembly Resolution 46/182, which led to the creation of the Department of Humanitarian Affairs, establishes the fundamentals of disaster assistance by articulating

---

[91] Article 4.3 of the Tampere Convention.

[92] V. Bannon & D. Fisher. Legal Lessons in Disaster Relief from the Tsunami, the Pakistan Earthquake and Hurricane Katrina, 10-6 *ASIL Insight* (2006), 2-3.

[93] See Background paper on International Disaster Response Laws, Rules and Principles for the FEMA Legal Round table on Foreign Assistance in the United States, Washington DC, November 28, 2005, 3.

[94] See Background paper on International Disaster Response Laws, Rules and Principles for the FEMA Legal Round table on Foreign Assistance in the United States, Washington DC, November 28, 2005, 3.

guiding principles to strengthen the coordination of humanitarian emergency assistance of the United Nations. Principle 3 of the Guiding Principles of UN General Assembly Resolution 46/182 states: "Humanitarian assistance should be provided with the consent of the affected country and in principle on the basis of an appeal by the affected country."[95] Furthermore, Principle 4 states that "each State has the responsibility first and foremost to take care of the victims of natural disasters and other emergencies occurring on its territory."[96] Hence, the affected state has the primary role in the organization, coordination, and implementation of humanitarian assistance within its territory.

Some instruments establish concrete rules dealing with responsibility and the standard of operations. They include principles and rules for Red Cross Movements and the Code of Conduct for the International Red Cross and Red Crescent Movement and Non-Governmental Organisations in Disaster Relief, the Sphere Project's Humanitarian Charter and Minimum Standards in Disaster Response,[97] and the Principle and Good Practice of Humanitarian Donorship[98] endorsed mainly by European nations.

The code of conduct of the Red Cross [99] – amongst others – clearly addresses the principles of rights and obligations. The Code of Conduct for the International Red Cross and Red Crescent Movement and NGOs in Disaster Relief states that "the right to receive humanitarian assistance, and to offer it, is a fundamental humanitarian principle which should be enjoyed by all citizens of all countries."[100] The Code of Conduct is accompanied by recommendations

---

[95] Principles 3 of Guiding Principles of UN General Assembly Resolution 46/182; UN Doc. A/RES/46/182(1991).

[96] Principles 4 of Guiding Principles of UN General Assembly Resolution 46/182; UN Doc. A/RES/46/182(1991).

[97] Sphere's Projects are the result of cooperation between various independent agencies engaged in humanitarian assistance, including most major NGOs and the Red Cross and Red Crescent. Its aim is to define basic practical standards that should govern humanitarian assistance; specifically, those immediately relevant to saving lives, such as water and sanitation, food, health care, and shelter.

[98] Endorsed in Stockholm, 17 June 2003 by Germany, Australia, Belgium, Canada, the European Commission, Denmark, the united States, Finland, France, Ireland, Japan, Luxemburg, and some provisions are concrete, such as Principle 14 of the Principle and Good Practice of Humanitarian Donorship states the requirement to: "Contribute responsibly, and on the basis of burden-sharing, to United Nations Consolidated Inter-Agency Appeals and to International Red Cross and Red Crescent Movement appeals, and actively support the formulation of Common Humanitarian Action Plans (CHAP) as the primary instrument for strategic planning, prioritization and co-ordination in complex emergencies."

[99] Six of the world's oldest and largest networks of NGOs came together in 1994 with the Red Cross and Red Crescent Movement to draw up a professional code of conduct.

[100] 310 *International Review of Red Cross*, January–February (1996), 120.

to the governments of disaster affected countries, donor governments, and intergovernmental organisations to create the framework for disaster assistance. Recommendations to the governments of countries affected by a disaster are advisory in nature and facilitate government actions based on goodwill.

The Code of Conduct is also voluntary and self-policing: that is, no NGO is going to force another to act in a certain way. Furthermore, there is no international body for disaster response that possesses any authority to sanction its members, which means that there is no enforcement mechanism. Even if the NGO or government does not respect the Code, it will not be sanctioned. The Code of Conduct and principles only bind the members of the Red Cross; therefore, it does not create a sufficient legal basis for an overall framework for disaster response.

Finally, the Principles and Rules for Red Cross and Red Crescent Disaster Relief are designed to balance the rights and responsibilities of donor and recipient, whilst taking into account the need for international coordination.[101] The Principles and Rules for Red Cross and Red Crescent Disaster Relief, as amended by the 26th International Conference of the Red Cross and Red Crescent, provide that the Red Cross and Red Crescent "has a fundamental duty to provide relief to all disaster victims."[102]

4.3.2.2. *National Legal Instruments*

At a national level, some nations have specific laws addressing disaster management. For instance, Japan has a Disaster Countermeasure Basic Act[103] that addresses disaster situations in general. It establishes obligations associated with disaster response, and clear mandates to, and responsibilities of, concerned persons and entities. It adopts both 'top-down' and 'bottom-up' approaches to maximize the effectiveness of disaster management. It has a 'top-down' approach as to the dissemination of warnings: upon receiving warning related to disasters, governors have the obligation to inform the local authorities and mayors,[104] whilst mayors need to inform potentially affected populations if they receive warnings related to the disasters.[105] The law also establishes that mayors have the mandate to implement the evacuation of populations. Japan takes a 'bottom-up' approach to the reporting obligation on the extent of the disaster beginning at the levels of villages, towns, and cities to a prefecture, and from a

---

[101] See P. Macalister-Smith, *International Humanitarian Assistance Disaster Relief Actions in International Law and Organization*, Martinus Nijhoff Publishers, Geneva, 1985, 165.

[102] 310 *International Review of Red Cross* (January-February 1996), 119–127.

[103] Disaster Countermeasure Basic Act, document no. 223, adopted on 15 November 1961. [Hereafter: Japanese Disaster Law]

[104] Article 55 of the Japanese Disaster Law.

[105] Article 56 of the Japanese Disaster Law.

prefecture and/or local authorities to the prime minister.[106] The Japanese law addressing disaster management is concrete and specific enough to be applied to many different disaster situations.

The trend is that – even in nations without a solid general disaster management law – adoption of national legal instruments related to disaster management is increasing, particularly following deadly disasters. Governments have sought to address problems arising from the application of general laws in an *ad hoc* fashion, frequently under a state of emergency regime.[107] For instance, Indonesia proposed a new Disaster Management Bill to its parliament, and the governments of Sri Lanka and India both adopted new disaster bills; however, whilst the Indonesian and Indian bills refer to the possibility of receiving international assistance, neither these nor the new Sri Lankan law contain specifics of when and how this should occur.[108] Furthermore, national legislation is designed for each individual disaster and adopted without prior planning, often serving to meet short term objectives intrinsic for the specific disaster, rather than as a regime that can be a solid reference for future disaster response. For example, the Philippines Congress and the Office of the President promulgated a series of laws and regulations to govern the country's reconstruction and rehabilitation by creating successive task forces for the rehabilitation of areas affected by the eruption of Mount Pinatubo.[109] These were intended to guide and coordinate the rehabilitation efforts of the government as well as the implementation of programs and projects for the specific purpose of recovery and reconstruction from the eruption.[110] Such laws are effective in implementing specific operations following disasters, but do not regulate relief operations in anticipation of future disasters.

Looking at various legal instruments on disaster management that exist today, one can conclude that certain issues, such as customs clearance, – particularly giving rise to the operational problems as mentioned previously – are addressed relatively well; however, critical issues that affect disaster response and relief

---

[106] Article 60 of the Japanese Disaster Law.

[107] See Background paper on International Disaster Response Laws, Rules and Principles for the FEMA Legal Round table on Foreign Assistance in the United States, Washington DC, November 28, 2005, 2.

[108] See V. Bannon & D. Fisher. Legal Lessons in Disaster Relief from the Tsunami, the Pakistan Earthquake and Hurricane Katrina, 10-6 *ASIL Insight* (2006), 6.

[109] See A.K. Sinha & S. Srivastava, Survey Report on the Reconstruction following the Earthquake in Marmara, *Report: Recovery and reconstruction reports of the worldwide disasters and their comparative study*, Asian Disaster Reduction Center, Kobe, 2005, 14. Legislation includes Memorandum order no. 369, of 26 June 1991, Republic Act 7637 of 24 September 1992, and Presidential Proclamation 1201 of 19 March 1998.

[110] See A.K. Sinha & S. Srivastava, Survey Report on the Reconstruction following the Earthquake in Marmara, *Report: Recovery and reconstruction reports of the worldwide disasters and their comparative study*, Asian Disaster Reduction Center, Kobe, 2005, 8.

operations – such as concrete rules to create a predictable mechanism for disaster response, and clarification of liability involved in provision of aid – are not addressed sufficiently under the current legal framework. Furthermore, multilateral conventions are not necessarily effective to facilitate international disaster response, as demonstrated by the Tampere Convention and the ASEAN agreement. The current legal framework surrounding disaster management is therefore inadequate and in need of an appropriate legal model that facilitates effective disaster relief.

4.4. The Appropriate Legal Model that Fosters Effective Disaster Management

Effective disaster management is potentially fostered by taking two distinct approaches. One is through establishing predictable mechanisms, where operations are regulated and rules are followed in such way that the behaviour of states and organisations – both donors and beneficiaries – in response to disasters, is easily foreseen and, as a result, relief operations are galvanized. Another approach is quite the opposite: to encourage and boost voluntary assistance by creating incentives to help, and by minimizing the burden on donors. In current practice, disaster relief is geared toward assistance on a voluntary basis; however, it is necessary to explore the humanitarian and human rights framework to ascertain whether these can serve at all as a viable model to facilitate disaster management.

4.4.1. *The Approach from Human Rights and Humanitarian Law*

Disaster relief is traditionally a humanitarian operation because it serves to satisfy the fundamental needs of people in peril who need effective protection and assistance from outside, as they cannot overcome their problems themselves.[111] Humanitarian and human rights law systems are instruments recognizing and granting rights to victims with humanitarian needs; the issue is to determine clearly whether these instruments could serve as models for promoting international disaster relief through their binding force.

The first thing to explore is the question of protection for the victims of natural disasters within the current legal framework surrounding human rights. This section examines whether the victims of natural disasters have the right to humanitarian assistance as a part of their human rights. The explicit right to humanitarian assistance to victims of natural disasters is not found under

[111] See B. Jakovljevic, International Disaster Relief Law, 34 *Israel Yearbook on Human Rights* (2004), 273.

the human rights legal system;[112] hence, the question is whether such right is inferred through the human rights that have already been granted. Article 3 of the Universal Declaration on Human Rights states that "everyone has the right to life, liberty and security of person."[113] When one considers the legal protection for a population affected by natural disasters, the issue becomes whether it is possible to construe that non-provision of aid constitutes infringement of a fundamental human right – specifically, the right to life. Human rights treaties commonly prohibit arbitrary deprivation of life.[114] The European Convention on Human Rights is specific on the interpretation of such deprival: "Everyone's right to life shall be protected by law and no one shall be deprived of his life intentionally save in the execution of a sentence of a court following his conviction of a crime for which this penalty is provided by law."[115] From the quoted texts it can be interpreted that the right to life can only be enforced when it is taken away unlawfully and arbitrarily. Therefore, the right of an affected population to receive humanitarian assistance may not be inferred or extended from the 'right to life'.

Past attempts to apply the human rights principle to establish a universal right to receive assistance in case of disasters have been unsuccessful. As previously mentioned, the 1988 UN General Assembly Resolution[116] was originally intended to establish international rights and obligations regarding humanitarian assistance associated with disaster management; more specifically, the human rights of victims to receive assistance, the duty of international governmental organizations and non-governmental organizations to provide such assistance, and the obligation of recipient governments to facilitate the relief workers' access to the victims. Before being presented to the Committee, however, the references to the Universal Declaration of Human Rights and other international instruments relating to human rights were deleted from the Resolution.[117] It is thus questionable whether establishing a law to ensure a

[112] See B. Jakovljevic, International Disaster Relief Law, 34 *Israel Yearbook on Human Rights* (2004), 257.

[113] UN Doc A/810 at 71 (1948), Universal Declaration of Human Rights.

[114] For instance, Article 6(1) of the International Covenant on Civil and Political Rights states: "Every human being has the inherent right to life. This right shall be protected by law. No one shall be arbitrarily deprived of his life". 1966, Article 4(1) of the American Convention on Human Rights, 1969, states that: "Every person has the right to have his life respected. This right shall be protected by law and, in general, from the moment of conception. No one shall be arbitrarily deprived of his life."

[115] Article 2 of the European Convention on Human Rights.

[116] UN Doc. A/RES/43/131(1988), Humanitarian Assistance to victims of natural disasters and similar emergency situations. The originator of the resolution was Bernard Kouchner, founder of Médecins Sans Frontières and Médecins du Monde.

[117] See Y. Beigbeder, *The role and status of international humanitarian volunteers and organizations*, Martinus Nijhoff Publishers, Dordrecht, 1991, 380.

right to humanitarian assistance following disasters would boost international disaster response.

More specific situations exist where humanitarian assistance is rendered in compliance with existing law. A leading example is a protection regime for victims of armed conflicts and refugees. The basic rules on protection for victims of armed conflicts are codified in the 1949 Geneva Conventions applicable for different categories of victims, which consists of Geneva Convention I for the Amelioration of the Condition of the Wounded and Sick in Armed Forces in the Field;[118] Geneva Convention II for the Amelioration of the Condition of Wounded, Sick and Shipwrecked Members of Armed Forces at Sea;[119] Geneva Convention III Relative to the Treatment of Prisoners of War;[120] and Geneva Convention IV Relative to the Protection of Civilian Persons in Time of War;[121] with their additional Protocols: the Protocol additional to the Geneva Conventions of 12 August 1949, and Relating to the Protection of Victims of International Armed Conflicts (Protocol I);[122] and the Protocol Additional to the Geneva Conventions and relating to the protection of victims of Non-International Armed Conflicts (Protocol II).[123]

Victims of natural disasters within a particular afflicted state could be associated more with victims of non-international armed conflicts than with those of international armed conflicts. The reason is that a disaster often strikes a specific part of a single state, and even if the impact of a disaster is transboundary, victims are usually taken care of by the government of the afflicted state within the respective domestic jurisdiction. Hence, any problem faced by a natural disaster victim in the course of aid operations is generally of a domestic nature. For the above reason, the analysis of this section will be limited to armed conflict situations of non-international character. In case of non-international armed conflicts, Article 3(1)a of Geneva Conventions I, II, III, and IV, respectively, prohibit, "violence to life and person, in particular murder of all

[118] Geneva Convention I for the Amelioration of the Condition of the Wounded and Sick in Armed Forces in the Field Aug. 12, 1949, 6 *UST* 3114, 75 *UNTS* 31. [Hereafter: Geneva Convention I].

[119] Geneva Convention II for the Amelioration of the Condition of Wounded, Sick and Shipwrecked Members of Armed Forces at Sea Aug. 12, 1949, 6 *UST* 3217, 75 *UNTS* 85. [Hereafter: Geneva Convention II].

[120] Geneva Convention III Relative to the Treatment of Prisoners of War, Aug. 12, 1949, 6 *UST* 3316, 75 *UNTS* 135. [Hereafter: Geneva Convention III].

[121] Geneva Convention IV Relative to the Protection of Civilian Persons in Time of War, Aug. 12, 1949, 6 *UST* 3516, 75 *UNTS* 287. [Hereafter: Geneva Convention IV].

[122] Protocol additional to the Geneva Conventions of 12 August 1949, and Relating to the Protection of Victims of International Armed Conflicts June 8, 1977, 1125 *UNTS* 3.

[123] Protocol Additional to the Geneva Conventions and relating to the protection of victims of Non-International Armed Conflicts, Dec. 7, 1978, 1125 *UNTS* 609.

kinds, mutilation, cruel treatment and torture,"[124] therefore, inhumane acts to cause suffering of the affected populations are prohibited. Obligations not to commit violence as well as rights of war victims against violence are granted under the Convention.

The right to be cared for is granted to sick and wounded in all categories of war victims under the Geneva Convention. Article 3(2) of Geneva Conventions I, II, III, and IV each provides that in case of non-international armed conflict, "the wounded and sick shall be collected and cared for and by an impartial humanitarian body, such as the International Committee of the Red Cross, which may offer its services to the Parties to the conflict."[125] This 'right to be cared for' can be interpreted as the medical obligation to meet the needs of sick and wounded.[126] The reason this right is granted for these two types of victims is that their movement and possibility of obtaining relief is limited when under the control of an enemy power.[127] The right to receive medical aid as well as aid of other kinds is granted to certain categories of war victims, such as prisoners of war[128] and civilian internees.[129] For other categories of victims, including the

[124] Article 3 (1) of the Geneva Convention IV states that " In the case of armed conflict not of an international character occurring in the territory of one of the High Contracting Parties, each Party to the conflict shall be bound to apply, as a minimum, the following provisions: 1. Persons taking no active part in the hostilities, including members of armed forces who have laid down their arms and those placed hors de combat by sickness, wounds, detention, or any other cause, shall in all circumstances be treated humanely, without any adverse distinction founded on race, colour, religion or faith, sex, birth or wealth, or any other similar criteria. To this end, the following acts are and shall remain prohibited at any time and in any place whatsoever with respect to the above-mentioned persons: (a) Violence to life and person, in particular murder of all kinds, mutilation, cruel treatment and torture; (b) Taking of hostages; (c) Outrages upon personal dignity, in particular humiliating and degrading treatment; (d) The passing of sentences and the carrying out of executions without previous judgment pronounced by a regularly constituted court, affording all the judicial guarantees which are recognized as indispensable by civilized peoples."

[125] Article 3 (2) of the Geneva Convention I, II, III, and IV.

[126] See B. Jakovljevic, International Disaster Relief Law, 34 *Israel Yearbook on Human Rights* (2004), 255.

[127] See B. Jakovljevic, International Disaster Relief Law, 34 *Israel Yearbook on Human Rights* (2004), 255.

[128] Article 72 of Geneva Convention III states: "Prisoners of war shall be allowed to receive by post or by any other means individual parcels or collective shipments containing, in particular, foodstuffs, clothing, medical supplies and articles of a religious, educational or recreational character which may meet their needs, including books, devotional articles, scientific equipment, examination papers, musical instruments, sports outfits and materials allowing prisoners of war to pursue their studies or their cultural activities."

[129] Article 108 of the Geneva Convention IV states that "Internees shall be allowed to receive, by post or by any other means, individual parcels or collective shipments

victims of natural disasters, the conventions of international humanitarian law are silent as to defining such right.[130] Therefore, the victims of natural disasters are not the intended recipients of any type of aid, even medical.

Indeed, the circumstances surrounding natural disasters are fundamentally different from those of armed conflicts. The population affected by natural disasters should be distinguished from victims of armed conflicts, since the cause of disasters is natural and the occurrence is inevitable, as opposed to man-made, and relief is usually provided willingly to those who need it. The victims of natural disasters can benefit from assistance inside and outside the state unless their own government refuses such aid.

Furthermore, attempts to establish a law have been unsuccessful. Although a Geneva Convention on international disaster relief was envisaged by the International Committee of the Red Cross, it was perceived as challenging to achieve such a convention as it would require extensive commitments from both the donor and recipient states.[131] It does not appear to be feasible to establish the rights of victims of natural disasters in a sphere of humanitarian law at the same level that is granted to victims of armed conflicts.

Next, there is the possibility of the protection of victims of natural disasters on the basis of refugee status. UNHCR has the right to intervene and ensure refugee protection.[132] Whether victims of natural disasters can be classified as refugees should be explored. Article 1 of the 1954 Convention relating to the Status of Refugees[133] as amended by the 1967 Protocol[134] defines a refugee as "a person who, owing to a well-founded fear of being persecuted for reasons of race, religion, nationality, membership of a particular social group, or political opinion, is outside the country of their nationality and is unable to or, owing to such fear, is unwilling to avail him/herself of the protection of that country." Hence, the word 'refugee' refers to political refugees – in other words,

---

containing in particular foodstuffs, clothing, medical supplies, as well as books and objects of a devotional, educational or recreational character which may meet their needs. Such shipments shall in no way free the Detaining Power from the obligations imposed upon it by virtue of the present Convention".

130 See B. Jakovljevic, International Disaster Relief Law, 34 *Israel Yearbook on Human Rights* (2004), 258.

131 See P. Macalister-Smith, *International Humanitarian Assistance Disaster Relief Actions in International Law and Organization*, Martinus Nijhoff Publishers, Geneva, 1985, 157.

132 See B. Jakovljevic, International Disaster Relief Law, 34 *Israel Yearbook on Human Rights,* (2004), 262. See Statute of the Office of the United Nations High Commissioner for Refugees, UN General Assembly Resolution, A/RES/428 (V), 14, 14 December 1950.

133 1954 Convention relating to the Status of Refugees, 189 *UNTS* 150.

134 1967 Protocol to Convention relating to the Status of Refugees, 606 *UNTS* 267.

'man-made' refugees. Victims of natural disasters are different from political refugees, who flee abroad because their own government denies human rights protection either by persecuting them or by not protecting them against dangers they face from third parties.[135] Clearly, victims of natural disasters are neither persecuted nor deliberately abandoned to the risks of danger from others. Under the existing Convention, therefore, it is clear that victims of natural disasters are not amongst those victims who have the right to humanitarian assistance.

4.4.1.1. *Exercising the Right to Humanitarian Assistance: Issues of Humanitarian Intervention and Consideration of its Applicability to Victims of Natural Disasters*

The first thing to determine is: Under what circumstances can the right to humanitarian assistance associated with natural disasters be exercised? Jakovljevic contends that victims may exercise the right to humanitarian assistance when they have no other alternative to satisfy their basic needs, and when they have exhausted all the possibilities for satisfying those needs in normal ways and from regular sources at the local level.[136] It follows that as long as the afflicted state is providing its citizens with internal aid, albeit insufficient or unsuccessful, it is difficult to argue that refusing aid from a particular external source would amount to denial of the right to humanitarian assistance. The common view is that a state does not have an obligation to accept international aid,[137] so that has been the practice. On the basis of spontaneous (not caused by man's actions) humanitarian needs of victims of natural disasters alone, it seems difficult under humanitarian law to establish international obligations on every afflicted state to 1) actively seek international assistance or to 2) ensure victims access to any external assistance offered, especially if internal aid resources are not exhausted. Considering the political nature of providing and receiving assistance, attempting to give international assistance without the consent of the stricken countries is not likely to succeed.

Then the question is: Under what circumstances states or organizations can intervene and rescue people in peril, despite the attitude of the affected state's

[135] F.L. Kirgis, Hurricane Katrina and Internally Displaced Persons, *ASIL Insight* (September 21 2005), 1. International law protects refugees by means of the 1951 Convention Relating to the Status of Refugees and its 1967 Protocol.

[136] See B. Jakovljevic, International Disaster Relief Law, 34 *Israel Yearbook on Human Rights* (2004), 257.

[137] The Final Report on the Reappraisal of the Role of Red Cross states: "current international law, which is largely based on traditional practice, does not obligate a State in any way to accept emergency aid even when its population is in extremely grave danger." See D. Tansley, *Final Report: An Agenda for Red Cross*, Geneva, 1975, 80. Cf. Hardcastle, R.J., and Chua, A.T.L., 325 *International Review of the Red Cross* (1998), 604.

government? Humanitarian intervention has taken place in extreme situations when states did not protect their own citizens as a result of civil wars, humanitarian crisis, or crimes. Indeed, historical evidence of humanitarian intervention is associated with conflicts and not with natural disasters.[138] In reality, intervention took place in a limited context where human rights were seriously violated through both deliberate and arbitrary actions to cause suffering of civilians, such as during the Kosovo Crisis.[139] In 1999, the NATO intervention in the Kosovo Crisis, through the bombing of the Federal Republic of Yugoslavia in an attempt to end the repression of ethnic Albanians population by Serbian armed forces, was considered a humanitarian necessity.[140] Humanitarian intervention for victims of natural disasters can only be justified in a narrow context. Such intervention can be justified on the ground of protection for refugees, if the affected population in a particular region is deliberately abandoned to the risk of great suffering as the result of the affected state government's reluctance to promptly allow aid to a specified group of the population for political reasons (as in the case of the previous examples of the 1998 Earthquake in Afghanistan, and the 2001 Earthquake in El Salvador).

One can conclude that serious inaccessibility to aid faced by an affected population because of political reasons may be addressed under the current protection regime for refugees; it is generally not appropriate to apply human rights and humanitarian law to ensure the protection of victims of natural disasters. Establishing a new law based upon humanitarian principles devoted to the protection of natural disaster victims does not appear to be either feasible or beneficial. Therefore, another type of approach – the Good Samaritan Principle – may facilitate rendering assistance on a voluntary basis.

4.4.2. *The Approach from the Good Samaritan Principle*

In the absence of a binding regime that supports an international response to disasters, it is essential to maximise incentives whilst minimizing the hesitation to assist people in peril. One of the major reasons for hesitation in offering assistance is liability risk. The Good Samaritan Principle derives from a New

[138] Examples include humanitarian intervention by the UK, France, and Russia in a conflict between Turkey and Greece to protect Christians; and French Intervention in Zaire in 1978 to stop the massacre of civilians. See Y. Beigbeder, *The role and status of international humanitarian volunteers and organizations*, Martinus Nijhoff Publishers, Dordrecht, 1991, 362–369.

[139] See M. Shaw, *International Law*, Fifth Edition, Cambridge University Press, Cambridge, 2003, 1046–1047.

[140] See M. Shaw, *International Law*, Fifth Edition, Cambridge University Press, Cambridge, 2003, 1046–1047.

Testament parable, referring to Samaritan, who was the only passer-by to stop and render assistance to a man who had been left half-dead.[141] It is a legal principle that prevents a rescuer who has voluntarily helped a victim in distress from being successfully sued for 'wrongdoing.' The Good Samaritan law is a "statute that exempts from liability a person (such as an off-duty physician) who voluntarily renders aid to another in imminent danger but negligently causes injury while rendering the aid."[142] It is a protection mechanism that waives or limits liability for people who aid others with the best of intentions, but inadvertently cause some harm in the course of giving aid.

The Good Samaritan doctrine has been applied most notably to medical service providers.[143] In the United States, medical volunteers in emergency situations have been exonerated from liability for an unsuccessful attempt to render medical assistance. In *Boccasile v. Cajun Music Limited,*[144] a doctor who volunteered to staff a first-aid station at the 1989 Cajun Music Festival in Escoheag, Rhode Island, conducted mouth-to-mouth resuscitation on a man suffering from food poisoning who died the next day; the doctor was not liable for the death. In another case, *Willingham v. Hudson,*[145] an off-duty doctor who was called in to help tornado victims at the hospital was not liable for his treatment that eventually led to the amputation of a victim's leg.

There are certain conditions to be fulfilled for the Good Samaritan Principle to be applied. As shown in the above examples, the primary condition is the absence of an obligation to help. People with no special obligation to prevent harm can be called 'bystanders' and bystanders who prevent harm are called 'Good Samaritans'. The Good Samaritan laws apply when someone is considered to be a 'Good Samaritan';[146] if the person giving aid has a duty to assist, then responsibility is clear and the Good Samaritan Principle does not apply. For instance, if two states conclude a bilateral agreement to render assistance to each other, then they owe a duty under the agreement and will not be considered as 'Good Samaritans' when one provides assistance to the other.

Obligations to help people in distress are framed differently in common law and civil law jurisdictions. In most of the civil-law EU nations, except for

---

[141] See Luke 10: 25–37.

[142] See Good Samaritan law, Black's Law Dictionary, Ninth Edition, 2009, 763.

[143] Most Canadian provinces and territories have Good Samaritan laws to protect a rescuer from liability. In the United States, all states and the District of Columbia have a Good Samaritan law that applies to all citizens.

[144] *Boccasile v Cajun Music Limited,* 694 A2d 686.

[145] *Willingham v. Hudson*, 274 Ga App. 200, 617 S.E. 2d 192 Ga. App. 2005.

[146] F. Schauer & W. Sinnott-Armstrong, *The Philosophy of Law: Classic and Contemporary Readings with Commentary,* Harcourt Brace College Publishers, Oxford, 1996, 816.

Finland and Sweden, such a general duty to come to the rescue of strangers is set forth in the national penal code.[147] In common law countries, such as the United States, the general duty to rescue is absent, but a specific duty exists – limited to special relationships such as those between parents and child, or in the case that a person occupies a position that requires him to act, or in the case of a contract between an individual and a potential rescuer.[148] Reflecting this, it is in the common law jurisdictions such as the United States and parts of Canada, where the Good Samaritan Principle is specifically codified.[149]

In practice, there are ambiguities as to how the Good Samaritan Principle is applied. For instance, it has been applied in the United States in a manner that whenever a person voluntarily comes to the aid of another and the latter relies upon such undertaking, there is imposed duty of care upon the former.[150] The Good Samaritan Principle has been applied in this manner, amongst others, to cases involving air traffic control – such as *Ingham v. Eastern Airline*.[151] Originally the Good Samaritan Principle is applied to waive or limit the liability of voluntary rescuers; in reality – most notably, in the United States – the principle is often applied to the benefit of the ones rescued. These gaps of application of principles to real cases need to be bridged for the Good Samaritan Principle to serve as a catalyst for the active provision of aid following natural disasters.

---

[147] See J. Smits, *European Private Law On the Perils of Principles without a programme and a Programme for the Future*, Kluwer Publishers, Deventer, 2000, 4. National penal codes, which incorporated a general duty to rescue of strangers, include the Netherlands (art. 450 Wetboek van Strafrecht 1886), Norway (art. 387 Straffeloeven 1902), Italy (art. 593 codice penale 1930 on Ommissione di soccorso; the zanardelli Code of 1889), Denmark (art. 253 Straffeloven 1930).

[148] See J. Smits, *European Private Law On the Perils of Principles without a Programme and a Programme for the Future*, Kluwer Publishers, Deventer, 2000, 6.

[149] Enacted laws are, amongst others, the 2000 Emergency Medical Aid Act of Alberta R.S.A. 2000, c. E-7, the 2001 Good Samaritan Act of Ontario S.O.2001, c. 2 for Canada; 1996 Bill Emerson Good Samaritan Food Donation Act P.L. 104-210; and the 1997 Volunteer Protection Act 111 Stat. 218 for the Untied States.

[150] Sections 323 and 324 of A Restatement of Tort, Second, vol. 2, American Institute Publishers, 1965. Cf. B.D.K. Henaku, Expanding Global Navigation Services: Selected Legal Issues, *Proceedings of Workshop on Space Law in the Twenty-first Century*, United Nations Publications, New York, 2000, 172. See Sections 323 and 324 A Restatement of the law second torts, vol. 2, American Institute Publishers, 1965.

[151] *Ingham v. Eastern Airlines*, 373 F. 2d 227 at 236 (2nd Cir.,). The case involved the crash of an Eastern Airlines DC-7 passenger airplane on 30 Nov. 1962 at Kennedy Airport in NYC. The crash was caused by the negligence of an air traffic controller to provide accurate advice to the pilot on the weather conditions. The US government was held liable on the grounds that it was responsible for voluntarily providing the control service when it was not required.

A binding multilateral regime would create specific obligations that would force state parties to respond, so a voluntary approach is a model better suited to promote disaster relief. Whilst the Good Samaritan type of approach has the potential to encourage voluntary assistance, the desired result is not likely to be achieved unless laws ensuring the protection of volunteers are implemented universally. Such protection can be achieved in the sphere of satellite remote sensing by clearly defining the applicability of the Good Samaritan Principle to the use of Earth observation data during disasters and incorporating it into a legal regime. The next section will examine how remote sensing can support disaster management, including the analysis of successful operational programmes based on voluntary assistance.

4.5. Satellite Remote Sensing Applications to Disaster Management and Associated Legal Issues

Remote sensing is a key tool in supporting disaster-related operations from both inside and outside the affected state by providing critical information about the disaster and/or the devastated area. The afflicted state can benefit from a satellite system in their possession, or from other states or entities that can utilize their systems to provide assistance to the afflicted state. Satellite images provide critical information on disaster-prone areas, progress of disasters, and the aftermath of disasters, which cannot be obtained by other means. With increased availability, particularly through constellations of satellites or a combination of different satellites operating under a coordinated framework, satellite remote sensing permits users to gain understanding of the precise situations related to disasters in time to help the governments and other actors make critical decisions, and to provide them with up-to-date information on the rapidly changing developments during disasters.

Satellite images serve as a source of critical information to aid disaster operations and the use of satellites for disaster management dates back several decades. Weather monitoring satellites have monitored hurricanes and tornados since the 1960s, and scientists have used satellite data, such as Landsat data, to monitor patterns of flood, drought, coastal erosion, and volcanic activities.[152] Satellite remote sensing supports different phases of disaster management and has proved to be particularly beneficial in large-scale disasters, where the extent of the damage is not immediately known.

[152] See K.J. Markowitz, Legal Challenge and Market Rewards to the Use and Acceptance of Remote Sensing and Digital Information as Evidence, 12 *Duke Environmental Law and Policy Forum* (Sept. 2002), 219–264.

4.5.1. *The Application of Satellite Remote Sensing in Different Phases of Disaster Management*

An examination of the diverse operations involved in the different phases of disaster management has shown that states with low capability to respond to disaster can aggravate the impact of disasters. Capacities are strongly linked to the ability to access timely and accurate information. In every phase of disaster management, satellite images can serve as a significant source of information upon which to base actions or decisions related to disaster operations, and satellite images are utilized to support a variety of operations.

4.5.1.1. *Prevention*

Risk assessment is a major operation for disaster prevention and satellite images are useful for hazard mapping, such as tectonic and flood plain maps. Tectonic maps are used in combination with other geological and geophysical data to identify earthquake hazards, and to supplement missing information on seismic history to plot the boundaries of various flood levels and the probability for recurrence.[153]

4.5.1.2. *Preparedness*

Satellite images can assist in preparing for impending disasters, or may be used to prepare for possible future disasters. There are several areas where satellite data can support disaster management, including warning and prediction.[154] The development of a disaster warning capability must be accompanied by the development of a disaster warning delivery system.[155] For instance, active microwave satellites are useful in facilitating flood warnings during storm surges through the determination of wind velocities over the oceans.[156] Furthermore,

---

[153] See Office of the United Nations Disaster Relief Coordinator (UNDRO), *Space Application and Dissemination of Disaster-Related Data*, Expert Meeting Geneva, 14–17 June 1983, 22.

[154] See Office of the United Nations Disaster Relief Coordinator (UNDRO), *Space Application and Dissemination of Disaster-Related Data*, Expert Meeting Geneva, 14–17 June 1983, 20.

[155] See Office of the United Nations Disaster Relief Coordinator (UNDRO), *Space Application and Dissemination of Disaster-Related Data*, Expert Meeting Geneva, 14–17 June 1983, 20. See further R. Holdaway, Is space global disaster warning and monitoring now nearing reality?, 17 *Space Policy* (2001), 127–132.

[156] See Office of the United Nations Disaster Relief Coordinator (UNDRO), *Space Application and Dissemination of Disaster-Related Data*, Expert Meeting Geneva, 14–17 June 1983, 32

data on surface wind velocities may also help enhance capabilities in prediction of tropical cyclones.[157]

4.5.1.3. *Response*

The extent of disasters and damage can be assessed effectively using satellite images, which can capture large-scale disasters across boundaries or provide detailed scenes of specific area of interest hit by a disaster. Images taken before and after the disasters are compared to identify damaged areas effectively. Appropriate direction for aid can be given based on such assessment. Satellite images can be combined with other sources of information to produce useful knowledge. Following Hurricane Katrina, the combination of satellite images and GPS technology was used to provide the necessary GIS information to the aid workers for rescue operations. It was particularly helpful in this case because the affected population asking for rescue indicated their location using their street address; however, since the land was completely flooded and many of the rescue services were from out of state, there were no visible landmarks so there was no way to locate people. Satellite images combined with GPS data helped identifying the exact location of the people in peril. Furthermore, high resolution images such as Quickbird were used for detailed assessment of the conditions in the aftermath, such as flooding and levee break locations. Maps indicating locations of flooded areas provided critical assistance to search and rescue operations.

Beneficiaries of satellite images are not just space agencies and companies that operate systems and generate satellite imagery. Many international actors, including UN organizations, benefit from using satellite images for their relief efforts. Other examples include UNICEF, UNHCR, UNDP, WHO,[158] and FAO – to name a few.[159] Specifically, UNICEF has used IRS and Landsat imagery of areas exposed to earthquakes in order to plan special relief and assistance programs directed toward children in affected areas.[160]

---

[157] See Office of the United Nations Disaster Relief Coordinator (UNDRO), *Space Application and Dissemination of Disaster-Related Data*, Expert Meeting Geneva, 14–17 June 1983, 32.

[158] A satellite image map was made to support WHO field teams assessing the current Marburg virus situation. This virus is a deadly hemorrhagic fever and a relative of the Ebola virus. Bjorgo, E., Space Aid: Current and Potential Uses of Satellite Imagery in UN Humanitarian Organizations, *Virtual Diplomacy*, United States Institute of Peace, 30 April 2002, 11.

[159] For utilization of satellite images by UN organizations, see Bjorgo, E., Space Aid: Current and Potential Uses of Satellite Imagery in UN Humanitarian Organizations, *Virtual Diplomacy*, United States Institute of Peace, 30 April 2002.

[160] See E. Bjorgo, Space Aid: Current and Potential Uses of Satellite Imagery in UN Humanitarian Organizations, *Virtual Diplomacy*, United States Institute of Peace, 30 April 2002, 10.

4.5.1.4. *Recovery and Reconstruction*

Satellite-derived information can be extensively used for the purpose of recovery and reconstruction. The images are used to map affected populations and to facilitate effective planning for reconstruction activities such as rebuilding bridges and schools. Following Hurricane Katrina, images from Landsat, Spot-4 and 5, and NigeriaSat-1 (DMC) were provided; these were used to support clean-up efforts, and to estimate the impact on public health. Images have been used for reconstruction efforts by relief entities such as the Red Cross.

Satellite data and derived information have been used for both national disaster management[161] and to assist other countries. Increasing numbers of space agencies and other EO entities voluntarily participate or plan to take part in international or regional disaster management programmes and initiatives in which they provide assistance under preferential conditions. Some entities have voluntarily provided images free of charge in response to particular disasters, without adherence to specific programmes.

The foregoing examples show that data uses have been dedicated more to post-disaster response and recovery phases – mainly assessment of damage and aftermath – than to the pre-disaster phase. Still, there are increasing uses of satellite remote sensing to assist with operations during prevention, preparation, and real time monitoring. Insufficient international assistance and cooperation in these phases of general disaster management can be corrected with the use of satellite remote sensing, because assistance is achieved more easily through data provision and sharing, without much intrusion. Applications in these phases are still limited: the accuracy of prediction using satellite data is not always reliable, or the monitoring is in need of more frequent information updates of disaster situations – at minimum, a revisit of approximately every three hours would suffice.[162] Utilization of data in these phases will be more active and widespread if the accuracy of the data is ensured and data is willingly and generously supplied by the data generators.

4.5.2. *Operational EO Based Programmes for Disaster Management*

There are an increasing number of programmes to utilize satellite data for disaster management, such as the International Charter on Space and Major

[161] For example, use of IRS satellite data for the Indian Ocean Tsunami and use of Landsat data for Hurricane Katrina.

[162] Constellations of satellites are providing an improvement to this problem. For instance, participating entities to the DMC are increasing; if ten to twelve satellites comprises a constellation, then a revisit time of a few hours will be achieved.

Disasters[163] (hereafter: the Disaster Charter), and the Disaster Monitoring Constellation (hereafter: the DMC) – both EO applications are coordinated mechanisms for disaster response. The key advantage of these programmes is that the beneficiaries can gain access to necessary information more promptly to make appropriate decisions thanks to a combination of data from different satellites that each entity contributes.

The Disaster Charter establishes extensive international cooperation amongst remote sensing satellite operators to provide satellite images free-of-charge to the countries affected by disasters.[164] This programme strengthens the legal regime of disaster management through establishing a predictable mechanism of assistance to afflicted states through effective information distribution. The Disaster Charter was initiated by ESA and CNES following the Third United Nations Conference on the Exploration and Peaceful Use of Outer Space (UNISPACE) III[165] in 1999, and has been operational since November 2000. Satellite imagery has been provided to afflicted states in response to diverse disasters – from floods, fires, and earthquakes, to rescuing ships stranded in ice. The Charter now embraces 10 partners: CNES, ESA, the Canadian Space Agency (CSA), NOAA, the Indian Space Research Organisation (ISRO), Comision Nacional de Actvidades Espaciales (CONAE), the Japanese Exploration Agency (JAXA), the United States Geological Survey (USGS) – on behalf of US partners such as Digital Globe and GeoEye, the British National Space Centre (BNSC) acting on behalf of the international Disaster Monitoring Constellation, and the China National Space Administration (CNSA). There are also other non-partner organizations serving as intermediaries. In response to authorized requests, partner space agencies provide data from their satellites free-of-charge to the states affected by natural or man-made disasters.

The Disaster Charter is triggered by certain intermediaries, referred to as Authorized Users. The unique feature of the Disaster Charter is that neither the Charter partners nor afflicted states can initiate the process for activation. It is only the pre-designated Authorized Users – being the relief agencies of the countries whose jurisdiction covers the member space agencies or other authorized entities – who can request the activation of the Charter, and it is the appointed Project Manager that organises the tasking of satellites, and directs

[163] Charter on cooperation to achieve the coordinated use of space facilities in the event of natural or technological disasters Rev. 3 (2/4/2000), 2. See Annex 2.

[164] For details, see A. Ito, Issues in the implementation of the International Charter on Space and Major Disasters, 21-2 *Space Policy* (May 2005), 141–149, A. Ito, Legal Aspects of the International Charter on Space and Major Disasters, *Proceedings of the Forty-Seventh Colloquium on the Law of Outer Space*, IISL, AIAA, 2005, 233–241.

[165] UNISPACE is a UN-organised international meeting where UN members and space agencies gather. At UNISPACE III the use of space technology for solving regional and world problems was discussed along with the need for international cooperation and use of space applications amongst developing countries.

data acquisition to member space entities. There are three major ways to activate the Charter: 1) direct activation- activation by the Authorized User of a member country; 2) activation via an Authorized User on behalf of a user from a non-member country; 3) activation via the UN for the UN users. In this manner, the Disaster Charter provides a mechanism to make critical space assets available to communities affected by disasters under a single point of contact. The afflicted states can use the data to monitor their disasters, assess the course of the disasters, and then respond to the aftermath of these disasters. The Charter has been operating successfully since November 2000 and the Charter activation takes place on a regular basis: the average number of calls per month requesting Charter activation is 1.9.

The DMC, which contributes to the Disaster Charter, is the result of extensive international cooperation to share EO space assets leading to the creation of a constellation of EO micro-satellites led by Surrey Satellite Technology Limited (SSTL) in the United Kingdom. With the capability of providing 32 m spatial resolution over a 600 km swath of any location in the world within 24 hours, DMC is designed to serve both national EO needs and global humanitarian needs for disaster relief.[166] The constellation was initiated with the launch of Alsat-1 for Algeria in 2002, followed by the launches in 2003 of BILSAT- 1 for Turkey, NigeriaSat-1 and UK-DMC, and then, in 2005, the launch of Beijing-1 for China. Each of these DMC partners – Algeria's Centre National des Technicques Spatiales, Turkey's Scientific and Technological Research Council, Nigeria's Research Council, China's Ministry of Science and Technology, and the British National Space Centre – owns its satellite for its own purposes and exchanges images within the consortium; however, they also provide images to non-partners free of charge in response to declared disasters. DMC is distinctive in that the constellation of five satellites can provide daily image revisit worldwide and thus provide fresh data to governments, as well as to relief and aid agencies. Whilst a certain capacity of the DMC is provided to aid agencies (including the framework of the Disaster Charter), other capacities are made available to commercial users through DMCii, which distributes the data on a commercial basis.

4.5.3. *Case Study: The 2004 Indian Ocean Tsunami*

This section discusses further the operational services of the Disaster Charter, which includes contribution from DMC, provided in response to the 2004

[166] M. Sweeting et al., Evaluation of the Utility of the Disaster Monitoring Constellation in support of Earth Observation Applications, *Selected Proceedings of the 5th International Symposium of the International Academy of Astronautics*, Berlin, April 4-8, 2005, 436.

Indian Ocean Tsunami and how assistance based on satellite remote sensing can be coordinated effectively in terms of an institutional framework.

On 26 December 2004, an undersea earthquake of magnitude 9 occurred 150 kilometres off the west coast of the northern Sumatra island in Indonesia.[167] The resulting Tsunami devastated the shores of a large number of countries in Southeast Asia, including: Indonesia, Sri Lanka, India, Thailand, the Maldives, Somalia, Myanmar, and Malaysia[168] and resulted in the loss of more than 200,000 lives.[169] There were no effective warning systems[170] in the Indian Ocean to detect tsunamis, and there was no effective communications system to disseminate warnings to the endangered populations.

Unfortunately, space assets were not helpful to predict or to prepare for the tsunami this time and did not contribute very much to real-time monitoring of the tsunami, although by coincidence the satellites TOPEX/Poseidon and Jason-1 happened to pass over the tsunami as it was crossing the ocean.

Satellite images were helpful in providing immediate assessment and direction of aid efforts in response to the Tsunami; three days after the tsunami hit, the Disaster Charter was activated and extensive amounts of data were provided. The way satellite imagery contributes to the disaster response effort merits a discussion. An immediate assessment of the scale of the disaster effectively took place within the following day or two through satellite images such as Landsat and those from DMC. Immediately after the tsunami, nobody knew precisely the scale of the disaster: contrary to the original predictions, the death tolls climbed higher and higher. The DMC supplied satellite images from its archives for Sri Lanka and the Maldives, showing the area before the disaster occurred for comparison with post-disaster images, so that damaged areas could be detected easily. The DMC data was very effective for immediate assessment because with the medium resolution and large-scale swath, it could cover very long stretches of coastline, so that damaged areas can be identified quickly. The DMC provided comprehensive coverage for the initial stage of damage assessment and facilitated a smooth transition to directing delivery of aid through effective targeting by high-resolution satellites. The high resolution satellites could detect at street level where the resources should be allocated; many

---

[167] Sharing Information for tsunami recovery in South Asia, *World Disaster Report 2005*, International Federation of Red Cross and Red Crescent Societies, 2005, 105.

[168] Sharing Information for tsunami recovery in South Asia, *World Disaster Report 2005*, International Federation of Red Cross and Red Crescent Societies, 2005, 105.

[169] Sharing Information for tsunami recovery in South Asia, *World Disaster Report 2005*, International Federation of Red Cross and Red Crescent Societies, 2005, 105.

[170] In a United Nations conference held in January 2005 in Kobe, Japan it was decided that, as an initial step towards an International Early Warning Programme, the UN should establish an Indian Ocean Tsunami Warning System.

high-resolution images including SPOT 5, Quickbird, and IKONOS were provided.

Satellite images were also used during the reconstruction period. For instance, high resolution Digital Globe satellite imagery helped relief workers in Indonesia establish refugee camps and medical teleconferencing Images were provided to various UN and NGO organizations for the rebuilding effort. A French technical engineering group, Geosciences Consultants, volunteered its expertise to non-governmental organisations, providing maps of the devastated areas: they had a data-base of world maps and a data-base of all infrastructures – such as airports, populated areas, and harbours, which allowed them to identify the spread of disasters and the exact areas affected.[171] The Indian Ocean Tsunami case has demonstrated clearly how space technology contributes to assessment and recovery.

4.5.4. *Coordination of Operations Based on Satellite Remote Sensing*

Various organizations are involved in provision of assistance; however, operations are fragmented and not well coordinated. In the field of satellite remote sensing, coordination is increasingly taking place to provide effective assistance. There are successfully operating programmes providing information based on satellite remote sensing, such as the Disaster Charter – a good example of coordinated disaster relief under a single point of contact. Institutionally, a network for information on space technology-based disaster management is being implemented. During the 43rd session of the UN Scientific and Technical Subcommittee in 2006, the group of experts presented a study on the possibility of creating a Disaster Management International Space Coordination Entity.[172] A more concrete step was taken on December 2006 by the adoption of the General Assembly Resolution to establish United Nations Platform for Space-based Information for Disaster Management and Emergency Response (UN-SPIDER) as a UN Programme, in order to "ensure that all countries and international and regional organizations have access to and develop the capacity to use all types of space-based information to support the full disaster management cycle."[173] The UN-SPIDER program aims to serve as central access point

[171] Satellite Images Boosting Asian Tsunami Aid Efforts, *Space Daily*, Paris, France (SPX), Jan. 4, 2005.

[172] See UN Doc. A/AC.105/C.1/L.285(2006), Study on the possibility of creating an international entity to provide for coordination and the means of realistically optimizing the effectiveness of space-based services for use in disaster management.

[173] See UN Doc. A/RES/61/110, 14 December, 2006, United Nations Platform for Space-based Information for Disaster Management and Emergency Response.

to space information for disaster management support, as a bridge to connect the disaster management and space communities. and to facilitate capacity-building and institutional strengthening. The anticipated role of UN-SPIDER is to coordinate space-based disaster management information – not just from remote sensing, but also derived from different space technologies such as GPS and telecommunications. The successful examples of disaster response in the field of satellite remote sensing can serve as a good model for information coordination in general disaster management.

4.5.5. *Revisit of Legal Principles Surrounding Satellite Remote Sensing*

The use of remote sensing for mitigating the impact from disasters is supported by the UN Remote Sensing Principles. A review of the provisions of the UN Remote Sensing Principles is necessary for further examination of the rights and obligations associated with disaster management. There is only one principle addressing natural disasters. Principle XI encourages the supply of disaster related information to affected or potentially afflicted states: "Remote sensing shall promote the protection of mankind from natural disasters. To this end, States participating in remote sensing activities that have identified processed data and analysed information in their possession that may be useful to States affected by natural disasters, or likely to be affected by impending natural disasters, shall transmit such data and information to States concerned as promptly as possible."[174] This provision stands on the presumption that imaging of states takes place without prior consent.

The major reason remote sensing can contribute to disaster management is that it enables disaster response unaffected by political impediment, at least in terms of information gathering. Since remote sensing permits data collection of afflicted states without their consent, case-by-case response from outside is improved. Foreign states and external entities that wish to render assistance do not have to wait for the approval of the afflicted state to sense and, even in case of refusal of such offer, the stricken country can still be sensed from space without intrusion. On the other hand, even in case of a deadly disaster, sovereignty concerns by affected countries are still of great importance. Following the 2004 Indian Ocean Tsunami, India was sensed only by its own satellite IRS series.

Critical information can be obtained externally about the extent of disasters as well as identification of priorities of affected areas, regardless of the capacity of the afflicted state. On the basis of such information, useful knowledge can be provided, and international bodies and NGOs can send in relief teams more

[174] Principle XI of the UN Remote Sensing Principles.

effectively and efficiently. Constant updates of data collection covering all of the damaged areas through effective coordination minimizes the overlapping services and reduces the level of disparities in served areas.

4.5.6. *Legal Analysis of a Successful Model Based on the Disaster Charter*

As discussed earlier, the Disaster Charter is a successful model for the effective coordination of disaster relief based on supply of satellite images. At this point, the stance of operations will be analyzed from the legal standpoint including the legal nature, and the implications for the responsibility and liability associated with the provision of aid.

The Disaster Charter is a non-binding legal instrument based on goodwill and best endeavours; the cooperation is voluntary.[175] Throughout the provisions of the text, the voluntary nature is confirmed by the language: "shall use best endeavours." For instance, Article 4.5 states: "The parties shall use their best endeavours, in accordance with the identified crisis scenarios, to supply associated bodies and, where appropriate, beneficiary bodies with data, and if necessary associated information and services, gathered by the space facilities."[176] No liability is involved in its whole course of operations; indeed, any liability is clearly waived under the Charter. Article 5.4 of the Disaster Charter clearly states that: "The parties shall ensure that associated bodies which, at the request of the country or countries affected by disaster, call on assistance of parties undertakes to confirm that no legal action will be taken against the parties in the event of bodily injury, damage or financial loss arising from the execution or non-execution of activities, services, or supplies arising out of the Charter."[177] In short, the services are not guaranteed: members will not be held liable either in the case in which the Charter was not activated for a particular disaster, nor where actual provisions of images did not lead to successful rescue operations. The intention of those who drafted the Disaster Charter was clearly that they do not wish to be held responsible for their favours; however, it is questionable whether the free provision of service is sufficient to waive any potential liability. The waiver of liability is not agreed upon by all potential afflicted states or concerned parties, such as families of aid workers. It is uncertain whether a unilateral waiver of liability is enforceable, and this situation emphasizes the need for clarification of liability.

175 Article 3.1 of the Disaster Charter states that "The parties shall develop their cooperation on a voluntary basis, no funds being exchanged between them".

176 Article 4.5 of the Disaster Charter.

177 Article 5.4 of the Disaster Charter.

4.6. The Need for an Improved Regime

There is an urgent need for a clear legal regime governing disaster relief to enhance effectiveness, as well as an unambiguous liability regime associated with the supply of satellite data. A better coordination with the division of responsibility will improve effectiveness. Furthermore, the liability associated with disasters (such as liability of volunteers) needs clarification.

4.6.1. *How a Better Coordination with Division of Responsibilities will Enhance Effectiveness*

The earlier section demonstrated that, in most disasters, the competence to provide assistance to the victims is shared amongst a number of actors; however, such operations are often uncoordinated. For instance, following the flash floods and landslide that hit Venezuela in December 1999, the issue of division of responsibilities and decision-making between central and decentralized powers became a source of conflict.[178] The coordination and division of responsibilities is also a key issue in provision of disaster relief using satellite remote sensing as various actors are involved before the delivery of final products. A better coordination is critical to minimize possible conflicts that may arise amongst diverse actors and enhance the effectiveness of operations.

Disaster response in the past has occasionally involved inappropriate assistance and/or unwanted side effects. For instance, following the Indian Ocean Tsunami, an unknown NGO had vaccinated some of the children without leaving any vaccination records, thus generating confusion as to who had and had not been served.[179] Obviously, there is the need for establishing clear mandates and responsibilities for each organization and/or specific tasks to enhance effectiveness and avoid serious mistakes.

The Disaster Charter demonstrates a well-coordinated disaster response that includes a predictable mechanism with the division of responsibility and effective direction of aid based on the provision of disaster information. Since a project manager appointed for each case is responsible for directing specific satellite operators to provide data to the appropriate Charter partner for imaging a particular location, data collection is effectively arranged, overlaps are minimized, and response time is improved.

---

[178] See C. Villacis, Latin American Cases Hurricane Mitch, Flash Floods and Landslides in Venezuela, El Salvador Earthquakes, *[Report] Recovery and reconstruction reports of the worldwide disasters and their comparative study*, Asian Disaster Reduction Center, Kobe, 2005, 15.

[179] See V. Bannon & D. Fisher. Legal Lessons in Disaster Relief from the Tsunami, the Pakistan Earthquake and Hurricane Katrina, 10-6 *ASIL Insight* (2006), 4.

Minimizing errors is crucial to achieve effective operations. A clear mandate and division of responsibilities of the authorized users and Charter partners would contribute to error reduction; however, minimizing errors involves cost. Providing assistance is already costly and covering the cost of accidents can place a severe burden on donors, who already bear the costs for liability risks. The aid provided is often not sufficient both in terms of quantity and quality, particularly following deadly large scale and prolonged disasters, where the need usually exceeds the offers of assistance.[180]

The provision of aid should be something appreciated and praised; however, it is also true that the victims sometimes do not look at the efforts made or the surrounding circumstances; rather they can only see the loss they or their family have actually incurred.

4.6.2. *Issues of Clarification of Liability: Analysis of Lawsuits following Disasters against Volunteers*

Lawsuits have been brought on unreasonable grounds following disasters. Volunteers who have rendered assistance in good faith are not exceptions, and they, too, have become the target of lawsuits.[181] For instance, Mark Morice, a local New Orleans lawyer who used boats to rescue reportedly more than 200 Hurricane Katrina victims, was sued by the boat owner although the lawsuit was not pursued.[182]

A large scale lawsuit that arose in relation to the Indian Ocean Tsunami is striking and worth examining in detail. A group of Austrian and German victims have since brought action against the NOAA Pacific Early Warning Centre, together with the Thai government and the French hotel chain Accor for not responding properly to the disaster.[183] The case is notable, as this type of lawsuit is the first of its kind. The NOAA Pacific Center is accused of having registered the earthquake but failing to alert Indian Ocean countries of the impending tsunami.[184] NOAA is a member of the Disaster Charter. Although the lawsuit is irrelevant to the operations of the Disaster Charter, it raises an important

180 See B. Jakovljevic, International Disaster Relief Law, 34 *Israel Yearbook on Human Rights* (2004), 264.

181 See J.R. Wible, Indeed, in the past volunteers have been sued for their rescue operations following disasters. See Volunteer's Liability Issues, Medical Reserve Corps National Leadership and Training Conference, April 17–20, 2007.

182 See L. Williams, Boat Owner drops lawsuit- Man took it to rescue victims of Katrina, *New Orleans Times- Picayune*, September 8, 2006.

183 See W. Nicholson, Legal Issues: Warning Systems, Early Waning Systems- A public Entity Risk Institute Symposium, 2005, 1.

184 See W. Nicholson, Legal Issues: Warning Systems, Early Waning Systems- A public Entity Risk Institute Symposium, 2005, 1.

question as to risks of exposure to liability and the uncertain outcome of claims. Another question is whether the Good Samaritan doctrine applies in this case.

The Pacific Tsunami Warning Center of NOAA had no obligations to warn, since it was covering only the Pacific Region, but it attempted to contact the affected countries in vain,[185] so it can be considered that the Center was a 'Good Samaritan'. In contrast, the Charter partners may *not* be considered as Good Samaritans, particularly taking into account the fact that the relationship of dependence does not necessarily require the existence of a contract or a personal relationship between the rescuer and rescued – insofar as the one aided counts on the aid by the rescuer. The primary reasons for the status of the Charter partners in this scenario are:

1) The initiators of the Disaster Charter actually set up the mechanism deliberately in advance to help the potential crisis victims, and followers appreciate this philosophy;
2) The Disaster Charter service takes place after the request from the afflicted states;
3) Under the Disaster Charter, there is no contract; but the degree of dependency by the crisis victims is high;
4) A certain degree of duty of care is assumed; and
5) Waiver of liability is unilateral; since there is no agreement between the afflicted states and partners or potential claimant(s), a waiver may not be enforceable.

Compared to NOAA's case in the Indian Ocean Tsunami, the Disaster Charter service appears to have a greater exposure to liability. In fact, as with any operation, assistance cannot be 100% successful. With respect to the Disaster Charter, services provided were not completely successful due to late response, unavailability of data, and the limited usefulness of the products that were supplied. According to the 6th Annual Report of the Disaster Charter, following the Suriname flood of early May 2006, it was found that the delivered products were of limited use because they focused on the initial area of interest and did not cover the main flood areas that were later discovered.[186] The Report shows that end users of the Disaster Charter were by and large appreciative of and benefited from the information provided, even if it was not completely useful. It is not possible to completely eradicate the risk that inaccuracies will lead to inappropriate decisions or actions that may result in unexpected loss. It is less likely that beneficiaries would bring actions against Disaster Charter partners

[185] See W. Nicholson, Legal Issues: Warning Systems, Early Waning Systems-A public Entity Risk Institute Symposium, 2005, 1

[186] Disaster Charter, Executive Secretariat, 6th Annual Report January-December 2006, 36–37.

for their aid that turn out to be unsuccessful; however, Disaster Charter partners may be at risk, for example, with regard to claims brought by families of aid workers against Charter partners for providing inaccurate images leading to loss of the aid workers' lives. In case of deadly disasters, the possibility of aid workers becoming victims is high. Rescue operations can be dangerous and aid workers are vulnerable to numerous risks, including endemic diseases.[187] In fact, infectious diseases accounted for over 75% of the emergency evacuations of humanitarian workers.[188] If the families of victims sue data suppliers for inaccurate data that indirectly led to their deaths, the outcome of possible lawsuits is not easily predicted under the current uncertain legal regime.

In the Indian Ocean Tsunami lawsuit case, since it was brought in the US court, the application of the Good Samaritan doctrine is foreseeable. Since the Good Samaritan Principle is well recognized,[189] the US Congress has considered a Samaritan act specifically addressing natural disasters. A bill entitled "The Good Samaritan Liability Improvement and Volunteer Encouragement Act of 2005" has been introduced[190] to grant certain immunities from liability to both national and international disaster relief volunteers, as well as to governmental and intergovernmental organizations.[191] Many countries lack comprehensive Good Samaritan laws and, consequently, foreign relief personnel and organizations are exposed to liability risks arising as a result of unsuccessful aid.[192] Considering the international nature of disaster management, particularly assistance based on satellite remote sensing, clarification and universal applicability of the Good Samaritan Principle is essential.

States or organizations would become more confident about providing or receiving aid if they know precisely the scope of their responsibility and that they will not be held liable, even if assistance is unsuccessful. Indeed, concern about vicarious liability was one of the factors contributing to the decision by

[187] Relief workers are vulnerable to endemic disease as they work in regions where sanitary systems are substandard, damaged or destroyed by disasters. J. Gerlach & J. Irwin, Insurance for relief and development workers, Safe Passage International, Lakewood, 3.

[188] See J. Gerlach & J. Irwin, Insurance for relief and development workers, Safe Passage International, Lakewood, 3.

[189] According to Charitable Immunity Manual: A Review of US Charitable Immunity Legislation for Volunteer Hearth Care Providers, 43 states and the District of Colombia have some sort of charitable immunity legislation.

[190] The Good Samaritan Liability Improvement and Volunteer Encouragement Act of 2005, S. 1747, 109th Congress (2005).

[191] See V. Bannon & D. Fisher. Legal Lessons in Disaster Relief from the Tsunami, the Pakistan Earthquake and Hurricane Katrina, 10-6 *ASIL Insight* (2006), 2–3.

[192] See Background paper on International Disaster Response Laws, Rules and Principles for the FEMA Legal Round table on Foreign Assistance in the United States, Washington DC, November 28, 2005, 2.

FEMA and other United States agencies to refuse any international offers of certain types of aid, including blood products, medical kits, and most search and rescue personnel arriving to assist the people affected by Hurricane Katrina.[193] The fear of being sued in association with providing aid has also generated reluctance to accept aid.[194]

Reluctance to accept aid may arise with respect to assistance based on satellite remote sensing as well. The close examination of the Good Samaritan Principle in association with the Indian Ocean Tsunami lawsuit case and the Disaster Charter has reaffirmed the need for establishing a clear liability regime applicable to the operation of systems and data provision. Establishing a more comprehensive legal regime that clarifies responsibility and liability would mitigate unwillingness and hesitation, and facilitate data supply as both donors and beneficiaries would then be able to accept or provide data without concern for the outcome of the assistance.

4.7. Conclusion

Today the contribution of satellite remote sensing to disaster management is evident in the extensive use of images, particularly in the post-disaster phases and the various decision-making processes related to relief and reconstruction that are based upon such information. The problems associated with the use of remote sensing, despite its proven helpfulness, illustrate the need for a more robust legal regime that also harmonizes existing rules.

Looking at the historical development of the legal framework on disaster management, the reluctance of government and private actors to enter into a binding regime, the nature of existing legal frameworks that are limited in scope, and the examination of appropriate models for disaster management, it can be concluded that:

1. The general legal regime of disaster management should be left as it is whilst effective coordination should be achieved in a narrower sphere, such as aid based on space technology and aid based on medical supply;
2. The application of the Good Samaritan Principle to remote sensing data should be established clearly and incorporated into the regime of satellite remote sensing law with clarification of liability.

---

[193] See V. Bannon & D. Fisher. Legal Lessons in Disaster Relief from the Tsunami, the Pakistan Earthquake and Hurricane Katrina, 10-6 *ASIL Insight* (2006), 2–3.

[194] See the preamble of the US Volunteer Protection Act that refers to impediments to the willingness to render assistance due to uncertain liability.

It appears to be most productive to leave the general regime as it is, since the codification of unified rules that require governments and organizations adhere to a binding regime will impose obligations that are most likely to discourage the rendering of assistance, instead of facilitating effective operations. Rather than establishing an overall regime that specifies the rights and obligations of donors and recipients of aid, it is more realistic to achieve rationalization of operations in a non-binding mechanism.

Successful aid provided by volunteers and voluntary organizations – as well as a viable model of voluntary yet coordinated disaster relief based on satellite data, as represented by the Disaster Charter, – illustrate that the voluntary approach is a better suited approach. The Disaster Charter is an interesting vehicle for the Good Samaritan (voluntary assistance without liability) model of disaster response. Taking into account the operational mechanism of the Disaster Charter, effective coordination should be achieved by a single entity, which should act as a clear-cut point of contact for afflicted states. Such an entity should have the mandate of delegating tasks and determining the associated responsibilities, but details can be determined on an ad hoc basis according to the particular circumstances of each disaster.

Although encouraging states, organizations, and private entities or individuals to render assistance on a voluntary basis appears, in the end, to be more useful for potential beneficiaries, lawsuits following disasters demonstrated that liability risks faced by volunteers are significant enough. Voluntary assistance, including that based on satellite images, can be provided more readily if good faith volunteers are not subject to lawsuits. To these ends, a Good Samaritan law protecting voluntary helpers for victims of natural disasters should be established globally. This should address the current uncertainty regarding liability, which continues to impede the effective delivery of aid, or, in the worst case discourages the provision of aid altogether.

Part II

Improvements to the Current Framework

Chapter Five

Data Policy and Verification of Accuracy and Authenticity

5.1. Introduction

Today there is a need for a large amount of reliable data to enable a variety of environmental assessments, as well as for large-scale and site-specific disaster monitoring. However, the need is not satisfied fully because the current legal framework does not address the issues sufficiently: for example, how data is accessed, charged, distributed, and used; and how the authenticity and accuracy of data is verified.

As illustrated in Chapter 2, there are no guidelines for the data policies. The only Principle related to data policy is Principle XII of the UN Remote Sensing Principles that addresses issues in a limited fashion: "As soon as the primary data and the processed data become available, Sensed States shall have access to the data on a non-discriminatory basis and on reasonable cost terms."[1] This Principle deals with the data availability regarding both sensed states and other states that wish to access to the data, and both are treated equally in terms of conditions for access. The provision limits the scope of data availability to only sensing states, sensed states, and other states interested in the data. Furthermore, it does not serve as a basis to allow free access to data for public benefit service, particularly for disaster relief.

[1] Principle XII of the UN Remote Sensing Principles.

The only provision concerning the pricing policy, viz. 'reasonable cost terms', is ambiguous and open to different interpretations: it could mean either extremes of marginal costs, or a commercial price, insofar as it is reasonable for the particular image in question. It in no way serves as a general guideline for the price setting expected for the different types of data or a variety of uses by the different data generators. Although Principles X and XI encourage the disclosure of information for environmental protection and disaster management, it does not state in what terms data are supplied in such cases;[2] therefore, it does not serve as a legal basis to allow favourable treatment for such uses. The Principles are far from sufficient to govern the supply and use of data currently taking place world-wide.

The UN Remote Sensing Principles are completely silent as to the rights of data generators, such as intellectual property rights or the conditions for the use of the data. The conditions of the applicability of the intellectual property rights are not spelled out (e.g., the categorisation of different types of products subject for protection and the rights associated with each of them). There are no provisions regarding the accuracy, integrity, and reliability of data, or how data should be validated and authenticated. Hence, the UN Remote Sensing Principles do not serve as a guideline that governs the activities of data suppliers; neither do they provide any expectations for the data recipients in dealing with the data.

At this time, access, distribution, and use of remote sensing data are governed largely by the varying policies set by an individual supplier and the implementation of any data validation also depends on the respective entity. Under the current environment, none of the interests of stakeholders are safeguarded properly. The rights associated with the data are not guaranteed to data suppliers; users may face limited access to the data as well as having low confidence in the data because of the uncertainties regarding its accuracy and authenticity. The current legal regime neither fosters easy access to data for the benefit of users, nor offers suitable protection of the different types of data to the appropriate parties in the data supplier chain.

There is an urgent need to consider new regulations in line with the current situation. A balance of interests of the different stakeholders needs to be achieved, the integrity of the data must be ensured, and the guidelines for

---

[2] Principle X of the UN Remote Sensing Principles states that: "States participating in remote sensing activities that have identified information in their possession that is capable of averting any phenomenon harmful to the Earth's natural environment shall disclose such information to States concerned." Principle XI states that: "States participating in remote sensing activities that have identified processed data and analysed information on their possession that may be useful to States affected by natural disasters, or likely to be affected by impending natural disasters shall transmit such data and information to States concerned as promptly as possible."

the assessment of reliability of the resulting information products must be established. The aim of this Chapter is to reveal the reality associated with data policies, data accuracy, and authenticity validation – analysing the tendency for each, to identify gaps to be filled and then to propose improvements.

5.2. Data Policy Associated with Satellite Images and Derived Products

The recent trend in the field of remote sensing saw the proliferation of data use and growth of the data market. Nowadays it is a minority of governmental space agencies that are keeping their data for internal use and/or giving out the data at marginal cost, but most of the time significant capacity is allotted for distribution of data on a commercial basis. With the emergence of different funding models for satellites and numerous satellite-based programmes based on varying forms of cooperation, the issues of data policy have gained more and more attention. It is critical not only for users but for all remote sensing stakeholders to understand precisely what data policy is and its impact on accessibility and utility of data by users.

Data policies determine the rights and obligations for data providers and recipients.[3] Effective data policies are essential to ensure that the right information can be made available to those who need it, in the right form and at the right time.[4] Ideally, data policy should be established in a manner that users can enjoy the full benefits of data whilst suppliers can retain and exercise rights associated with data.

As discussed in Chapter 2, whilst government may have an overall framework to implement the law and policies to exercise control on domestic actors – particularly as to data dissemination – data policies are usually set independently for each satellite. A variety of data policies that exist worldwide have differing impact on accessibility to the data and the utilisation of the data, all of which affect the interests of the data suppliers, end-users, and the third parties.

5.2.1. *Access Policies*

Existing data policies, adopted by the different entities in terms of access and pricing, can primarily be categorised to two types. The clear distinction is made

[3] For a condensed version of this Chapter, see A. Ito, Improvements to the Legal Regime for the Effective Use of Satellite Remote Sensing Data for Disaster Management and Protection of the Environment, 34-1 *Journal of Space Law* (2008), 45–65.

[4] *GMES Issues and Need for Action Report from the 1st GMES forum Brussels*, 15–17 July 2002, 24.

between governmental agencies and commercial entities that operate satellites and generate data/products. As far as the governmental agencies are concerned, they appear to be further categorised into: those entities, most notably those of the United States, which adopt an open access policy; and those – particularly European entities – that tend to have more regulated access.

Open access commonly allows users to access the data freely at least at marginal cost irrespective of the types of users and the purpose for which the data is utilised. For instance, Orthorectified Landsat Data is available free-of-charge via FTP.[5] NASA makes QuickSCAT data and MODIS data available free of charge for scientific and educational use, with virtually no restrictions.[6] This type of access policy is reflective of US federal policy, applying the concept of freedom of information for the data collected at the federal level or using federal funds, and consequently, no government copyright is claimed and no fee is charged for the spatial data.[7]

On the other hand, space agencies, including ESA, and the Canadian Space Agency, adopt two (or multiple) tier systems encompassing different categories of users. ENVISAT distinguishes between category 1 and 2 uses: category-1 is for research and application development use; category-2 is for other uses, including operational and commercial use.[8] For ERS, there are five categories: (A) internal use; (B) research and demonstration use; (C) national meteorological services belonging to WMO members; (D) those organizations of participating states engaged in operational services for public utility; and (E) commercial use.[9]

The private entities operating remote sensing systems such as GeoEye and Digital Globe primarily distribute their data on a commercial basis. The commercial data generators only hold a single category of commercial users, unless a special arrangement is made so that the data is supplied free-of-charge for disaster relief in certain cases, such as within the framework of the Disaster Charter.

Finally, there is access policy based on the programme rather than based on each satellite or its owner. The DMC Consortium has a unique data policy.

---

[5] See http://www.landsat.org. Last accessed on 20 April 2010.

[6] See R. Harris, and R. Browning, *Data Policy Assessment for GMES Final Reports*, EVK2-CT-2002-80012-DPAG., University of London, London, 2004, 35.

[7] See R. Longhorn, and M. Blakemore, Re-visiting the Valuing and Pricing of Digital Geographic Information, 4-2 *Journal of Digital Information*, 2004, 4. The root of the open access policy lies in numerous relevant domestic laws including the 1976 Copyright Act, the Freedom of Information Act, the Paperwork Reduction Act, Electronic FOIA Amendments of 1996, Office of Management and Budget Circular No. A-130.

[8] See F.G. von der Dunk, Non-discriminatory data dissemination in practice, *Earth Observation Policy and Europe*, A.A. Balkema Publishers, Lisse, 2002, 44.

[9] See ESA/PB-EO(90)57, rev.6, Principles of the Provision of ERS data to users 1994.

Part of the data capacity of each satellite in the constellation is reserved for national governmental use by the owner of the individual satellite and there is a collective commitment to supply data for disaster relief purposes, whilst the remaining capacity of the satellites is released to the commercial market. Within the total capacity of the DMC, approximately 5% to 10% of the data capacity is provided free of charge to disaster relief agencies. Each partner determines how the rest of the capacities are allocated – such as its own national use or commercial sale released onto the market place through the commercial distributor, Disaster Monitoring Constellation International Imaging (DMCii).

As discussed in Chapter 2, entities are subject to constraints imposed by the government through national legislation and policies, including shutter controls and priority access. Entities that fall under the jurisdiction of the United States, Canada, Germany, and France (amongst others) need to adhere strictly to the obligations they owe under the laws of the respective state. Other states, such as India and China, have framed specific policies that place restrictions on data dissemination in terms of spatial resolution. Consequently, users may be inevitably restricted by law and/or policy in accessing data under the framework of any of these states. In the near future, these types of restrictions may extend to other states in consideration of the launch or planned launch of their high resolution satellites.

The access policies of divergent entities are not uniform; however, analysis highlights the trend in clear distinction between public and commercial use. For governmental agencies, data access for public benefit purposes is reserved as just one category amongst many others, whereas for commercial entities it is a special arrangement. That means that the same data can be distributed both for the public good and for use in commercial products.[10] For those governmental agencies that adopt more than one category of users, the distinction is made not by a type of data, but by a purpose for which the data is used, or the nature of the data recipients. Indeed, it is a common practice now that data from governmental space agencies are distributed on a commercial basis[11] and vice versa: data from commercial entities is being supplied as a public service in support of disaster relief.[12] In such cases, the policies and conditions for supply are not

---

[10] For instance, the flood mapping requested for insurance claim can be commercial, whereas the data supply requested by aid agency is public.

[11] It needs to be restated that space activities had been purely governmental, as were remote sensing activities; thus, the resulting data remained within governmental use rather than commercially available to the public. Marketing of the data is relatively recent, beginning in the 1980s. Now data is distributed globally by more than one distributor.

[12] A number of high resolution IKONOS images of the aftermath of the Tsunami have been distributed.

necessarily clear to the external parties. If easier access for the purposes of public benefit is to be achieved, it is necessary to establish a clear overall policy for public service applications.

5.2.2. *Pricing Policies*

Generally speaking, the pricing policies for different satellites and their derived products are non-uniform. The different access policies examined in the previous section are directly linked to variable pricing. A fundamental distinction is made between those entities operating under an open access policy adopting free or marginal cost pricing, and those entities operating under regulated access with cost recovery pricing.

5.2.2.1. *Pricing for Open Access*

For the open access policy, which is typically adopted by US governmental suppliers, the price is set at a marginal cost. The marginal cost is regarded as "the price that recovers the costs incurred in providing data beyond the costs of the basic ground infrastructure."[13] For instance, NASA adopts a 'cost of fulfilling user request' mechanism for Landsat;[14] it implies that all marginal costs that are necessary to fulfil the request of the user – which may include the programming, the satellite, and/or the retrieval of the data from the archive[15] – are charged. The Landsat data is set at a marginal cost, where the cost of fulfilling the user request is assessed as approximately US$ 600 per scene[16] for newly collected data, and US$ 25-50 for archived data.[17] Furthermore, NASA's ASTER data was made available free of charge until August 2002; currently, the charge is approximately US$ 55 per scene.[18]

---

[13] See R. Harris, Earth Observation Data Pricing Policies, *Earth Observation Policy and Europe,* A.A. Balkema Publishers, Lisse, 2002, 117.

[14] Land Remote Sensing Policy Act 1992, Section 5615.

[15] See G. Schreier, Data Policy Implications on Archive Design, *Earth Observation Policy and Europe,* A.A. Balkema Publishers, Lisse, 2002, 175.

[16] See R. Harris, Comparison of regime: ESA, NASA and NOAA, *Proceedings of the International Conference: Satellite Remote sensing in aid of development: legal considerations,* Tunis 26-27 September 2002, European Centre for Space Law, 2003, 37.

[17] Each Landsat 7 ETM+ scene in the Landsat.org Archive is 50 USD. Each Landsat 4/5 TM scene in the Landsat.org Archive is 25 USD.

[18] See R. Harris, Comparison of regime: ESA, NASA and NOAA, *Proceedings of the International Conference: Satellite Remote sensing in aid of development: legal considerations,* Tunis 26-27 September 2002, European Centre for Space Law, 2003, 37.

5.2.2.2. *Pricing for Regulated Access*

The regulated access pricing environment is further categorized into (A) the pricing for entities that adopt categories of users and/or uses, and (B) commercial entities that basically have one single commercial user. Amongst those having more than one category of users, only data regarded as for public use – including internal, educational, and humanitarian use – are distributed under preferential conditions, either with a price waiver or distributed at a marginal cost. For example, ERS of ESA distinguishes between category 1 and 2 use: category 1 is for research and application development use, and category 2 is for other uses, including operational and commercial use.[19] The marginal cost, referred to as 'at or near the cost of reproduction' is applied to category 1.[20]

When a governmental space agency creates a commercial use data category, the data will then be distributed on a commercial basis. Governmental space agencies do not distribute data themselves but delegate to authorized entities to market the data. More often than not, more than one supplier undertakes the data sale, most notably to cover the distribution in different regions. An increasing number of data distributors is independent from the governmental space agency that originates the data.

Private commercial data generators operate satellites and distribute their data on a commercial basis. The data generators of very high resolution satellites, such as GeoEye's Orbview, sell their data at a full commercial price; for instance, data with spatial resolution of 60 cm costs US$ 10 per square km.[21] The examples make it clear that open access and regulated access, particularly by commercial entities, present a striking contrast in pricing models.

Finally, price waiver arrangements adopted by certain data generators for public benefit warrant a brief discussion here. In operational programmes using remote sensing, the participating data generators often compromises their individual access and pricing policy. Under the Disaster Charter, data from different systems are supplied free of charge, compromising the data policy of each system to the benefit of all potential afflicted states, provided that they follow the designated process. Such arrangements are increasingly adopted for a coordinated service.

GMES aims to provide integrated information services for monitoring the environment through contributions from different sources to the benefit of users in both public and private sectors. Precise data policy has not been defined for GMES, but a certain harmonization will be achieved, ensuring a

[19] ESA/PB-EO(97)57 rev.3, 4–6.

[20] ESA/PB-EO(97)57 rev.3, 4–6.

[21] That means that the image of 60 km × 60 km would cost US$ 36,000.

user-friendly service and cost effective access to the standardised data.[22] It is envisaged that a considerable part of GMES information should be available free of charge.[23] Indeed, all EUMETSAT data and products will be made available free of charge to the pre-operational services of GMES, provided that every GMES user contacts EUMETSAT directly to receive the data/products and signs a simple license.[24] First, however, a balance must be established between non-discriminatory access and the incentive for investment in satellite development and its financial sustainability.

5.2.3. *Observation of Factors Determining Pricing*

The factors affecting price settings for divergent satellite images and derived products merit close analysis. Obviously, the quality and characteristics of the data determine the pricing of both the data and derived products. Such factors as spatial resolution, image size, different levels of processing, and added value all contribute to varying prices associated with each product. Naturally, the data is more expensive if it has higher spatial resolution, if its image coverage is wide, and if it has value-added attributes – for instance, a product in the form of a digital elevation model[25] – and if it is newly tasked data, in contrast to archived data.[26]

The complexities of the pricing of satellite images often lie in the factors external to their quality and type of data; specifically in broader dimensions such as national policies, the type of data generator, and the financing for building and operating the satellites. National policies, as previously mentioned,

[22] Communication from the Commission to the European Parliament and the Council Global Monitoring for Environment and Security (GMES): Establishing a GMES capacity by 2008 – Action Plan 2004–2008, COM(2004) 65 final, Commission of the European Communities, 6.

[23] Communication from the Commission to the European Parliament and the Council Global Monitoring for Environment and Security (GMES): Establishing a GMES capacity by 2008 – Action Plan 2004–2008, COM(2004) 65 final, Commission of the European Communities, 13.

[24] EUMETSAT, Eumetsat agrees to provide data for GMES, *Press Releases*, July 3, 2008.

[25] DEM is a file containing a uniform grid of terrain elevation values of an area of interest, generated from a pair of automatically correlated scenes. See *Product Guide*, Spot Image, Toulouse, 2003, 10.

[26] Most suppliers set higher prices for tasking/programming data than for archive data, such as Spot Image charging 1,000 Euros more for programmed data. Some entities, such as DMCii, set different pricing within their archived data for recent data and older data. DMC charges 614 Euros for 160 km × 160 km plus 0.024 Euros/square km, whilst archived data is only 0.04 Euros/square km.

affect the overall pricing extensively, with the difference between the governmental US providers and European providers being particularly notable. The funding source is also a major pricing factor. The fundamental distinction is whether the satellite is financed fully or partially by public funding, or financed solely by private funding. As discussed in Chapter 2, the 2006 US Licensing for Private Systems holds three distinct pricing policies depending on the percentage of funding from the US government. Commercial entities commonly seek to retrieve their initial investment for manufacturing and launching satellites, as well as attempt to cover operational cost through data sales. Consequently, they are driven to set the price at fair market value.

In short, various factors determining the access policy and pricing of satellite remote sensing products are so complex that it is not feasible to achieve a single access policy with a uniform price setting for all. What can be feasible, however, is to establish a common access and pricing policy for public benefit services, as already achieved under the Disaster Charter and the DMC.

Common access and pricing policy for the benefit of users is increasingly recognized and implemented to foster environmental protection, in addition to emergency response. In Europe, public access to environmental information is specifically written into a Directive. As discussed previously, the Directive on Access to Environmental Information establishes the right of any requestor to access environmental information held by or for public authorities at reasonable cost and without having to state his/her interest.[27] The Environmental Directive contains basic data policy for broad categories of environmental information defined as "any information in written, visual, aural, electronic or any other material from."[28] It is important to note that the Directive is applicable to environmental information insofar as the information is held by or on behalf of public authorities: a) government, other public administration; b) any natural or legal person performing public administrative functions under national law; and c) having public responsibilities or functions or providing public services.[29] This language reflects the approach that allows public access to the environmental information on the basis of the content; that is, information on the environment itself, rather than on how the information is produced.[30] The language of the Environmental Directive covers EO data produced by private

[27] Article 3 of the Environmental Directive states: "Public authorities may make a charge for supplying nay environmental information but such charge shall not exceed a reasonable amount."

[28] Article 2(1) of the Environmental Directive.

[29] Article 2(2) of the Environmental Directive.

[30] See G. Winter, Access of the Public to Environmental Data from Remote Sensing, *Droit télédétection et environnement: actes du colloque international: le droit face aux techniques de télédétections par satellite au service de l'environnement*, S. Courteix (Ed.), Strasbourg, 2-3-4 juin 1993, Sides, Antony, 1994, 250.

entities as long as it is used for a public service related to environmental operations.

The Environmental Directive sets a timeframe for access to be: "as soon as possible or at the latest, within one month after the receipt of the public authority of the request"[31] and sets forth a systematic mechanism to ensure easier access to environmental information.[32] The Directive has a solid enforcement mechanism whereby when an applicant's request for information has not been processed properly by the public authorities, that applicant has the right to review both administrative and legal procedures.[33] The Directive is thus a powerful instrument that has a significant impact on the access to satellite data for environmental purposes.

There is a movement within the European Community to establish a common infrastructure for spatial information, and data policy issues are amongst the considerations. The initiative commonly referred to as INSPIRE, became a Directive and was adopted by the European Commission on 14 March 2007.[34] The scope of INSPIRE is limited to Europe for the purpose of environmental applications[35] and covers "any data with a direct or indirect reference to a specific location or geographical area," which includes satellite data and products.[36] It is important to note that INSPIRE is applicable primarily to spatial data held by public authorities for its use by those authorities in executing public tasks; however, it can also be applied to persons other than public authorities if they request it, thus extending its applicability to individuals and private entities.[37] INSPIRE clearly establishes the pricing policy as follows: (1) the availability of

---

[31] Article 3(2)(a) of the Environmental Directive.

[32] Article 3(5) states that: "For the purpose of this Article, Member States shall ensure that: (a)officials are required to support the public in seeking access to information; (b) lists of public authorities are publicly accessible; and (c) the practical arrangement are defined for ensuring that the right of access to environmental information can be effectively exercised."

[33] See Article 6 of the Environmental Directive.

[34] Directive 2007/2/EC of the European Parliament and the Council of 14 March 2007 establishing an infrastructure for spatial information in the European Community. [Hereafter: INSPIRE Directive]

[35] Article 1 of the INSPIRE Directive states that it "lays down general rules for the establishment of an infrastructure for spatial information in Europe for the purposes of Community environmental policies and policies or activities which may have an impact on the environment."

[36] Furthermore, Preamble 10 of the INSPIRE Directive states that "the establishment of INSPIRE will represent significant added value for and benefit from GMES".

[37] Preamble 12 of the INSPIRE Directive states that "This Directive should apply to spatial data held by or on behalf of public authorities in the performance of their public tasks. Subject to certain conditions however, it would also apply to spatial data held by natural or legal persons other than public authorities, provided that those natural or legal persons request this."

the spatial data sets to public authorities at marginal costs; (2) availability of network services,[38] and the data sets specifically to fulfil the reporting obligations under Community legislation relating to the environment, free of charge.[39] The INSPIRE Directive is a major step toward a common access and pricing policy and toward establishing mechanisms for data sharing and coordination at a regional level; it can also serve as a catalyst for creating an operational satellite-derived information service on a global basis, particularly through GMES. It is imperative to facilitate data access on a global level beyond the existing frameworks for both disaster management and protection of the environment.

5.3. Distribution Agreements and Conditions Associated with the Use of Data

The practice of most data generators is that data supply and use is regulated through contractual agreements. They do not just release their data in an uncontrolled manner, but retain control over the data and products after distribution to the users. The contractual agreements between providers and users are valid only amongst the contracting parties.[40] In practice, the majority of data generators provide data in a form of license agreements.[41] In most cases, the pricing levels set for products are the fees for granting a license to use the products – just as in the case of software products. In the license issued to users

[38] See Article 11(1) of the INSPIRE Directive, which states that "Member States shall establish and operate a network of the following services for the spatial data sets an services for which metadata have been created in accordance with this directive." Services include (a) discovery services, (b) view services, (c) download services, (d) transformation services, and (e) services allowing spatial data services to be invoked.

[39] Article 17 of the INSPIRE Directive states that "Where charges are made, these shall be kept to the minimum required to ensure the necessary quality and supply of spatial data sets and services together with a reasonable return on investment, whilst respecting the self-financing requirements of public authorities supplying spatial data sets and services, where applicable. Spatial data sets an services provided by Member States to Community institutions and bodies in order to fulfil their reporting obligations under Community legislation relating to the environment shall not be subject to any charging."

[40] M. Mejia-Kaiser, Satellite Remote Sensing Data in Databases Copyright or *sui generis* Protection in Europe?. XXII-I *Annals of Air and Space Law* (1997), 496. See further M. Ferrazzani & S. Courteix.(Eds.), International Agreements and Contractual Practice for Collecting and Distributing Remote Sensing Data, *Droit télédétection et environnement: actes du colloque international: le droit face aux techniques de télédétections par satellite au service de l'environnement,* Strasbourg, 2-3-4 juin 1993, Sides, Antony, 1994, 181–184.

[41] Examples include but are not limited to: IKONOS, SPOT, ENVISAT, ERS, and DMC.

upon distribution of data, data distributors set a variety of conditions associated with the data.

5.3.1. *Concept of License Agreement*

The term license originates in *licentia*, a Latin word meaning 'permission' and 'agreement.'[42] License is defined as "a permission, usually revocable, to commit some act that would otherwise be unlawful."[43] A license permits a third party to conduct certain activities and exercise privileges associated with products which would be illegal without the license. A license associated with satellite data and products clarifies and enforces the rights of data suppliers and controls the activities of data recipients, thus serving as a means to implement and enforce the terms and conditions associated with products. A license normally involves a fee, but, theoretically, the license agreement exists even in the case of free of charge licenses.[44] The supply of data under the Disaster Charter is a good example of free of charge license-based distribution with attached conditions.

5.3.2. *Rights and Obligations with Regard to Licenses*

A major obligation of a licensee is to pay a license fee (also referred to as a royalty) to a licensor in return for permission to obtain rights that belong to the licensor. In case of satellite images, there is a distinction between a license fee and a royalty: a license fee is when users pay for access to the data, whereas royalty is a fee paid to the owner/ operator of a satellite by distributors and value added resellers upon the sale of products.[45] For instance, the individual DMC satellite owners receive royalties for use of their copyright from DMCii when the data from their satellites are sold by DMCii and the customers actually make payment.[46]

Spot Image charges a royalty fee to the value-added resellers upon the subsequent sale of value-added products rather, than when the data is sold for the first time.[47] In case of ESA's ERS data, a royalty is not applicable only when

[42] See H. Kenoko, Principles of License, *Business Law Practice Licensing Contracts*, Nihon Hyoron Publisher, Tokyo, 2007, 2.

[43] Black's Law Dictionary, Ninth Edition, 2009, 1002.

[44] See H. Kenoko, Principles of License, *Business Law Practice Licensing Contracts*, Nihon Hyoron Publisher, Tokyo, 2007, 7.

[45] A license fee is applied to the end users, while a royalty fee is charged for the other entities of the supplier side, such as a value added reseller or the satellite owner.

[46] Communication to Martin Sweeting of SSTL on 22 December 2006.

[47] Interview conducted with Stéphane Dupont of Spot Image on 12 January 2006. The royalty fee for the first time sale appears to be included in the initial licensee fee

products for further distribution have been modified significantly so that the resulting products do not retain the original pattern.[48] According to ESA personnel, whether or not products retain the original pattern is determined on a basis of 'reversibility'; that is, whether resulting products can be re-constituted and reformatted back to the original data.[49]

In general, a license gives rise to implicit obligations of a licensor not to prohibit a licensee from using the associated intellectual property rights.[50] On the other hand, a licensee is constrained by a set of obligations imposed by a licensor. Licenses are issued under conditions set by the licensor. Such requirements are imposed as conditions for use in a license agreement. Licenses are classified into those that are exclusive, where the use of the intellectual property granted under the license is valid solely for the specific licensee; and those that are non-exclusive, where rights given under the license are also granted to those other than the licensee.[51] In general, licenses for satellite images are non-exclusive and non-transferable.[52]

5.3.3. *Conditions Associated with the Use of Data*

This section examines the typical obligations expected of a licensee in a license agreement with the satellite remote sensing data generators in terms of conditions associated with the data. Licensing conditions are commonly specified as either permitted activities or prohibited activities. Although licensing conditions vary in detail from one data generator to another, the overall tendency clearly indicates the intention of data generators to retain certain control over the data.

---

paid by value added resellers, as the fee is set higher than that for the standard end-user license.

[48] Whenever the users of an ERS primary or processed product wish to sub-license the product to a third party, ESA shall be credited and a royalty fee is due as long as products for onward distribution retain the original pattern ESA/PB-EO(90) 57 rev, p. 3.

[49] Interview conducted with Marco Ferrazzani of ESA on 14 February 2004.

[50] See H. Kenoko, *Principles of License. Business Law Practice Licensing Contracts*, Nihon Hyoron Publisher, Tokyo, 2007, 7. A licensor does not owe explicit obligation over a licensee.

[51] See H. Kenoko, *Principles of License. Business Law Practice Licensing Contracts*, Nihon Hyoron Publisher, Tokyo, 2007, 4.

[52] For instance, Spot Image issues a Non-Exclusive Licence to use Spot Satellite Products between Spot Image and the end-user, and Macdonald Dettwiler grants non-exclusive licenses for a single user. See respectively, Article 2.1 of the Standard License of Spot Image, 2007, and Article 9 of the Single User License, Radarsat-2 End User License Agreement, L1-11525-607NOV15.

Permitted activities include: the reproduction of products and use of the products for internal use; permission for subcontractors to use products on behalf of the licensee; and the generation of derived and value-added products. Prohibited activities include: further distribution of products to a third party;[53] sales of products; and sub-licensing without notification. In any case, the commercial resale of the data is prohibited unless the licensee is an authorized value-added reseller. Prohibited activities vary with the individual supplier and affect the degree of pervasiveness and utility of the data.

Some entities are very restrictive as to the further distribution and the treatment of the data, whilst others are less restrictive. In the case of Digital Globe, end-users can share the data with affiliated entities by means of a joint project and only if they agree to be bound by the conditions of the license and shared products may not be retained by affiliated entities after the completion of the project. On the other hand, IRS data can be disseminated for non-commercial purposes insofar as they are in a non-downloadable format.[54]

A single use is typically permitted under a standard license. Overall, data generators restrict the further distribution and sharing of data with external parties. This is clear from the higher charges levied by data generators in the case of multiple users sharing the same data.[55] This serves as an impediment to the effective implementation of programmes related to environmental protection and disaster management, as they benefit extensively from data sharing, e.g., sharing the data under the different projects requiring overlapping datasets. As identified by the preamble of the INSPIRE Directive, it should be possible to combine data with other sources; to share data between several users and applications; for data collected at one level of public authority to be shared between other public authorities; to make finding data availability easy; and to make data available under conditions which do not unduly restrict their extensive use.[56] Under an operational programme such as the Disaster Charter – although each partner provides the data free of charge – the intellectual property rights and restrictions associated with data that are set by each partner, such as prohibition on further distribution to the third party, are not waived. Indeed, according to non-disclosure agreements, the ownership is retained and copyright is claimed by the data generator; further distribution of data to the third party is prohibited.[57] Under the INSPIRE Directive, the intellectual property

[53] This is the case with RADARSAT, EROS, and DMC. For IRS images, distribution for non-commercial purpose is allowed.

[54] See 2(C) of End User License Agreement for IRS product, Euromap Satellitendaten-Vertriebsgesellschaft, 2008.

[55] Spot Image and DMCii place additional fees for multiple uses, depending on number of entities.

[56] Preamble 6 of the INSPIRE Directive.

[57] See *Non-disclosure agreement*, RSCSA-LS0076-N/C.

rights associated with data remain with data suppliers;[58] therefore, despite certain frameworks that have included steps for the establishment of a common access and pricing policy for public service, the challenge remains to improve the restrictive conditions associated with the data and facilitate effective data sharing for the benefit of the public. These conditions need to be relaxed so as to enhance the exploitability of data, as well as to allow effective data sharing amongst relevant communities for public service purposes.

Many conditions associated with the data provided in the distribution agreements are the result of efforts by the data generators to protect their intellectual property rights.[59] The conditions associated with data and intellectual property rights are interrelated and inseparable. In order to understand the control and restriction imposed on the data recipients in terms of intellectual property, and to examine the legitimacy of claimed rights as well as to consider suitable protection for satellite images and derived products, it is necessary to further explore the issues of intellectual property rights.

5.4. Intellectual Property Rights Associated with Satellite Images

Data generators attempt to seek protection for their products, which are the results of significant financial investment, and their technical achievements. In fact, it is common that the ownership of data remains with the data generators and copyright is claimed by a majority of data generators – with some exceptions, such as Landsat 7.[60] Despite this widespread practice of claiming such rights, it is doubtful as to whether those rights are actually safeguarded. Although access and pricing policies can be exercised largely at the discretion of entities on a contractual basis, absolute security of intellectual property rights depends upon whether the legal regime permits such practices. It is hence necessary to explore and clarify the intellectual property regime, to determine how the rules may be applied to satellite images, and to identify the gaps between the regime and the current practices of the data generators.

---

[58] Article 2 (2) of the INSPIRE Directive states that "this Directive does not affect the existence or ownership of public authorities' intellectual property rights."

[59] Intellectual property such as copyright and patent is the primary subject for licensing.

[60] See R. Harris, Legal Approaches: contractual and regulatory; the European commission directive, *Proceedings of the International Conference; Satellite remote sensing in aid of development: legal considerations*, Tunis 26–27 September 2002, European Centre for Space Law, 2003, 37.

5.4.1. *Examination of the Current Legal Framework for Intellectual Property Rights*

Intellectual property rights protect businesses wishing to exploit them.[61] Historically, the need for copyright protection arose from the large amount of pirated copies of world renowned books by Dickens and Victor Hugo.[62] Victor Hugo, having incurred losses because his works such as "Les Miserables" had been extensively printed and circulated at a low price in Belgium and Germany, established a union to appeal for the importance of copyright.[63] This union led to the creation of the Berne Convention for the Protection of Literary and Artistic Works of 1886,[64] which is the oldest treaty and primary source of international law regarding intellectual property rights.[65] The aim of the Convention is to provide international protection for works of literature, science, or art and to establish an overall framework to harmonize the varying rules on intellectual property. It has established the rules for the protection of work without registration,[66] a mechanism for mutual protection,[67] and independent protection.[68] Although it has created an overall framework, the Berne Convention leaves the implementation and enforcement to the national laws.[69]

---

[61] See S. Norman, *Practical Copyright for Information Professionals,* Facet Publishing, 2004, London, 1.

[62] See H. Miyatake, *Intellectual Property and Creativity,* Misuzu Publisher, Tokyo, 2007, 8–9.

[63] See H. Miyatake, *Intellectual Property and Creativity,* Misuzu Publisher, Tokyo, 2007, 8–9.

[64] Berne Convention for the Protection of Literary and Artistic Works of 1886, *UNTS* 11850 vol.828, September 9 1886. 161 States are parties to the Convention. [Hereafter: Berne Convention]

[65] For an overview of Berne Convention, see S. Ricketson, *Berne Convention for the Protection of literary and artistic works,* Deventer, Kluwer Publisher, 1987.

[66] It provides that copyrights are recognised automatically when work is created and does not require registration as in the case of patents.

[67] Article 5(1) of the Berne Convention provides that the author enjoys his right in the country of origin, where copyright can be claimed as well in other countries that are parties to the Berne Convention. One can argue that the Berne Convention recognises national and territorial jurisdiction, as it covers both authors who are and are not nationals of member states but physically publish their works in territories of the member states.

[68] Article 5(1) and (2) provides that the enjoyment and exercise of the rights granted are independent of the existence of protection in the country of origin of the work.

[69] Article 5(3) states that with respect to the enjoyment and the exercise of these rights, authors can enjoy the protection shall be governed exclusively by the laws of the country where protection is claimed.

The Universal Copyright Convention (UCC) of 1952[70] is a compromise of the Berne Convention. The UCC is complex, and originally was not accepted by many states for signature.[71] The main features include a provision with respect to artistic works from a given country that must carry a copyright notice to secure protection in other UCC countries – the internationally recognised © copyright symbol – and it has concrete provisions with regard to certain measures, such as requiring a license for translation rights.[72]

A global agreement on Trade Related Aspects of Intellectual Property Rights[73] (TRIPS Agreement) of 1994, which follows the trend of recognizing works as equivalent to trade goods, obliges over 120 states to implement the protection provided. In contrast to the Berne Convention, which was intended to provide protection for persons, TRIPS Agreement was intended to set rules between governments.[74] It provides international protection for different kinds of artistic works, including computer programs, satellite broadcasting, and databases. Although TRIPS does not mention databases specifically, the compilations of data or other material, which constitute intellectual creations by reason of the selection or arrangement of their contents, shall be protected as such.[75]

The Copyright Treaty of the World Intellectual Property Organisation (WIPO)[76] covers the internet environment and emerged as an extension[77] of the Berne Convention to deal with computer programs, compilation of data, and the right to communicate with the public.[78] States signatories to the Berne Convention can choose whether they wish to adhere to the WIPO Copyright Treaty or not. This Treaty serves as a significant bridge from the Berne

---

[70] The Universal Copyright Convention of 1952, 943 *UNTS* 178. [Hereafter: UCC] See further Bogsch, A., *The law of copyright under the Universal Convention*, Sijthoff publisher, Leiden, 1968.

[71] See S. Norman, *Practical Copyright for Information Professionals,* Facet Publishing, 2004, London, 140.

[72] For instance, an author's translation rights may be subjected to compulsory licence. See Article V of the UCC.

[73] Global agreement on the Trade Related Aspects of Intellectual Property Rights 33 *ILM* 1140-1272(1994). [Hereafter: TRIPS Agreement]

[74] See K. Nawa, *Digital Copyright*, Misuzu Publisher, Tokyo, 2004, 39.

[75] See M. Mejia-Kaiser, Proprietary Rights in Remote Sensing Image, *Proceedings of the Thirty-Eighth Colloquium on the Law of Outer Space*, IISL, 1995, 2.

[76] The World Intellectual Property Organisation (WIPO) Copyright Treaty 1996 1 36 *ILM* 65 (1997). See further World Intellectual Property Organization.

[77] It is a special agreement under Article 20 of the Berne Convention and it does not prejudice any rights and obligations under any other treaty apart from the Berne Convention. See A. Ilardi, A., and M. Blakeney, *International Encyclopaedia of Intellectual Property Treaties*, Oxford University Press, New York, 2004, 1112.

[78] T. Miyamoto, Intellectual Property Law and Space Activities, Communication to *UN workshop on Capacity Building in Space Law*, Hague, November 2002.

Convention protecting primarily literary and artistic works to database law. Article 5 of the WIPO Copyright Treaty is applicable specifically to compilations of data; as a result of this Treaty the European Community adopted the Council Directive on the legal protection of databases of 1996[79] to achieve a European-wide database law. Amongst the relevant treaties, this section will focus primarily on the Berne Convention and EC Database Directive in analysing the protection for satellite images through examination of the protections for literary and artistic works, and the protection for databases.

5.4.1.1. *The Analysis of Protection for Literary and Artistic Works*

Intellectual property can be broadly categorised into two types: (1) protection for artistic and literary works such as copyright; and (2) protection for industrial property such as designs, patent, and trademarks. This Chapter will only examine the first category, as it is relevant to satellite images and their derived products. The fundamental concepts of protection for literary and artistic works include the concept of copyright and author's rights, the conditions for application, the measure for exercise of rights, and the consequences for infringement.

It is important to understand the two distinct types of rights recognized for the protection of literary and artistic works. One is an economic right, which is typical of copyright; the other is an author's right known as droit d'auteur, which is characterized as a legal or moral right or claim.

An author is given opportunities to profit from trading economic rights[80] from reproduction, public performance, and broadcasting rights, to rental rights. Authors profit from their work through selling and licensing these rights, whereas moral rights provide the author with a continuing control over the work, despite its exploitation.[81] The moral right allows an author to claim paternity of the work independent of any economic right, and to object to distortions, mutilations, and other modifications in relation to the work.[82] Under the moral rights doctrine, the author alone decides when the work is ready for release and how it may be exploited; the work must be properly

---

[79] Directive of European Parliament and of Council on the legal protection of databases, 96/9/EC, of March 1996; OJ L77/20(1996). [Hereafter: EC Database Directive]

[80] See S. Norman, *Practical Copyright for Information Professionals,* Facet Publishing, 2004, London, 1.

[81] See D. Vaver, The copyright Mixture in a Mixed Legal System: fit for Human Consumption? 5-2 *Electric Journal of Comparative Law* (May 2001), 6.

[82] Art. 6 bis 1 of the Berne Convention states that independently of the author's economic rights, and even after the transfer of the said rights, the author has the right to claim authorship of the work and object to certain modifications and other derogatory actions. See A. Ilardi & M. Blakeny, *International Encyclopaedia of Intellectual Property Treaties,* Oxford University Press, New York, 2004, 1066.

credited and it must not be modified contrary to the author's wishes – at least not in ways prejudicial to the author's honour or reputation.[83]

The term 'copyright' originally meant a 'right to copy' the literary works of another, to encourage the publication of a diversity of written works.[84] Copyright is a property right, which may be bought or sold.[85] It protects creativity in literary and artistic works that involve original expressions. Although the Convention does not introduce 'copyright' specifically, it establishes that an author has exclusive rights over work he created.[86] Copyright is a measure against unauthorised reproduction, distribution, use, and sale of one's creative expressions.

There are certain conditions to be fulfilled in order for the work to benefit from copyright protection. The most fundamental specific limits to copyright applicability is that copyright is applicable to expressions and not to ideas and methods[87] and neither protects facts, bibliographical citations, nor headlines.[88] As to more detailed conditions, copyright requirements mandated under the domestic law need to be fulfilled.[89] This section provides an analysis of the trend in domestic laws with regards to the conditions for protection.

Sterling states that "The individual who produces a work by his or her creative activity is normally regarded as the initial beneficiary of the copyright or author's right in the work."[90] The Berne Convention does not define 'author'; however, it implies a natural person, that is, a human being, and does not refer to a juridical person, which is an entity, such as a corporation, created by law and given certain legal rights and duties of a human being.[91] The existing domestic law must be referenced with regard to the definition of author.

[83] See D. Vaver, The copyright Mixture in a Mixed Legal System: fit for Human Consumption?, 5-2 *Electric Journal of Comparative Law* (May 2001), 6.

[84] See A. Longhorn, Legal Issues in Use of Agricultural and Environmental Geospatial Data and Tools, *Special Workshop in the framework of Geospatial Applications to Support Sustainable International Agriculture*, 28 May 2002, South Dakota, 12.

[85] See S. Norman, *Practical Copyright for Information Professionals*, Facet Publishing, 2004, London, 1.

[86] Exclusive rights granted by the Berne Convention include the right of translation (Article 8), reproduction (Article 9), public performance (Article 11), adaptation (Article 12), the right of broadcasting (Article 11 bis), and the right of public recitation (Article 11 ter).

[87] See website of World Intellectual Property Rights. www.wipo.int. Last accessed on 17 April 2010.

[88] See S. Norman, *Practical Copyright for Information Professionals*, Facet publishing, London, 2004, 7.

[89] See M. Mejia-Kaiser, Satellite Remote Sensing Data in Databases Copyright or *sui generis* Protection in Europe?, XXII-I *Annals of Air and Space Law* (1997), 496.

[90] J.A.L, Sterling, *World Copyright Law*, Sweet & Maxwell, London, 2003, 184.

[91] For example, Article 3(2) states that "Authors who are not nationals of the countries of the Union but who have their habitual residence in one of them shall, for the

It is interesting to look at the definitional differences between the civil law system and common law system on 'author' and the concept of a 'creator'. The initial ownership or right is given either to the individual who produces a work or to the legal entity responsible for its production, such as an employer.[92] In the common law system, the general trend is that the employer is the first owner of the copyright when the work is made for hire.[93] UK Copyright Act 1988, for instance, states that "Where a literary, dramatic, musical or artistic work is made by an employee in the course of his employment, his employer is the first owner of any copyright in the work subject to any agreement to the contrary."[94] Likewise, in the US, Section 201(b) of the Title 17 of the United Code stipulates that "In the case of a work made for hire, the employer or other person for whom the work was prepared is considered the author for purposes of this title, and, unless the parties have expressly agreed otherwise in a written instrument signed by them, owns all of the rights comprised in the copyright."[95]

On the other hand, in civil law jurisdictions, when an employee creates a work in the course of his employment, he is the initial owner of the author's right; nevertheless, contractual provisions, whether expressed or implied, can transfer or license the economic rights associated with the work to the employer.[96]

The Berne Convention does not explicitly state that originality and intellectual creation are required in order for a work to be protected under copyright; however, most domestic legal systems require works to involve certain degrees of originality and creativity as distinctive human intervention in order to merit protection.[97] The application of such conditions and the interpretation of the terms vary by jurisdiction.

---

purposes of this Convention, be assimilated to nationals of that country." Article 7 states that "the term of protection granted by this Convention shall be the life of the author and fifty years after his death." From the above terms, namely, habitual residence, life of the author, and his death, one can construe that the 'author' under the Berne convention refers to a natural person.

[92] See J.A.L, Sterling, *World Copyright Law,* Sweet & Maxwell, London, 2003, 185.

[93] See J.A.L, Sterling, *World Copyright Law,* Sweet & Maxwell, London, 2003, 201–203.

[94] Section 11(2) of the UK Copyright, Designs and Patent Act, 1988 Chapter 48.

[95] § 201(b) of the Title 17 of the United States Code.

[96] See J.A.L, Sterling, *World Copyright Law,* Sweet & Maxwell, London, 2003, 200. For instance, Article L.111-1 of the French Code provides that the existence or conclusion of a contract of employment for service by the author of a work does not imply any derogation from his author's right.

[97] See M. Mejia-Kaiser, Satellite Remote Sensing Data in Databases Copyright or *sui generis* Protection in Europe?, XXII-I *Annals of Air and Space Law* (1997), 496.

Generally, the civil law countries take the conditions of creativity and intellectual creation into significant account.[98] In contrast, originality is not linked very closely to creativity in common law countries. For instance, in the British tradition 'originality' has been treated as a low threshold: in most cases the traditional approach has been to refer to "skill, judgement and labour" in assessing the originality.[99] The basic concept is that the work must be the result of the investment of individual skill, judgement, or labour. Skill and judgement imply the application of a personal choice and some intellectual contribution, whilst mere labour does not.[100] In practice, the element of labour alone does not appear to meet the condition for protection; in fact, it was demonstrated in the case of *Cramp v. Smythson*,[101] that not only must the work not be a copy of previous work, but judgement has to be original based on individual skill.

The well-known US Supreme Court case, *Feist Publications Inc. v. Rural Telephone Service*,[102] which concerned the alleged copyright infringement on use of a white page listing, demonstrated that the copyright protection for factual compilations is not granted if no originality is involved. This case clearly illustrates that copyright does not protect facts, and that works within the scope of protection need to be a product of intellectual activity – not merely a product of time, labour, and money invested.[103]

The application of the conditions for protection varies, but the protection seems to require certain degrees of intellectual input with the involvement of subjective choices.[104] Following this rule, there is doubt as to whether satellite

---

[98] In Germany, Article 2(2) of the 1965 German Author's Right Law states that only personal intellectual creations are works in the sense of the Law. In France, the 1992 Code does not mention originality as a criterion of protection, however, the concept of criteria has established in jurisprudence. With regards to technological productions, *cour de cassation* accepted the criterion of 'intellectual contribution,' rather than 'mark of personality.' See J.A.L. Sterling, *World Copyright Law*, Sweet & Maxwell, London, 2003, 298.

[99] See J.A.L. Sterling, *World Copyright Law*, Sweet & Maxwell, London, 2003, 292.

[100] See J.A.L. Sterling, *World Copyright Law*, Sweet & Maxwell, London, 2003, 306.

[101] *G.A. Cramp & Sons Ltd. v. Frank Smythoson Ltd.*, [1944] A.C. 329, H.L. The plaintiff published pocket diaries containing calendars, postal information, important dates, tables of weights and measures, but it was found that the material was not sufficiently original.

[102] *Feist Publications Inc. v. Rural Telephone Service*, 111 S. Ct 1282 (1991).

[103] See D. Karjala, Copyright in Electronic maps, 35 *Jurimetrics Journal* (1995), 395.

[104] See E. Back-Impallomeni, Legal Protection through agreements, contracts of scientific information and data, *Proceedings of the International Conference: Satellite Remote sensing in aid of development: legal considerations*, Tunis 26-27 September 2002, European Centre for Space Law, 2003, 58.

data without human intellectual contribution, such as the raw data derived from a system where an operator makes no intellectual choice in taking such imagery (e.g., a decision on the best angle for image acquisition), merit copyright protection – even when the minimum standards of creativity requirements are applied.

Infringement of the creator's rights occurs when reproduction, communication, display, or distribution of works to the public is done without the transfer of right or permission with respect to the protected subject matters. The analysis under this Chapter will focus on the case of reproduction.

The consequence of infringement is that "the infringing copies of a work shall be liable to seizure in any member States where the work enjoys legal protection"[105] and "the seizure shall take place in accordance with the legislation of each country."[106] The consequences of infringement may result in litigation and/or involve compensation by the infringing parties subject to the respective national legislation. The concrete remedies for infringement depend upon domestic legislation.

The majority of disputes concerning copyright are settled out of court;[107] however, when an infringement action is brought, the magnitude of consequences can be serious. For instance, the UK Automobile Association was fined 20 million pounds in 2001 after copying maps of Ordnance Survey, the National Mapping Agency of the United Kingdom.[108] Any person that is involved in the supply and use of satellite data faces such risks. The copyrights and associated entitlement are unaddressed anywhere, and consequences for infringement may be detrimental to all parties concerned.

5.4.1.2. *The Analysis of Protection for Database*

As has been discussed, time, labour, and money invested to create works do not count as elements toward copyright protection; but these elements are subject for protection under the database category. This part examines the concept of database protection, conditions for protection, and infringement.

The EC Database Directive protects databases on two grounds: the protection based on copyright for databases that constitutes the author's own intellectual creation in terms of the selection and arrangement of their contents,[109]

---

[105] Article 16(1) of the Berne Convention.

[106] Article 16(2) of the Berne Convention.

[107] See S. Norman, *Practical Copyright for Information Professionals,* Facet Publishing, London, 2004, 129.

[108] A. Clarke, Copying maps costs AA £20m, *The Guardian,* Tuesday March 6, 2001. For more cases regarding copyright infringement, Cf. S. Norman, *Practical Copyright for Information Professionals,* Facet Publishing, London, 2004, 130.

[109] See Article 3(1) of the Database Directive.

and the protection on the basis of *sui generis* right, which protects the contents of the database.[110] *Sui generis* means "of its own kind" in Latin and the term is used to describe a regime designated to protect right that fall outside the traditional doctrines such as copyright.[111] The right of *sui generis* is clearly contrasted from the copyright, which protects the "form" or how the database is presented.[112] *Sui generis* applies "irrespective of the eligibility of that database for protection by copyright or by other rights."[113] Thus it can be interpreted that a creative database may enjoy protection in terms of both copyright and *sui generis* rights. The extraction right permits authors to conduct or authorize acts including reproduction and distribution of the contents;[114] this right prevents the unauthorized retrieval and use of all or substantial parts of database.[115]

Unlike copyright, protection on the basis of the *sui generis* right does not require intellectual creation but only requires that substantial qualitative or quantitative investment be made in obtaining, verifying, or presenting the contents.[116] Time and money invested in making a database are solid grounds for protection under the *sui generis* right.

The database right applies to "a collection of independent works, data or other materials arranged in a systematic or methodical way and individually accessible by electronic or other means."[117] The natural person or group of natural persons who created the base, or the legal person designated as the right holder by legislation, can be the database right holder.[118] This is more explicit than the regime of protection for literary and artistic works under the Berne Convention, which infers that the natural persons or a group of natural persons are primarily the beneficiary of such right. In these ways, database protection, in theory, is more readily applicable to satellite images and derived products than copyright.

There have been a number of cases that involve database rights. The first French case-law application *sui generis* over a database concerned a telephone directory. It was held that the unauthorised down-loading of the France Télécom

---

[110] See Article 7(5) of the Database Directive.

[111] Black's Law Dictionary, Ninth Edition, 2009, 1572.

[112] See L. Dufresne, Protection of space data products under the European Directive on the legal protection of databases, *Proceedings of the International Conference: satellite remote sensing in aid of development: legal considerations*, Tunis 26–27 September 2002, European Centre for Space Law, 43.

[113] Article 7 (4) of the Database Directive.

[114] See Article 5 of the Database Directive.

[115] See Article 7(5) of the Database Directive.

[116] See Article 7(1) of the Database Directive.

[117] Article 1(3) of the Database Directive.

[118] See Article 4(1) of the Database Directive.

electronic directory by a third party for the purposes of creating an inverse directory was an infringement of the *sui generis* right.[119]

In the case of *British Horseracing Board (BHB) v. William Hill*[120] the BHB objected to William Hill, a bookmaker, using information in its Internet betting service derived from horse race meetings conducted by BHB. William Hill extracted data without obtaining a licence from Satellite Information Services Ltd. licensed by the BHB to transmit much of the Board's data to its subscribers. William Hill used the information on the Internet betting site and the court found that, as William Hill had downloaded a substantial amount of data, they had therefore infringed database rights.[121] This case is highly relevant to satellite images, as many of them are downloadable from the Internet. A third party can easily infringe the rights of the satellite data generator; therefore, the rights associated with the products should be clarified.

5.4.2. *Applying Rules of Intellectual Property Rights to Satellite Remote Sensing Data*

As far as the protection of the satellite data is concerned, although certain characteristics of the author's right, such as prohibition on modification, are inherent in the license itself, the primary purpose is to get economic benefits associated with the products.[122] Economic rights are claimed by the data generators, most notably in the form of copyright.

The clarification of the differences between the regimes for copyright and database rights shows that database rights, by nature, are a more suitable basis for the protection of satellite images. However, depending on their characteristics, some data may be protected under copyright law. The following sections closely examine how different types of satellite data would fit under the regime of copyright and database right, respectively.

---

[119] See L. Dufresne, Protection of space data products under the European Directive on the legal protection of databases, *Proceedings of the International Conference: satellite remote sensing in aid of development: legal considerations*, Tunis 26-27 September 2002, European Centre for Space Law, 54.

[120] *British Horseracing Board (BHB) v. William Hill*, HC 2000 1335, judgement on 9 February 2001.

[121] See S. Norman, *Practical Copyright for Information Professionals*, Facet Publishing, 2004, London, 135.

[122] In general, see A. Patrick, Proprietary Aspects of Remote Sensing Imagery, 13 *Northwestern Journal of International Law and Business* (1992), 349–373. See also P. Gaudrat, La Protection des Données de télédétection par recours aux conventions internationales, *Droit télédétection et environnement : actes du colloque international : le droit face aux techniques de télédétections par satellite au service de l'environnement*, Strasbourg, 2-3-4 juin 1993, Sides, Antony, 1994, 227–235.

5.4.2.1. *The Copyright Protection for Satellite Images*
With regard to the protection of satellite images, three types of data should be considered: 1) raw data, 2) processed data, and 3) analysed information, value-added products, and derived products. Applying copyright protection to digital information is not an uncommon practice nowadays, since the scope of its applicability has widened in accordance with increased variety of works. Initially intended primarily for protection of literary and artistic works, many media have been recognized and now subject to copyright protection.[123] This section explores how copyright law applies to raw data, processed data, and analysed information.

Primary data (often referred as raw data) is defined under the UN Remote Sensing Principles as "the data that are acquired by remote sensors borne by a space object and that are transmitted or delivered to the ground from space by telemetry in the form of electromagnetic signals, by photographic film, magnetic tape or any other means." [124] Raw data from satellite imagery is simply a reflection of Earth phenomena; in other words, mere representation of facts. Theoretically, raw data does not precisely fit within the regime of copyright protection since human intervention is involved to a minimal extent. The development and operation of a satellite involves significant investment of time and money; however, it is questionable as to whether or not the raw data derived from such systems can be regarded as original expression.

If they are to be protected at all under copyright, the elements that would count toward protection can be found in how the satellite image is taken and the degree of creativity involved in the process of data collection. It seems difficult to find creativity in the operation of data collection alone from a sensor, such as Landsat, since it is continuously taking imagery simply of nadir, whereas with Spot images can be taken from different angles[125] requested through human intervention. In this regard, it is more justifiable to argue that raw data from the Spot satellites involves subjective judgement and thus merits protection under copyright.

Processed data are defined under the Principles as "the product resulting from the processing of the primary data needed to make them useable."[126] It is more difficult in this instance to deny copyright applicability. Processed

---

[123] See L. Lessig, *Creative Commons Intellectual Property in the Digital Era*, NTT Publisher, 2005, Tokyo, 123. For instance, the US Office of Technology Assessment (OTA) - 86 Report provides a variety of works subject to copyright protection, classified into three different categories: works of art, works of function, and works of fact.

[124] Principle I(b) of the UN Remote Sensing Principles.

[125] Indeed, the scanner for SPOT satellites is able to tilt up to 22 degrees off nadir – the point on the ground directly under the satellite – and this tilting capability makes it possible to acquire data from either side, and also to create digital elevation models(DEM) using two off-nadir SPOT datasets.

[126] Principle I(c) of the UN Remote Sensing Principles.

data is generated after data handling operations such as standardization, classification, visualization, editing, and correction of data. If no creativity is involved in these data handling operations, processed data does not have the element of creativity and vice versa.[127] Thus, it can be argued that if the processing is done only by machine, then it is difficult to justify copyright protection for the resulting processed data. On the other hand, if the processing is done manually by an individual making subjective choices and judgements, then the resulting data involves an intellectual contribution and, consequently, is likely to merit copyright protection.

One aspect of creativity which could be found in data handling operations is a colour assignment of an image. An image can be colour assigned in different ways to produce different images. The fundamental difference lies between true colour and false colour images. Multi-band remote sensing images do not have the visual spectrum, but in order to see these in a colour monitor, one needs to assign red, blue, and green to bands outside the visual spectrum. The decisions to assign certain colours to produce the particular image can be considered to involve a certain degree of creativity, although the subsequent application is generally fully automatic and may indeed be based on accepted classifications requiring no individual input to a specific satellite image.

Processed satellite data is perhaps most closely analogous to photos, in the sense that it is capturing facts but may involve certain degrees of choice or creativity. Article 2(1) of the Berne Convention states that "literary and artistic works shall include every production in the literary, scientific and artistic domain, whatever may be the method or form of its expression, such as (…) photographic works to which are assimilated works expressed by a process analogous to photography."[128] Therefore, photographs are, in general, within the scope of copyright protection; however, the degree of creativity is a critical element with respect to photographs, and a photograph without creativity is not protected.

The fact that photographs without originality are not protected under copyright law is well illustrated in the case of *Bridgeman Art Library Ltd. v. Corel Corporation.*[129] In this case, the plaintiff claimed copyright regarding

[127] See K. Nawa, *Digital Copyright*, Misuzu Publisher, Tokyo, 2004, 145.

[128] Article 2 of the Berne Convention. Furthermore, Article 6 of the EC Database Directive states that "photographs, which are original in the sense that they are the author's creation, are to be protected in accordance with Article 1 of the Directive," That is, a creative photograph can be protected by copyright.

[129] *Bridgeman Art Library Ltd. v. Corel Corporation*, 36 F. Supp. 2d 191(S.D.N.Y. 1999). That no copyright is claimed for non-original photo can also be seen, e.g., in the comparison of 1) functional aerial photography by IGN mentioning no copyright, and 2) the copyrighted photographs by a French artist, Yann Arthus-Bertrand, who has used his own initiative and artistic insight to take pictures from an airplane flying over a number of place around the globe from different angles.

photographic transparencies incorporating photographs of paintings in the public domain; however, the defendant denied that the photographs concerned were protected by copyright. It was decided that no copyright in drawings, which were merely copies, is recognised even if skilfully made of previous drawings.[130] The plaintiff appealed, but ultimately – relying on the US case of *Batlin and Son Inc. v. Snyder* – the Court concluded that the photographs were not original.[131]

Analysed information is defined as "the information resulting from the interpretation of processed data, inputs of data, and knowledge from another source."[132] Derived products or value-added products are also considered to be analyzed information. There are different types of analyzed information including digital elevation models, maps, satellite data combined with metadata, population data, vector data, and navigation data. As previously discussed, satellite imagery is increasingly provided to end-users in a format that is ready to support decision-making such as maps of affected areas.

Whatever the form of value-added products derived from satellite data, they seem to fall into the category of protected work under Article 2 of the Berne Convention, as they include "illustrations, maps, plans, sketches and three-dimensional works relative to geography, topography, architecture or science."[133] Indeed, a map, which is a typical form of analysed information, is within the scope; however, not all maps are protected. Whether a particular map is protected or not hinges on both its originality and the intellectual activities involved in producing it.

A map was considered to be protected under copyright in the case of *Compagnie des Courtiers v. Cellier des Halles*,[134] where a map of wines of France was held to be original, in part because of the choice of disposition of evaluations, typography, and colours.[135] Protection for maps on the basis of copyright is applicable to digital maps as well.

Mason v. Montgomery Data Inc.[136] concerned real estate tract maps showing the location, tract size and shape, deeds, abstract numbers, acreage, and tract

---

130 *Interlego Ag v. Tyco Industries,* 1989 1 A.C. 217;1988 3 ALL E.R. 949[1988].

131 *Batlin and Son Inc. v. Snyder*, 516 F. 2d (2nd Cir. 1976) cert. Denied 429 U.S. 857. It was held that a plastic reproduction of an uncopyrighted mechanical cast-iron coin bank could not be protected by copyright for there was no distinguishable variation – something beyond technical skill in the plaintiff's reproduction which would make it original. See J.A.L. Sterling, *World Copyright Law,* Sweet & Maxwell, London, 2003, 313.

132 Principle I(d) of the UN Remote Sensing Principles.

133 Article 2 of the Berne Convention.

134 *Compagnie des Courtiers v. Cellier des Halles,* CA Paris, 4 Ch., March 26, 1991, (1991) 150 R.I.D.A. 148.

135 See J.A.L. Sterling, *World Copyright Law,* Sweet & Maxwell, London, 2003, 297.

136 *Mason v. Montgomery Data Inc,* 967 F. 2d 135 (5th Cir. 1992).

ownership, along with other topographical features.[137] The US Court of Appeals for the Fifth Circuit held that maps created by independent mapmakers meet the standard of creativity necessary for copyright protection on the ground of "selection of sources, interpretation of those sources and skill and judgement in depicting information."

In *Kern River Gas Trans. Co. v. Coastal Corp*,[138] the plaintiff mapped proposed gas transmission lines on 1:24,000 scale US Geological Survey topographic maps. The court held that the map was not protected by copyright, as the mapping of transmission lines created the only effective way of expressing the idea of the location of the pipeline. Karjala argues that a map should be copyright protected when it is the product of its maker's intellectual effort – that is, when it is not copied from another map.[139] Whether a particular work merits protection hinges on the intellectual activities involved to produce maps; thus, non-original maps are not protected.

Another important issue is the copyright of analysed information in terms of joint authorship. As previously noted, it is possible that more than one author is involved in analysing and producing information, such as maps, based on satellite images. The satellite images are often processed, interpreted, and analyzed by more than one person who belong either to the same organisation or who belong to different organisations. In such cases, it is important to clarify the nature of the joint authorship and how the total benefits are shared amongst the authors.

The Berne Convention provides recognition for joint authors in Article 7 *bis*;[140] however, it is silent as to how the work of each author is assessed and how the benefits of each author are allocated as a result of being a joint author. Hence, determination of what joint authorship entails has to be made by consulting the national laws that contain more concrete provisions. A work of joint authorship means either that the separate contributions to the work are indistinguishable in some cases, or in other cases the term may be used generally to describe works with more than one author.[141]

Joint work is recognized amongst European jurisdictions. Article 8(1) of the 1965 German Copyright Law stipulates that "if more than one person has

[137] See D. Karjala, Copyright in Electronic maps, 35 *Jurimetrics Journal* (1995), 395.

[138] *Kern River Gas Trans. Co.v. Coastal Corp*, 899 F.2d 1458 (5th Cir. 1990).

[139] See D. Karjala, Copyright in Electronic maps, 35 *Jurimetrics Journal* (1995), 395.

[140] Article 7 bis of the Berne Convention states that "The provision of the preceding Article shall also apply in the case of a work of joint authorship, provided that terms measured from the death of the author be calculated from the death of the last surviving author."

[141] See J.A.L. Sterling, *World Copyright Law*, Sweet & Maxwell, London, 2003, 188.

created a work together, and their respective contributions cannot be separately exploited, they are joint authors of the work."[142] Section 10(1), Chapter I of the 1988 UK Copyright, Designs, and Patents Act provides that "a work of joint authorship means a work produced by the collaboration of two or more authors in which the contribution of each author is not distinct from that of the other author or authors."[143] Whilst what 'joint' means is not defined, this provision infers that contributions by each party are not distinguishable from one another in terms of type and degree of each contribution. The German and UK laws seem to be inadequate to address a satellite product, which is, by nature, the result of contributions of different kinds and of unequal degrees, which may be exploited separately.

On the other hand, the 1992 French Intellectual Property Code[144] recognizes joint work where contributions are of different types, with differing degrees, and can be separately exploitable. The Code recognizes three types of joint works: work of collaboration, composite work, and collective work. Work of collaboration means a work in the creation of which several individuals ('plusieurs personnes physiques') have participated.[145] Composite work means a new work in which a pre-existing work is incorporated without the collaboration of the author of the earlier work.[146] Collective work means a work created on the initiative of an individual or a legal person, who edits it, publishes it, and divulges it under his/its direction and name, and in which the personal contribution of several authors participating in its elaboration merge in the ensemble for which the work is conceived, it being not possible to attribute to each of the contributors a distinct right on the work as a whole.[147]

Different types of joint works could lead to various interpretations amongst the concerned parties, and these differences have given rise to litigation.[148]

---

[142] Law Dealing with Copyright and Related Rights (Copyright Law) of September 9, 1965, as last amended by the Law of June 23, 1995. See J.A.L. Sterling, *World Copyright Law,* Sweet & Maxwell, London, 2003, 191.

[143] Section 10 (1) of the1988 UK Copyright, Designs, Patents Act, Chapter 48 1988.

[144] Loi n° 92-597 du 1er juillet 1992 relative au code de la propriété intellectuelle. [Hereafter: French Intellectual Property Code]

[145] Article L 113-2 of the French Intellectual Property Code.

[146] A composite work is the property of the author who made it, subject to the rights of the author of the pre-existing work. See Article L 113-4 of the French Intellectual Property Code.

[147] Article L 113-2 of the French Intellectual Property Code.

[148] See J.A.L. Sterling, *World Copyright Law,* Sweet & Maxwell, London, 2003, 190. *Duchier v. Gueroult,* CA Paris, 4 Ch.; (1995) 166, RIDA 240; photographs of flowers held to be the work of collaboration of flower arranger and photographer, who had collaborated to arrange the banquets for photographing; *Euro Vacancies v. Froment* Cass. I civ., Oct 24, 1995(1996) 168 RIDA 276: update of pre-existing guide by new original material was considered as composite work.

The French Intellectual Property Code appears to contain language that could be suitably applied to differentiate amongst the many types of value-added satellite-derived products. Applying the Code, a map constructed from two different satellites may be associated with a work of collaboration. On the other hand, satellite-derived products, in which the data generator and information extractor are different, can either be composite work or collective work.

It is critical to establish categories for the different types of joint work associated with satellite images and their derived products and to clarify what each of them involve. This highly relevant issue will be examined more thoroughly later in this Chapter.

5.4.2.2. *The Database Protection for Satellite Images*

The right of *sui generis* can be applied to all types of satellite products – raw data, processed data, and analysed information – all of which may also be protected simultaneously under copyright. Such application is possible according to Recital 17 of the EC Directive, which reads that "(...) whereas the database covers collections of independent works, data or other materials which are systematically or methodically arranged and can be individually accessed." Since a satellite image is a set of geographically located measurements, it is considered to meet the condition of autonomy of elements because its measurements are not only individually accessible, but are also independent of the other in the file.[149] Since the right of *sui generis* is applicable to non-electronic databases, it is also applicable to printed satellite images.

The database rights regime is much more concrete as to joint work than to the copyright regime. Article 4(3) provides that "In respect of a database created by a group of natural persons jointly, the exclusive rights shall be owned jointly."[150] Therefore, in the case of a product made using the data from more than one satellite, a claim for the raw data on the basis of the right of *sui generis* will be shared jointly amongst the data generators.

It is possible to have different entities holding the copyright and database rights for satellite-derived products when the data generation and the information extraction have been done by different entities.

---

[149] See L. Dufresne, Protection of space data products under the European Directive on the legal protection of databases, *Proceedings of the International Conference: satellite remote sensing in aid of development: legal considerations*, Tunis 26–27 September 2002, European Centre for Space Law, 41. Each of the pixels in a Spot scene represents a separate radiometry measurement, which can be found by reading each pixel with an appropriate programme.

[150] Article 4(3) of the EC Database Directive.

5.4.3. *The Practice of the EO Data Generator with Regard to Copyright and Database Right and Identification of Gaps and Problems*

Currently, the majority of data generators,[151] whether they are governmental agencies or private entities, explicitly claim copyright in their distribution contracts/licensing agreements. It is common practice that the copyright notice © followed by the name of the data generator is included along with the image. If the owner and suppliers are different, the copyright holders of data, rather than the suppliers of the data, are normally the owners of the satellites (e.g., CNES holds the copyright for Spot data from Spot 1 to Spot 5 even if Spot Image distributes it).

The current practice of claims over intellectual property rights of satellite data and derived products has both problems and gaps that can be highlighted through further analysis. There are discrepancies between the legal regime and practice, and varying practices in exercising and enforcing intellectual property rights create uncertainties and confusion, most notably: 1) problems related to varying domestic laws; 2) protection for value added products; 3) the issues of joint authorship; and 4) the lack of universal applicability of database rights.

5.4.3.1. *The Litigation over Copyright Infringement*

Varying domestic laws applicable for the protection of satellite data[152] can be a source of dispute. In the past, the problem concerning authorship of work has resulted in litigation related to satellite images. The dispute concerned published processed Meteosat data marketed by an ESA distributor in Germany without a copyright notice. Following the action brought by the distributor along with ESA, it was held that, according to German law, only a natural person could be recognised as an author of work; however, ESA could not prove before the court which employee was the author of the image, so the case was dismissed.[153] This case invokes the need to understand the differences in the

[151] Exceptions include the US with respect to Landsat 7 and MODIS. See further L.R. Shaffer, The American Distribution System for Satellite Remote Sensing Data, *Droit télédétection et environnement: actes du colloque international: le droit face aux techniques de télédétections par satellite au service de l'environnement,* 2-3-4 juin 1993, Sides, Antony, 1994, 123–132.

[152] Data generators normally provide terms and conditions of sales, including the specifics on intellectual property rights, and they state the applicable laws. For instance, the terms for use of IKONOS state that the product of GeoEye is protected under US copyright law; for ESA products, clauses on intellectual property are covered under the French legislation law.

[153] See M. Mejia-Kaiser, Propriety Rights in Remote Sensing Image, *Proceedings of the Thirty-Eighth Colloquium on the Law of Outer Space*, IISL, AIAA, Washington, 1995, 31.

intellectual property regimes under different jurisdictions and the necessity to make suitable arrangements in case of copyright infringement of a cross border nature.[154]

5.4.3.2. *Protection for Value Added Products and Issues of Joint Authorship*

A problem exists as to the types of products for which copyright is claimed. The practice is not uniform. Most data generators claim copyright over their products, irrespective of the type of data. Others do not claim copyright for the resulting products if they are modified to the extent that the original image is no longer identifiable. For instance, the Advanced Land Observation Satellite (ALOS) distinguishes between products retaining original patterns and those which do not, and copyright is not claimed for the latter, in which copyright is held by the persons who conducted the value-adding operations.[155]

Who has copyright over a value-added product is a subject for debate. It is useful to approach the issue by applying the definition of three types of joint work from the French Intellectual Property Code discussed above. The joint authorship in terms of work of collaboration is adopted for a map made of data collected from distinct satellites.[156] In this case, there are no critical problems because authors of collaborative works exercise their rights under the French Intellectual Property Code either by common accord or by separately exploiting their personal contribution.[157] The question arises as to the issue of joint authorship of two types of value-added products: collective work and composite work, respectively.

[154] Such adjustment is necessary, particularly amongst civil law jurisdictions, as the initial ownership of work tends to be employees rather than employers; hence it is necessary to disclose the names of individuals that transfer copyright under the terms of employment work.

[155] ALOS data policy, Available on the Internet at http://www.alos-restec.jp/pdf/ALOS_End_User_License_Agreement_J_rev1.pdf Last accessed 22 April 2010.

[156] A number of examples of joint authorship can be found for products supplied within the framework of the Disaster Charter. If the information comes from more than one satellite, then the resulting product is regarded as jointly copyrighted. For example, in the Charter activation dated 30 March 2007 for flooding and landslides in the city of Gualeguay, Argentina, there are some jointly copyrighted products: One is a composite map from Landsat and Spot images, copyrighted by CNES and CSA; another is a map made from ALOS and Landsat 5 and copyrighted by JAXA and USGS.

[157] See Article L. 113-3 of the French Intellectual Property Code. When the material contribution of each of the authors is of different kinds, each may, if not otherwise agreed, separately exploit his personal contribution, without prejudicing the exploitation of the common work.

Copyrights of value-added products are largely held by the data generators, with a credit given to value-added entities as far as products delivered within the Disaster Charter are concerned. Examples of how the intellectual activity of the value-adding operation is credited include the statements: "image processing and information extraction by ...", "analysis and map created by ...", or "image processing, map created by ...",[158] which describe the specific contributions made by the various value-added entities. However, there are generally no joint copyright notices.

The copyright practice on value-added products from collective work generates a situation in which value-adding entities are not in a position to exercise rights independently and to pursue claims in case of copyright infringement of the derived products to which they contributed. The rights of the entities that provide added value are secured only through the satellite data generator's goodwill. It appears that the legitimate rights that should be enjoyed by those entities that add value by conducting analysis and information extraction through intellectual contribution cannot be exercised because they are not even recognized.

The practice of collective work seems to be common amongst data generators, and is to be differentiated from the relatively minor practice involving composite work on value-added products where the value-adding reseller, on the basis of raw data, processes and extracts information to publish new work under its direction.[159] In this case, copyright associated with pre-existing work is held by the data generator and such right can be exercised through a royalty paid by the value-adding reseller to the data generator or by exercising database rights with respect to the users of value-added products.

Following the conventional rules on copyright, the copyright should be claimed by the person who has conducted the value-adding process in transforming the raw data into useful knowledge, rather than the data generator who has invested in the development and operation of the satellite. It is necessary to clarify the resulting benefits relating to three different categories of joint works: 1) work of two or more data generators; 2) collective work published under a single data generator; and 3) composite work published under a value-adding reseller.

5.4.3.3. *Non-Universal Applicability of Database Rights*

The non-universal applicability of database rights also creates a problem. Currently, protection on the basis of *sui generis* or extraction right is limited to certain jurisdictions. The EC Database Directive is applicable only within the

---

[158] See in general http://www.disasterscharter.org/web/charter/activations. Last accessed 27 April 2010.

[159] An example is RESTEC publishing a map under its copyright but placing a copyright notice to JAXA that reads simply "included ©JAXA."

European Community and to its nationals.[160] Some legal systems, apart from that of the European Community, also recognize database rights – including Japan– but others do not. For instance, the United States has not introduced the *sui generis* right into its national copyright system[161] and, accordingly, satellite images are not protected under the US database rights. Data generators can refer to the inherence of their extraction right in their distribution agreement, but enforcement hinges on jurisdiction. Assertion of the extraction right is likely to serve as a safety net in case of court decisions against protection on the basis of copyright within jurisdictions that recognizes database rights.

The foregoing examination of the legal framework of intellectual property rights and the practice of EO data generators demonstrates that there is a gap between the goal of the data generators seeking protection for all types of data and what protection is actually available to them under present law. Furthermore, there are inconsistencies in practice amongst different data generators that create uncertainties and confusion, particularly with regard to copyrights for value-added products; this harms concerned parties, including data suppliers, users, and third parties. Considering the international nature of data generation and sale, the clarification of the rules on how intellectual property rights apply to satellite images is critical and there is an evident need for a common regime that provides that clarification.

5.5. Issues of Data Reliability: Data Authentication and Accuracy Validation

The final section of Chapter 5 explores the issues of data reliability and mechanisms for data authentication and validation. Digital geographic information data quality has been defined in ISO standard 8402 as "the totality of features and characteristics of a product or services that bear upon its ability to satisfy stated or implied needs."[162] This definition is limited: although it implies the ability of the final product to satisfy the need, it does not establish clear criteria

[160] Article 11(1) of the EC Database Directive states that "The right provided for in Article 7 shall apply to database whose makers or rightholders are nationals of a Member State or who have their habitual residence in the territory of the Community."

[161] See P. Weiss, Borders in Cyberspace: conflicting government information policies and their economic impacts, Summary Report, US Department of Commerce, NOAA, 2002 70. Originally the United States was interested in the idea of a *sui generis* right and even contributed to WIPO's work in the field; but the US has met criticism mainly from national research and science circles on the basis that a *sui generis* system would lead, if not to a monopoly of knowledge, then at least to a privatisation of scientific data.

[162] ISO 8402: 1986, note 1.

as to how to verify whether the final product actually meets the specific user needs. Furthermore, it applies only to Geographical Information Systems (GIS); that is, digital maps or digital components of maps such as roads, cities, and lakes. The definition does not apply to satellite-generated raw images or processed images unless some cartographic layers are derived from the satellite image; then the definition applies.

The accuracy of the final product depends on two components: the quality of the original data, and the quality of the image processing used to obtain the final result. The quality of the original satellite image dramatically affects the subsequent decision possibilities. Generally, satellite image data is processed and the user(s) rely on the results to make decisions or take actions. Therefore, the ability to satisfy user needs hinge largely on the authenticity and accuracy of the data and those of data processing. The authenticity and accuracy of the data and those of data processing affect decision-making to such an extent that, if the image provider has not fully disclosed details regarding the image fidelity and potential errors, and/or the processing methodology is not described precisely, users must use the image with caution or discard it completely.[163]

Satellite data generators and distributors do not provide any measurements for the accuracy of standard products. In some cases, if specifically requested, they might provide their best knowledge about the conditions of the satellite when the image was taken.[164] There is no mechanism that enables the objective assessment of data in order to identify or measure its quality.

5.5.1. *Data Authenticity and Accuracy*

The issues relating to data quality can be categorised into: 1) authenticity, and 2) accuracy. The first category addresses data fidelity; that is, whether the satellite data is authentic and that there has been no manipulation involved of which the user may be unaware, such as intentional or unintentional modification and/or misinterpretation of the original data. For verification purposes, it is important to be able to trace the complete history of the data from the original source to the final product delivered to the user, including the image acquisition, the subsequent processing, and analysis that has been conducted at different stages.

The second category addresses how accurately the data reflects the actual phenomena on the ground. There are several methods to enhance the accuracy of the data during processing stages such as geometric and radiometric

[163] See K.J. Markowitz, Legal Challenge and Market Rewards to the Use and Acceptance of Remote Sensing and Digital Information as Evidence, 12 *Duke Environmental Law and Policy Forum* (Sept. 2002), 247.

[164] Communication to Mario Hernandez of UNESCO, 11 August 2008.

correction, calibration, and ground truth measurements. All these methods depend also on the accuracy of the auxiliary data used to process the satellite image.[165] In this way, the assessment of the overall quality of data is complex and it is necessary to look closely at all aspects – authenticity of data, accuracy of data, and an analysis of what needs to be done to ensure data authenticity and accuracy.

5.5.1.1. *The Authenticity of Data*

In the aforementioned ICJ dispute resolution between Cameroon and Nigeria, the satellite images used in court generated confusion due to the conflicting interpretations of the image.[166] In this case, Cameroon is alleged to have misinterpreted the image;[167]however, in absence of a definitive mechanism for data authentication, it may be difficult for a court or other third party to judge.

It should be noted that the possibility of data manipulation exists in satellite images primarily because of its modifiable format, such as JPEG. If the processing is not documented, it might become extremely difficult, if not impossible, to notice if any processing was applied;[168] consequently, experts should reject cases where there is any doubt of an artificial artefact created by image processing. In such cases, the original imagery should be requested from the satellite provider and compared to processed imagery to verify whether the original image contained such artefacts or not.[169]

It is necessary to ensure that the processed and/or interpreted satellite data reflects the reality of what is on the ground using the method of ground truth – that is, to compare image data to real features and materials on the ground to verify the accuracy of interpretation. Certain laws of evidence and environmental laws[170]require data authentication for verification use. It is necessary to be able to establish clearly the authenticity of images, as well as to identify those images that may have been manipulated and/or misrepresented.

The DIC of Delaware case illustrated that lack of information on the submitted photograph – including date, place, and the party responsible for defects – hampered its admissibility as evidence. In order to conduct data authentication, thorough disclosure of information on the data is crucial.

---

[165] Communication to Mario Hernandez of UNESCO, 11 August 2008.

[166] See *Report on the legal aspects of the privatisation of space activities Remote Sensing and national space legislation*, Berlin Conference 2004, International Law Bar Association, 12–13.

[167] See NPA Group, *Final Report BNSC Sectors Studies Programme Applications of Earth Observation to Legal Sector*, BNSC, London, 2001, 29.

[168] Communication to Mario Hernandez of UNESCO, 11 August 2008.

[169] Communication to Mario Hernandez of UNESCO, 11 August 2008

[170] Annex 2 of Resolution IMO A 152 of 17 November 1983.

Information providing the context of the data is referred to as metadata,[171] which is complementary information such as name of the author, when the work is made, copyright information,[172] methodologies for image processing applied to image, and software packages used.

Generally, the practice of satellite data providers is to include fundamental metadata[173] information – such as the position of the satellite, geographical location of the centre of the satellite image, date of image capture, copyright, satellite description, and sensors description – in the data file.

Some metadata information is included in the contract/license and only available to data recipients. These data concern, for example, the distribution information and content of the information, such as technical specifications. Often, advanced products are provided with more comprehensive metadata;[174] however, not all metadata information is available currently, particularly for standard products. Most notably, the quality of the data processing methods is not available. Metadata defined by the EC INSPIRE Directive includes "quality and validity of spatial data sets,"[175] but does not explain further how they are determined and applied. Such information is critical for verification use and, in the future, a common standard should be provided for this.

5.5.1.2. *The Accuracy of Data*

Article 1(8) of the Environmental Directive refers to 'accuracy' in addressing the quality of environmental information, but does not explain this term in further detail. Before getting into the detailed discussion of accuracy, some attention should be given to the scope of satellite imagery as a suitable verification tool: the capability of satellite imagery in terms of the level of information that it can provide. If the court is interested to know whether an oil spill occurred on a certain date and time, remote sensing can assist. On the other hand, if the need is to measure such an oil slick at sub-metre precision, then satellite images have limitations and cannot be used alone to provide such accuracy.[176] In fact, the use of data for verification purposes needs to be treated with

---

171 Article 3(6) of the INSPIRE Directive defines metadata as "information describing spatial data sets and spatial data services and making it possible to discover, inventory and use them."

172 See L. Lessig, *Creative Commons,* NTT Publisher, Tokyo, 2005, 49.

173 Black's Law Dictionary defines 'metadata' as "secondary data that organize, manage, and facilitate the use and understanding of primary data." See Black's Law Dictionary, Ninth Edition, 2009, 1080.

174 For example, Spot Image provides highly accurate reference for 3D products consisting of high-resolution-satellite digital elevation model, high resolution orthoimage, and quality and traceability of metadata. See *Spot Image Product Guide,* 10.

175 Article 5(c) of the INSPIRE Directive.

176 Communication to Mario Hernandez of UNESCO, 11 August 2008.

caution. Today, satellite images can be used in court cases when the disputed point is within the level of accuracy that the satellite can provide with certainty. If the disputed matter requires accuracy at a very precise level, the imagery may not be completely reliable.

The question of accuracy is obviously whether the data contains errors. Potential errors may be caused not only from deliberate manipulation or negligence, but by the inevitable uncertainties inherent in satellite remote sensing techniques, inaccuracies, ground data used, and associated image processing activities. Indeed, the original remote sensing data has potential exposures to errors which may arise in the course of the chain of remote sensing operations. Potential errors may exist in each step of the data chain: data acquisition, storage, transformation, and use.[177]

The following discussion focuses on the errors that occur in the phases of data transformation and use. In these phases, errors may occur in image interpretation because of ambiguous interpretation of surface features.[178]

The image processing of remote sensing data is a relatively new technology with uncertainties, and both the experience and knowledge of the image interpreter affects the image analysis. For instance, oil slicks that appear dark on a radar image can be mistaken for the dark spots that result from wind, oceanographic and current effects, or weather conditions, and much will depend on the expertise of an interpreter to distinguish such differences.[179] Subjective judgement is inevitably involved in the interpretation of images and it is essential that the standards for interpretation are clarified. Therefore, when the data are interpreted by another entity (possibly a subcontractor), it is necessary to quantify the levels of confidence in such interpretations and make them available to the ultimate user.[180]

Currently, a parameter of validation of data accuracies is available to users in the form of technical adjustments conducted in pre-processing and ground verification. Pre-processing operations such as geometric and radiometric correction and calibration correct the data for distortions or degradations resulting

---

[177] See K.J. Markowitz, Legal Challenge and Market Rewards to the Use and Acceptance of Remote Sensing and Digital Information as Evidence, 12 *Duke Environmental Law and Policy Forum* (Sept. 2002), 247. Errors include incorrectly calibrated satellite data, inaccurate GIS digitalisation, lack of spatial precision, distorted models, and data misinterpretation. The errors may be attributed to registration differences between reference data and remotely sensed data, delineation errors in digitizing, data entry errors, errors in image classification, and errors involved in sampling, collection, and interpretation of reference data.

[178] NPA Group, *Final Report BNSC Sectors Studies Programme Applications of Earth Observation to Legal Sector*, BNSC, London, 2001, 37.

[179] NPA Group, *Final Report BNSC Sectors Studies Programme Applications of Earth Observation to Legal Sector*, BNSC, London, 2001, 37.

[180] Communication to Mario Hernandez of UNESCO, 11 August 2008.

from the image acquisition.[181] Usually there is already some pre-processing on board the satellite.[182] Using ground truth verification, it is possible to provide the data with a considerable degree of accuracy,[183] thus enabling an overall improvement of accuracy from a technical perspective; nevertheless, human error is not adequately addressed.

Data accuracy is indicated in two components, namely, 1) maximum magnitude of error in metres, and 2) probability of occurrence of such an error. For example, Spot Image provides highly accurate digital elevation model products with their absolute horizontal accuracy[184] within 30 m and absolute vertical accuracy[185] within 20 m; these accuracies are provided with a 90% confidence level.[186] This indicates that 9 out of 10 products are accurate and, if they are not accurate, the errors are less than 30 m for vertical dimension and 20 m for horizontal dimension. These are the main criteria that can be relied upon in distinguishing more accurate images from those that are less accurate; however, the accuracy and fidelity of the processing procedures is not assessed for each product.

There are no objective assessments for all the processes of data handling activities from the raw data to final products. It is important to clarify the data handling processes so that accuracy is ensured in a uniform manner.

5.5.1.3. *What Needs to be Done to Ensure Data Authenticity and Accuracy*

As previously discussed, currently basic metadata and pre-processing information is available. What is not readily available is full comprehensive metadata, particularly relating to the quality of information, data authenticity, and accuracy of data handling activities. Currently, it is difficult to determine the precise degree of data fidelity and accuracy because there is no systematic means to identify modifications and errors, and locate the sources thereof. The difficulty in the identification of where and how the errors occur lies largely in the changeable format and unrecorded history of data handling activities, which is not executed in a systematic way.

The fidelity and accuracy of the data processing procedures need to be ensured at the different stages of data handling activities, including data acquisition, pre-processing, processing, and interpretation. The fidelity and accuracy

---

[181] See NPA Group, *Final Report BNSC Sectors Studies Programme Applications of Earth Observation to Legal Sector*, BNSC, London, 2001, 32.

[182] Communication to Mario Hernandez of UNESCO, 11 August 2008.

[183] See NPA Group, *Final Report BNSC Sectors Studies Programme Applications of Earth Observation to Legal Sector*, BNSC, London, 2001, 72.

[184] Positional error in the DEM with respect to the ground truth.

[185] Elevation error in the DEM with respect to the ground truth. It depends mainly on the slope of the terrain.

[186] This means that nine out of ten products are statistically accurate.

of the data processing procedures should be complemented by comprehensive information on the data/products at each stage. Instead of random availability of different types of metadata, metadata for disclosure should follow certain standards; for this, metadata endorsed by INSPIRE[187] as well as the ISO standard on metadata of digital data – ISO 19115 – should be referenced.[188] In accordance with these standards, depending on the applications, mandatory elements need to be set and must be available to users as well as to the concerned parties in the supply chains.

The determination of fidelity requires the recording of activities involved in product generation and the tracing of all subsequent processes, specifying by whom, when, and how processing is conducted. Recording such information in an unchangeable format would serve as an effective mean to avoid data manipulation and misrepresentation, and would enhance user confidence in data. The validation must enable the maintenance of data integrity and minimize the possibility of the data to be discredited by a court.[189]

5.5.2. *Issues of Traceability*

Traceability is defined as the ability to follow activities associated with products through specified stages of processing and distribution.[190] Traceability refers to the completeness of information for each step involved in the chain of production,[191] and is used mainly in the context of food traceability, which is defined by Codex, the agency established by the FAO and WHO as follows: "to follow movement of food through specified stages of production, and processing."[192] Since the term 'traceability' is not yet adopted widely in the context of satellite images, it is practical to approach this topic by drawing an analogy with the existing traceability systems applied to other products – most notably, agricultural products.

---

[187] Article 5(1) of the INSPIRE Directive.

[188] ISO Technical Committee 211, Geographical Information/ Geomatics releases the metadata ISO 19115.

[189] See NPA Group, *Final Report BNSC Sectors Studies Programme Applications of Earth Observation to Legal Sector*, BNSC, London, 2001, 29.

[190] See K. Yamamoto, *Agricultural Products and Traceability*, Seibyundo Publisher, Tokyo, 2003, 10.

[191] *See* K. Yamamoto, *Agricultural Products and Traceability*, Seibyundo Publisher, Tokyo, 2003, 10.

[192] Joint FAO/WHO Food Standards Programme Codex Alimentations Commission, Proposed Definition of Traceability/ Product Tracing, Annex VI to Report of the Twentieth Session of the Codex Committee on General Principles, 2004, ALINORM 04/27/33A.

5.5.2.1. *The Consideration of Adopting Traceability for Satellite Images*

Traceability in the verification of each process in a chain of processes from generation to distribution of products is applied extensively to various food products. The European Union generated a list of foods that must be subject to traceability: starting from beef (with the outbreak of bovine spongiform encephalopathy around 2000), a number of agricultural products, such as genetically modified foods and eggs, have been added to the list.[193] The purpose of introducing traceability is to manage risks, to ensure reliability, and to disclose information.

There are two elements required in order for products to be verified: specification of a product and documentation of the history of the product; that is, how the product goes through various processes at different stages. Authentication requires the documenting of all major processes as well as proof that the documentation is reliable.[194]Although it is not necessary to impose an obligation for all remote sensing data to follow such comprehensive traceability guidelines, they can serve as a concrete model. The most critical point is the ability to ensure the image is free from errors or without manipulation.

Just as with agricultural products, if traceability is applied to satellite products, then higher reliability and quality can be achieved. Tracing of process requires assigning identification numbers to every product. A number of satellite data have identification numbers; thus the traceability of satellite images is already at a pre-operational stage. Now the question is: how can traceability be implemented in practice?

5.5.2.2. *How to Implement Tracing Satellite Products*

The reasons for implementing a traceability requirement for satellite images are to achieve a uniform and objective approach and to establish a mechanism in assessing data authenticity and fidelity. Such implementation requires a methodology that would permit provision of metadata and recording of data history, such as how the data is generated and the degree of confidence in interpretation of images. Methods for traceability can range from documents, labels, marks, and databases, to bar codes.[195] For satellite images, an electronic authenticity tool is appropriate. Such a tool should fulfil several functions. First, the electronic authenticity tool should facilitate integration and disclosure of metadata and other relevant information that is currently available under different locations or is not available at all. Second, the tool should convey the information recorded at each stage of data acquisition and data handling activities. Third, the electronic authenticity tool should connect the information contained

[193] See Y. Niyama, *Traceability in Food Chain*, Showado, Tokyo, 2005, 127.

[194] See K. Yamamoto, *Agricultural Products and Traceability*, Seibyundo Publisher, Tokyo, 2003, 14.

[195] See Y. Niyama, *Traceability in Food Chain*, Showado, Tokyo, 2005, 155.

therein to a central database where the authorized agency can identify the available information sources and ascertain the quality of each product. For implementation, two different periods for tracing are required for satellite images: pre-data supply and post-supply data.

As noted above, traceability before product distribution is already adopted for food products. A traceability system using a bar-code is common practice. By accessing a bar-code, consumers can get information such as who made the product and the history before its delivery to a consumer. Toshiba Tec uses three-step traceability for two-dimensional bar codes: the first step is to request the providers of the primary materials to disclose who the generators are and how the product is made – such as the use of a pesticide, and additives involved in production – and information is compiled into the database and then materials are labelled with bar codes.[196] The second step is to make a database of processed food in the same way; and the final step is to build the infrastructure for disclosing all the information on the products to the consumers.[197] Similar steps can be adopted for satellite-derived products combining multiple sets of data. Such a method allows the recording and integration of: 1) a set of metadata from a single source; and 2) more than one set of metadata from different sources.

Whilst traceability after sale is not necessary for food products, for satellite images, traceability after data supply may be critical. For example, data is supplied to a state representative that uses the data as evidence for a boundary dispute, and the court needs to know whether the data has remained free of manipulation after data supply. The method of traceability should also serve as a means for a data supplier to retain control after data distribution. It should permit the data generator such as a space agency to follow what happens to the data after distribution and to be able to prevent unauthorized reproduction and data manipulation. For this purpose, a digital watermark is a potential electronic authenticity tool. A digital watermark is a technique which allows an individual to add hidden copyright notices or other verification messages to digital audio, video, or image signals and documents without disturbing the information contained therein.[198] Through the use of a digital watermark, information such as intellectual property rights can be imbedded, and the history of reproduction and possible manipulation can be traced. Such an

[196] See Food Rink News, *All about Traceability of Food Products*, Japan Association of Efficiency Publisher, Tokyo, 2003, 182–183.

[197] See Food Rink News, *All about Traceability of Food Products*, Japan Association of Efficiency Publisher, Tokyo, 2003, 182–183.

[198] See K. Matsui, *Basics of Digital Watermarking*, Morikita Publisher, Tokyo, 1998, 6-9. See further J. Seitz, *Digital Watermarking for Digital Media*, Information Science Publishing, Hershey, 2005.

electronic authenticity tool can serve to protect the interests of both the provider and recipient of the data in the way that copyright protection is enforced for data providers. Ensuring data fidelity creates a disincentive for piracy and misuse, and also improves user confidence.

The need for verifying the accuracy and authenticity of data is urgent. In order to assess the quality of data objectively, transparency is required. In order to achieve transparency of processes before the final product, it is necessary for processing wholesalers and information extractors to: 1) document the data history, 2) indicate the level of certainty in interpretation and analysis, coupled with uncertainties and errors that may be involved; and 3) to provide the comprehensive metadata information at each stage. Ensuring the quality of data requires establishing common principles for the authentication and validation of data accuracy, as well as for traceability guidelines. Then, principles need to be implemented through use of the suggested electronic authenticity tool. Achieving the above described transparency measures using a suitable electronic authenticity tool will all contribute substantially to better data management. Consequently, the recommendation enables more effective assessment of the environment as well as response to disasters.

5.6. Conclusion

Improvements in the areas of data policies and data verification are urgent because satellite data is becoming an increasingly crucial source of information for disaster management and environmental protection applications, both of which require a large number of reliable data sets. Today, scattered data policies are an impediment to the full utilization and wide-spread use of data for public benefit purposes, and it is necessary to establish a clear regime with an exclusive set of rules to ensure availability, exploitability, and reliability.

Although the findings of this Chapter show difficulty in unifying all the existing data policies, they infer the possibility for establishing a common data policy designed for public benefit service. Certain harmonization should be achieved as to access and pricing policies for the purposes of humanitarian aid and environmental protection. Data used for such public benefit programmes should enjoy special treatment to relax the conditions usually placed on data. With respect to pricing, data should be available for urgent disaster response free of charge and for other non-urgent public uses at the cost of reproduction.

The improved regime to be proposed should serve to clarify current uncertainties involved in data use, specifically, intellectual property rights and lack of guidelines for data authentication and accuracy validation. It is recommended that data supply in both the public benefit service and commercial service categories come under a common intellectual property rights regime and data authentication mechanism. With regards to a common intellectual property rights regime, there are three primary recommendations:

1. Ownership of all types of satellite data will rest with the data generator, but protection on the basis of copyright will depend upon types of EO data.
2. Joint authorship for value-added products should be clarified.
3. The names of the persons who conduct data handling operations should be documented.

With respect to the first recommendation, manually processed data and value-added products that involve intellectual activity should be protected by copyright, whereas the raw data should be protected by database rights or merely credited without legal protection in those jurisdictions where such right is not recognized. Concerning the second recommendation, products generated by more than one individual or entity should be subject to protection under copyright when intellectual input is involved by each party, irrespective of types and degrees of contribution by each:

A. Value-added products of works of collaboration should be protected by copyright, then each data generator and associated rights share the associated rights equally.
B. For value-added products of collective work, copyright should be held by the data generator but a copyright notice should be provided that includes the name of the value-adding entity instead of simply crediting the entity.
C. For value-added products of composite work, copyright should be held by the value-adding reseller with database rights or credit given to the data generator.
D. For B and C, and other complex cases, intellectual rights should be determined on a case by case basis through the contractual agreements.

In any case, the names of individuals who conducted each operation should be documented. In case of copyright infringement, entities should be able to identify the originators of specific products. The responsibility and benefits of joint work should be allocated in accordance with the percentage of work obtained from each contributor to the final work, if necessary under the contract.

In order to enhance the reliability of data for verification, it is recommended that the metadata of all satellite images should adhere to common standards such as ISO, and mandatory elements for disclosure should be established for certain verification uses. Clear distinction should be made between authenticated and unauthenticated products, as well as between information with verified accuracy and that which has not been verified. The authenticated and verified products should be traceable with a suitable method

For products verified as both authentic and accurate, the following principles should be applied:

1. Finished products at each data handling operation should be saved in a format that cannot be modified further after the completion of each processing stage, and/or upon distribution to users.
2. The names of persons who conducted processing and value-adding activities should be recorded and referred to if necessary.
3. Comprehensive metadata as well as the degree of certainties in the processing and analysis of images should be made available to all parties through the supply chain to end-users.
4. Proof of data authentication and verification should be marked clearly with a common symbol that should be used for all the authenticated and verified products.

The products with validated accuracy should be authenticated and their accuracies should be assessed objectively by a checklist including:

1. pre-processing information, such as whether the sensor is calibrated and the data is geometrically and radiometric corrected;
2. whether the products have been through ground truth verification or compared with other datasets for the same location; and
3. whether the final products are verified by external parties.

This way, the standards and qualities expected of the different products are set objectively.

Chapter Six

Liability

6.1. Introduction

Whilst Part I clearly illustrated that satellite remote sensing is a powerful tool to support critical decision-making – from disaster management to environmental applications – it also highlighted the uncertain liability regime that creates barriers to the full utilization of remote sensing data. Chapter 2 addressing space law identified ambiguities in applying the general legal responsibility and liability associated with space activities to satellite remote sensing.

Although the issues surrounding the need for a new regime have been raised,[1] such as the Brazilian proposal calling for an international convention on remote sensing,[2] the liability aspects of satellite remote sensing are still relatively unexplored with no concrete outcome to improve the current regime. Discussions of the Indian Ocean Tsunami lawsuit, and of the liability of volunteers introduced in Chapter 4, have particularly highlighted the need for a clarification of liability associated with disaster management. There is an urgent need for an unambiguous and comprehensive regime.

[1] See R.B. Trinder, Remote Sensing Via Satellite: The Need for a New International regime, XXII (I) *Annals of Air and Space Law* (1997), 522.

[2] UN Doc. A/AC.105/C.2/L.244(2003), Why is an international convention on remote sensing of the Earth from outer space necessary?, Proposals to the Committee on the Peaceful Uses of Outer Space for new items to be considered by the Legal Subcommittee at its forty-third session. In the proposal, it emphasized that potential harm does not lie in the actual operations involving collection, storage, processing, and distribution of the processed data, but in the use of analysed information.

The purpose of a comprehensive liability regime is two-fold: first, to compensate victims of damage; second, to strengthen data use by clearly defining liabilities. An ideal regime would not only provide a clear regulatory framework, but also a set of rules and guidelines to address potential claims effectively. In considering an improved regime for remote sensing, a thorough study is required. The aim of this Chapter is to conduct an interdisciplinary review of liabilities that will highlight the essence of liability, and to propose effective recommendations for an improved regime that will regulate and stimulate remote sensing activities.[3]

This Chapter undertakes a detailed examination of liabilities from different standpoints in both national and international contexts, including: 1) the current legal uncertainties dealing with consequential damage arising from remote sensing; 2) the issue of liability from different branches of law; and 3) the analysis of different bases of liability – specifically, absolute/strict liability, fault-based liability, and liability for products and services.

6.2. Current Legal Uncertainties with Liability Arising from Remote Sensing

Chapter 2 examined the issues of responsibility, liability and registration, and showed that the application of these to remote sensing from space is ambiguous, and thus does not address anticipated legal challenges today. Chapter 2 demonstrated that neither the current remote sensing legal regime nor the relevant *corpus juris spatialis* – the Outer Space Treaty, the Liability Convention, and the UN Remote Sensing Principles – addresses consequential damage arising from the supply and use of data. This section revisits the Outer Space Treaty, the Liability Convention and the UN Remote Sensing Principles with close attention to the relevant provisions in each, and then examines the specific problems raised by the inadequate regime.

6.2.1. *The Examination of Liability from the Outer Space Treaty and the Liability Convention*

As noted previously, Article VII of the Outer Space Treaty, together with the Liability Convention, establishes the international liability of a launching state for the damage caused by a space object or component part incurred on the Earth; however they do not provide an adequate answer to the question of

[3] For a condensed version of this Chapter, see A. Ito, Improvements to the Legal Regime for the Effective Use of Satellite Remote Sensing Data for Disaster Management and Protection of the Environment, *34-1 Journal of Space Law* (2008), 45–65.

whether damage arising from use of remotely-sensed data is covered. The earlier analysis of the Liability Convention (see Chapter 2) pointed out that the language on the absolute liability of a launching state to pay compensation for damage caused by its space object[4] is commonly interpreted to refer to the identifiable physical damage caused by the space object.[5] This Chapter further examines that provision to clarify the ambit of 'damage caused by a space object' provided under the Liability Convention in association with remote sensing. The fundamental question concerns whether the liability extends to damage of an indirect nature. Whether indirect damage is covered has been a subject for debate. Spradling contends that the position of the United States is that "indirect damages are not covered in any way by the Convention."[6] On the other hand, Christol expresses the view that indirect damage may be recoverable insofar as such damage is covered as a result of a physical initial impact.[7] An example of this type of indirect damage would be nuclear damage resulting from the fall of a nuclear powered satellite.

Damage arising from the information obtained through space applications would be not only indirect in nature, but also would not involve an original physical impact to cause further indirect damage. Even applying a broad interpretation of 'damage caused by a space object,' it is difficult to argue that the Liability Convention covers types of damage that may arise either from the supply or use of inaccurate data in the course of data handling activities, or from damages arising from misuse of data. The interpretation that consequential damage falls outside the ambit of damage under the Convention appears to be the usual approach to addressing the damage associated with GNSS. Although a certain interpretation exists that the Liability Convention covers consequential damage from GNSS signal-related damage,[8] the basis of liability

[4] See Article II of the Liability Convention. For a detailed analysis, see A. Kerrest, Liability for Damage Caused by Space Activities, *Space Law : Current Problems and Perspectives for Future Regulation*, M. Benkö, and K.U. Schrogl, Eleven International Publishing, 2005, 91–111.

[5] See Article I(a) of the Liability Convention, which defines damage as loss of life, personal injury or other impairment of health or loss of or damage to property. The nature of "damage caused on the surface of the Earth" is easily interpreted to be direct damage, rather than consequential damage.

[6] K.K. Spradling, International Liability Ramifications of the US NAVSTAR Global Positioning System, *Proceedings of the Thirty-Sixth Colloquium on the Law of Outer Space*, AIAA, 1990, 98.

[7] See C.Q. Christol, Space law: past, present and future, Kluwer Law and Taxation Publishers, Deventer, 1991, 223.

[8] See B.D.K. Henaku, *The Law on Global Air Navigation by Satellite: an Analysis of Legal Aspects of the ICAO CNS/ATM System*, AST Monographs, Leiden 1998, 230. Larsen also contends that the satellite navigation damage falls within the Liability Convention. See P.B. Larsen, Legal Liability for Global Navigation Satellite Systems, *Proceedings of the Thirty-Sixth Colloquium on the Law of Outer Space* (1993), 69.

currently adopted in operational regimes surrounding GNSS rather denies this view;[9] therefore, coverage for consequential damages under the *corpus juris spatialis* is unlikely.

6.2.2. *The Examination of Liability from the Standpoint of the UN Remote Sensing Principles*

The UN Remote Sensing Principles – the only international document dealing with satellite remote sensing – do not offer an adequate answer. As previously stated, the primary focus of the Principles is on the delineation of rights and duties of sensing states and sensed states, as well as on the rules concerning data collection and dissemination; they do not address adequately the matters of supply, use, and application of the data itself.

The principles are also silent as to the types of damage that might be associated with satellite remote sensing activities, and which activities might fall under the scope of liability. Indeed, one cannot find any provisions that would be of help in settling claims, such as the responsibility to provide accurate information by remote sensing and/or the responsibility not to cause damage to others by use and misuse of information derived from remote sensing.

The language of the Principles is ambiguous regarding the international responsibility expected of states engaged in remote sensing activities and the types of remote sensing activities to which the responsibility extends. The scope of the remote sensing activities covered under the Principles is "the operation of remote sensing space systems, primary data collection and storage stations, and activities in processing, interpreting and disseminating the processed data."[10] From this, it could be interpreted that the responsibility extends to data handling activities and distribution, but would not likely extend to the use and application of data. These are the matters of growing concern within the user community.

Finally, it can be construed that Principle XIV does not impose a responsibility on the end-users or other third parties who use or exploit the data. Indeed, Principle XIV limits the extent of responsibility to only the supplier community. In reality, the stakeholders associated with remote sensing – including suppliers, users, and third parties – are in a position to be both potential claimants and defendants. The language of the Principles does not correspond to the diversified dimensions of remote sensing nowadays; it is not sufficiently comprehensive to cover the responsibilities of all potential defendants, as well as consequential liabilities arising from the use of data.

---

[9] A detail analysis of GNSS regime will be provided later in this Chapter.

[10] Principle I (e) of the UN Remote Sensing Principles.

The documents in the current *corpus juris spatialis* are the only international compensation channels available to potential claimants, which are far from sufficient. On the other hand, on the basis of what is provided at the moment, it is difficult for potential defendants to envisage predictable scenarios in the case of claims. Potential legal challenges are undoubtedly beyond the scope of the current regime surrounding remote sensing.

6.2.3. *How the Inadequacy of the Current Regime Raises Specific Problems*

The current ambiguity over responsibility and liability is a major problem, as the risks for damage as well as liability risks are entirely conceivable; thus, the consequences can be detrimental.

The causes of claims can vary, but damage is most likely to arise from the misinterpretation of data and/or the misuse of data. In fact, there are examples in the past of both problems. The media has interpreted satellite imagery poorly and communicated incorrect information to the public. When the Chernobyl accident occurred it was announced that two reactors were on fire instead of one.[11] Furthermore, there was a case of an intentional misrepresentation of data in the US State of Kansas. The US Postal Inspection Service sought to bring fraud charges against Paul Caldwell, chief executive officer of Psytep Corp., for giving Kansas authorities alleged satellite imagery for use as conclusive evidence in their case, which turned out to be an aerial photograph taken at an irrelevant time.[12]

The above are only a few examples of the conceivable damage. Indeed, any person, whether natural or juridical, and/or any state can be exposed to the risks of damage arising from the use of remote sensing data. The population affected by disasters, as well as aid workers, could suffer injury or death by relying on instructions given based on a misinterpreted image; ships could be wrongly charged for environmental contamination; users and data suppliers may incur damage from the misuse of the data.

With the current uncertainties, the court is given no guidance as to how to approach potential claims. Users may not be able to recover compensation for

[11] See R. Dalbello & L. Martinez, The Legal and Political Implications of Media newsgathering from Space, *Proceedings of the Thirtieth Colloquium on the Law of Outer Space*, IISL, AIAA, Washington, 1987, 180. See further A. Florini, and Y.A. Dehqanzada, Commercial Satellite Imagery Comes of Age, *Issues in Science and Technology*, fall 1999, 2. See also A. Florini, and Y.A. Dehqanzada, Policy Implications of Commercial Remote Sensing Satellites, Governance, Transparency and Civil Society Global Policy Program, Project on Transparency, Working Paper no. 1, Carnegie Endowment for International Peace, July 1999, 2.

[12] See W. Ferester, Firm Suspected of Misrepresenting Imagery, *Space News*, 16 January 1995.

the losses they incur. As a result, users are less willing to rely upon the data when making critical decisions that may have significant effects on their interests or the interest of third parties. In addition, suppliers remain unprepared for liability risks and may incur unexpected litigation costs in case of claims. Consequently, suppliers are prone to be reluctant to provide the data, particularly on a voluntary basis; this would inevitably hamper the widespread use of data. The inadequacy of the current legal framework needs to be rectified so that remote sensing data can be rigorously and fully exploited with legal support. From the next section onward, the Chapter will examine the issues of liability from the standpoint of different branches and bases of law governing liability. Although the general liability regime can be applicable to disputes between states, this Chapter only addresses cases between natural and/or juridical persons.

6.3. Different Branches of Law Governing Liability

To get an overall picture of the law governing liability, one should deconstruct it to the two components: contract law, and tort law that covers third party liability. It is helpful to briefly refer to the legal regime surrounding air navigation as an example of a structured regime for a global and complex operation. International air law includes a relatively well regulated liability regime comprised of different sets of conventions. With regard to the liability aspects, the Warsaw Convention for the Unification of Certain Rules Relating to International Transportation[13] addresses the contractual liability of carriers with respect to passengers; the Rome Convention on Damage caused by Foreign Aircraft to Third Parties on the Surface[14] deals with a third party loss caused on the ground by an aircraft. Furthermore, in the field of satellite navigation – although its legal regime is less comprehensively established than that of air navigation – certain rules specific for GNSS exist as to contractual and third-party liability, which will be discussed in depth later in this Chapter. By contrast, international space law only regulates third-party liability in a limited context through the Liability Convention, and does not regulate at all with respect to contractual liabilities. It is evident that liability arising from remote sensing activities is unregulated and potential claims are inadequately addressed.

The absence of international rules has resulted in a complex legal environment surrounding data supply and use that depends almost entirely upon on varying national laws. Claims are addressed through contractual agreements

[13] 1929 Warsaw Convention for the Unification of Certain Rules Relating to International Transportation, 137 LNTS 11. [Hereafter: Warsaw Convention]

[14] 1952 Rome Convention on Damage caused by Foreign Aircraft to Third Parties on the Surface, 310 UNTS 181.[Hereafter: Rome Convention]

provided by individual data suppliers under the national law governing the respective entities,[15] as well as under the general tort law regime of the particular jurisdiction.

If disputes associated with remote sensing arise, how they are likely to be addressed depends on the following settings:

1. Between the parties to the contractual relations – such as a) data generator/supplier and data recipient, and b) data generator and their sub-contractor;
2. Between the parties outside contractual relations – such as a) data suppliers and the third parties, and b) between the third parties.

To consider an appropriate regime for satellite remote sensing, it is necessary to investigate the current liability regimes in a general context to address the two settings above and gain understanding of the fundamental principles and rules in each, as well as to examine how such principles may be applied to remote sensing.

6.3.1. *Contract Law*

The nature of contract law is that it binds the parties in contractual relations on agreed terms. A contract is "an agreement which creates, or is intended to create, a legal obligation between the parties to it."[16] A contract has three essential elements: (1) consensus in the same agreement about the same thing, (2) at least two contracting parties and (3) legal obligations.[17]

6.3.1.1. *The Concept of Contractual Liability*

First of all, one needs to understand that contractual liability arises when a contract is breached. A contract is governed primarily by domestic laws, although common principles do exist and certain rules are codified in the European context, such as the European Principles of European Contract Law 1998; and in the international commercial context, such as the UNIDROIT Principles of International Commercial Contracts.[18]

[15] For example, Spot Image distribution contracts are governed by French law, GeoEye under US law, MacDonnald Dettwiler under Canadian law, and DMCii under UK law.

[16] E. Jenks, English Civil Law *Digest of English Civil Law*, Butterworth, London, 1947, 2.1.

[17] See A. Gordon, *Contract Law Second Edition,* Thomson W. Green Publisher, Edinburgh, 2003, 2–4.

[18] The UNIDROIT Principles of International Commercial Contracts [Hereafter: UNIDROIT Principles]. They are applicable when the parties have agreed their contract be governed by the Principles.

Contractual liability arises if the performance agreed between the parties is not undertaken or achieved (e.g., non-performance). The central concept is that non-performance, defined by the UNIDROIT Principles as "failure by a party to perform any of its obligation under the contract," gives rise to liability.[19] Defective performance and late performance are included in non-performance.[20] A defective performance can be associated with the quality of performance and partial non-performance – when the expected level of performance on the agreed terms is not met.

6.3.1.2. *Remedies*

There are several remedies available for breach of contract. The European Principles of European Contract Law 1998, which serve as a good basis for understanding the fundamental principles governing contracts, provide the remedies for non-performance as: the right to withhold performance;[21] the right to terminate the contract;[22] and the right to damages.[23]

The right to withhold performance means that a party who is to perform simultaneously with or after the other party may withhold performance until the other has tendered performance or has performed.[24] The first party may withhold the whole of his performance or a part of it, as may be reasonable in the circumstances.

The right to terminate the contract means a party may terminate the contract in case of serious non-performance by the other party.[25] Termination of a contract can only be justified if the non-performance is "material" or "fundamental," or amounts to a "substantial failure of performance."[26]

The right to damages is what is known as a monetary compensation: the aggrieved party is entitled to damages for loss, which is reasonably likely when non-performance of the other party is not excused.[27] The Warsaw Convention is a good example of the codification of passengers' right to damages of different

[19] Article 7.4.1. of the UNIDROIT Principles.

[20] Contractual liability is perceived differently under different jurisdictions. Under common law, breach is considered to arise through total non-performance, partial non-performance, or defective performance; under French law, contractual liability arises in the event of the non-performance of a contractual obligation if the other contracting party incurs damage.

[21] See also Article 7.1.3. of the UNIDROIT Principles.

[22] See also Article 7.3.1. of the UNIDROIT Principles.

[23] See also Article 7.4.1. of the UNIDROIT Principles.

[24] See Article 9.201 of the Principles of European Contract Law.

[25] See Article 9.301 of the Principles of European Contract Law.

[26] See H. Kotz, Contract Remedies in Civil Law and Common Law, *De tous horizons melanges Xavier Blanc-Jourvan, Société, de Législation comaprée*, Paris, 2005, 573.

[27] Article 9.501 of the Principles of European Contract Law.

types including damages for death or injury,[28] damages for loss of or damage to baggage as well as cargo,[29] and damage occasioned by delay.[30] Unlike the Warsaw Convention, the right to damages associated with satellite remote sensing is not established under any legal instrument; therefore, it is necessary to specify the type of damage recoverable and to establish a clear liability regime surrounding remote sensing.

6.3.1.3. *Defences Available to Contractual Liability*

There are different types of defences available against contractual liability, all of which serve to exonerate or limit liability. This section looks at the major defences that are relevant to remote sensing data supply: Act of God/*force majeure*; contributory negligence; *volenti non fit injuria*; and liability disclaimers.

Defences based on the Act of God/*force majeure* serve to exempt a defendant from liability for non-performance or defective performance caused by operation of forces beyond human anticipation or avoidance; however, there are slight differences between the two. An Act of God is an "overwhelming, unpreventable event caused exclusively by forces of nature, such as an earthquake, flood, and tornado."[31] *Force majeure* is "an event or effect that can be neither anticipated nor controlled."[32] Unlike the Act of God, which limits the event to be called by natural forces, *force majeure* includes both acts of nature (e.g., floods and hurricanes), and acts of people (e.g., riots, strikes, and war). They are commonly regarded as interruptions of causation, hence, excusing defendants' responsibility in case of mere accidents.[33] *Force majeure* clauses may be included in a contract when "allocating the risk of loss if performance becomes impossible or impracticable as a result of an event or effect that the parties could not have anticipated or controlled."[34] Exoneration based on Act of God/*force majeure* is stated in Article 18.3.c and inferred under Article of 20-1 of the Warsaw Convention.[35]

---

[28] See Article 17 of the Warsaw Convention.

[29] See Article 18 of the Warsaw Convention.

[30] See Article 19 of the Warsaw Convention.

[31] Black's Law Dictionary, Ninth Edition, 2009, 39.

[32] Black's Law Dictionary, Ninth Edition, 2009, 718.

[33] See F. Werro, Strict liability in European tort law: an introduction, *The boundaries of strict liability in European tort law*, Carolina Academic Press, Durham, 2004, 6. See further Draetta, U., Les clauses de force majeure et de hardship dans les contrats internationaux,3-4 *Revue de Droit des Affaires Internationales* (2002), 347–358.

[34] Black's Law Dictionary, Ninth Edition, 2009, 718.

[35] The carrier is not liable if he proves that he and his agents have taken all necessary measures to avoid the damage or that it was impossible for him or them to take such measures.

In reality, *force majeure* tends to be applied in a limited context. For instance, the Société Nationale des Chemins de fer Français (SNCF), the French railways, was found to be responsible for a derailment caused by an act of sabotage.[36] Since the threats of sabotage had been received twenty days before the accident, it was considered by the court to be foreseeable by SNCF and therefore, the SNCF could not plead *force majeure*.[37] It is unforeseeability and insuperability in the circumstances that are the key elements constituting *force majeure*. The mere fact that the incident is not imputable to the defendants does not qualify as *force majeure*.

Contributory negligence is defined as: "A plaintiff's own negligence that played a part in causing the plaintiff's injury and that is significant enough to bar the plaintiff from recovering the damage."[38] Exoneration of a carrier's liability based on contributory negligence is incorporated under Article 21 of the Warsaw Convention.[39] An example of contributory negligence is well-illustrated in a British case concerning food poisoning[40] where it was held that the claimant contributed to the damage by not cooking the raw pork properly.

Volenti non fit injuria is defined as: "The principle that a person who knowingly and voluntarily risks danger cannot recover for any resulting injury."[41] It means that no liability can be claimed by one who has assumed and accepted the risk of injury voluntarily. An assumption of risk may be express or implied, but it will be inferred only where the injured party had full knowledge of the existence and magnitude of the risk, and where he acted voluntarily without the pressures of an economic or equivalent constraint such as that experienced by employees.[42] A case example is *Kupetz v. Deere & Co*[43] concerning an injured bulldozer operator, in which it was held that the operator's assumption of the risk completely barred recovery of damages on the ground that he, the plaintiff, had knowledge of the danger.[44]

---

[36] *S.A. Electricité de Strasborg c. SNCF*, bull. I, No. 27 (Cass. Civ. Ire, 26 January 1971).

[37] See G. Miller, *Liability in International Air Transport*, Kluwer Publisher, Deventer, 1977, 55.

[38] Black's Law Dictionary, Ninth Edition, 2009, 1134.

[39] Article 21 states that: "If the carrier proves that the damage was caused by or contributed to by the negligence of the injured person the Court may, in accordance with the provisions of its own law, exonerate the carrier wholly or partly from his liability."

[40] See S. Whittaker, *Liability for Products*, Oxford University Press, Oxford, 2005, 249.

[41] Black's Law Dictionary, Ninth Edition, 2009, 1710.

[42] See C.J. Miller & R.S. Goldberg, *Product Liability*, Oxford University Press, New York, 2004, 756. See *Smith v. Baker & Sons*, [1891]AC325.

[43] *Kupetz v. Deere & Co*, 644A 2d 1213(1994).

[44] See C.J. Miller & R.S. Goldberg, *Product Liability*, Oxford University Press, New York, 2004, 757.

A major defence to contractual liability exists when parties have liability disclaimers written into their contracts. Disclaimers are contractual terms whereby one party excludes, restricts, or modifies its potential liability arising out of a breach of a contractual obligation.[45] In general, the international legal frameworks addressing contracts recognize the legality of a liability disclaimer;[46] however, exoneration clauses are not always valid and enforceable. The issues of enforceability of exoneration clauses will be further examined later in this Chapter.

6.3.1.4. *The Practice of the Remote Sensing Data Suppliers*

Since data supply is exclusively governed by contracts, positions of data suppliers with regards to contractual liabilities must be explored. As discussed in the previous Chapter, satellite data products are provided under the terms and conditions stipulated in contracts, which are governed by the national law of the states whose jurisdiction cover the data suppliers. Total harmonization is impractical; however, it should be noted that there is an overall trend regarding the distribution agreements of the data suppliers. It is necessary to achieve harmonization of certain rules and identify the necessary items that must be harmonized.

The content of a typical licensing agreement for the use of remote sensing data to data recipients includes:

1. Data recipients have limited or no right of recourse;
2. Remedies available are limited to either a replacement of a product or a refund;
3. Wide ranges of defences are available to suppliers, including *force majeure* and *volenti non fit injuria*; and
4. Liability disclaimers apply to data recipients as well as to third parties.

Distribution contracts are commonly made more favourable for data suppliers with limited rights for data recipients. Rights recognized under the European Principles of Contract Law or the UNIDROIT Principles are not granted sufficiently to data recipients.

---

[45] See K. Henaku, The International Liability of the GNSS Space Segment Provider, XXI-I *Annals of Air and Space Law* (1996), 155. See further J. Rajski, Les clauses limitatives exonératatoires de responsabilité, 3-4 *Revue de Droit des Affaires Internationales* (2002), 321–328.

[46] Article 3.109 of the European Principles of Contract Law states that "parties may agree in advance to limit or exclude their liability for non-performance except where the non-performance is intentional or the limitation or exclusion is unreasonable." See also Article 25 of the Warsaw Convention.

The right to terminate contracts is often provided only to the data supplier in the distribution contract,[47] which is silent as to whether the right to termination of contract is granted to the data recipient if the data supplier breaches the contract provisions.

The right to damages – specifically, the issues of damages and liability – is not dealt with properly under the distribution contracts. Provisions are primarily focused on permitted and prohibited activities for data recipients and, most of all, the liability disclaimers of the data suppliers. Data generators and distributors commonly waive their liability under the terms and conditions they establish for a supply of satellite data. Hence, remedies available to data recipients are significantly limited. The right of recourse to litigation is not granted to data recipients by some data suppliers such as Spot Image and DMCii.[48]

Remedies available to the data recipients are generally limited to the replacement of a product or a refund.[49] Evidently, a warranty typically provided by a data supplier does not provide much substance for users who may wish to rely on it to seek compensation if damage arises from the products they purchased.

Contractual agreements commonly have clauses for immunity of the data generator for any damage associated with the data delivered to the customer. Often an agreement incorporates defences such as Act of God/*force majeure*, and *volenti non fit injuria*.

Force majeure is concretely defined in the contracts. The Spot Image distribution contract incorporates such a clause and defines *force majeure* as: "all the

[47] For instance, a standard license of DMCii provides that "should all or part of the price not paid on the due date, or should any one of the general or specific conditions not be met, the supply and all other supplies being carried out or on written notice may be terminated." See Article 17 of Terms and Conditions for the sale of DMC satellite Data or products, DMCii, August 2007. Likewise, Spot Image Article 5.1 states that Spot Image may, in addition to all other remedies to which it may be entitled under this end user license agreement or at law, terminate immediately this agreement by notice in writing if the end user breaches any provision of this agreement." See Article 5.1 of Standard license, Spot Image, November 2007.

[48] Spot Image mandates that: "Under no circumstances can Spot Image and or its licensor be held responsible, on the basis of responsibility arising from the faulty products, for the damages caused to the client's goods which are useful for his profession." See Article 7.4 of General Supply Conditions of Satellite Imagery Products, March 2008. DMCii states that: "Customers shall not be able to bring action against DMCii on these grounds." Terms and Conditions for the sale of DMC satellite Data or products, DMCii, August 2007.

[49] See for instance boiler plate contract language for the standard license for McDonald Dettwiler: Licensee's exclusive remedy with respect to the Data is, at MDA Geospatial Services Inc.'s option, to either (a) return the purchase price paid by the Licensee directly to MDA Geospatial Services Inc. or distributor of the Data, whichever is less, or (b) replace the Data that does not meet this Limited Warranty.

occurrences which are beyond the reasonable control of Spot Image of any nature whatsoever."[50] Likewise, the terms and conditions of the DMCii specify a number of events that can be regarded as *force majeure,* including strikes and sabotage.[51] An example of the type of *force majeure* that is likely to be established could be where the ground facility that receives the data is affected by an earthquake, resulting in the suppliers' inability to supply satellite data. Furthermore, acts or regulations issued by governments that refer to shutter control security-related governmental constraints (introduced in Chapter 2) and interruption of commercial service could serve as *force majeure* for data providers.

But again, a court has high discretion on determining how *force majeure* is interpreted and to what extent it is allowed as a defence against liability, as illustrated in *Hughes Communications Galaxy Inc. v. United States.*[52] In this case, NASA was held responsible for the cost owed to its contractual party, Hughes, for having had to switch to foreign launchers following the Challenger disaster.[53] This case shows that even though the performance was made impossible by an event beyond the control of the party, it did not serve as a legal basis for defence against the liability. These cases raise issues as to the validity of using distribution agreements that contain the Act of God/*force majeure* language as a defence against liability.

---

[50] According to Article 10 of the standard license of Spot Image, general supply conditions of satellite imagery product of March 2008, *force majeure* is taken into account as "All breakdowns of Spot Satellite in activity and or of another satellite of which Spot Image distributes the product and /or of the Spot System, natural catastrophes, bad weather, fires, embargoes, interruptions, or delays in the transport or means of communication, war, acts or regulations issues by French government, by public, civil or military authorities, by the UN, which may occur as from the date of the order and would prevent its total or partial execution."

[51] According to Article 16 of the standard license of DMCii, Terms and Conditions for the sale of DMC satellite Data or products, August 2007: "All breakdowns of a DMCii and of another satellite of which DMCii distributes the data, distributes the product and /or of the Spot System, natural catastrophes, bad weather, fires, embargoes, interruptions, or collective work disputes, strikes, sabotage, delays in the transport or means of communication, war, acts or regulations issues by UK government, by public, civil or military authorities, by the UN, which may occur as from the date of the order and would prevent its total or partial execution."

[52] *Hughes Communications Galaxy Inc. v. United States*, United States Claims Court, Washington, DC Number 91-1032C.

[53] The claim was concerning NASA's failure to launch ten commercial satellites following the 1986 Challenger Disaster. Hughes incurred the expense of switching to the Ariane launcher, which was more expensive than the Space shuttle. In this case, at first the US Court of Federal Claims ruled that the government was not held liable but upon appeal, the decision was reversed: "(…) it does not prevent the President or Congress from implementing the space policy, but NASA still has to cover the cost Hughes incurred by switching to Ariane."

Clauses specifically dealing with contributory negligence generally do not exist under the terms and conditions of contracts provided by data suppliers; however, from the phrasing of a typical exemption clause, it is evident that such clauses reflect the intention of data providers to waive any liability, obviously including language about the negligence of the end-user so as to add the defence of contributory negligence. An example of circumstances where contributory negligence may be invoked would be that damage resulted from the end users' employing the data for purposes other than those for which the image was requested. Defence on the basis of contributory negligence, and the resulting apportionment of liability between the contractual parties, need to be clarified.

A defence on the basis of voluntary assumption of the risk, *volenti non fit injuria*, is also used by the satellite data suppliers, and is often incorporated in the contracts. For instance, the DMCii agreement stipulates that "the risks pertaining to the data or products sent shall be transferred to the customer as from the placing of the data or products on the medium of transport."[54] The stance taken by data suppliers is that it is for consumers to assume the risks for any damage associated with their products and to accept them through contract, thereby exonerating suppliers from any potential liability. As seen in the case of *Kupetz v. Deere & Co.*[55] the successful establishment of voluntary assumption of risk depends upon the claimant's knowledge of the specific danger. Unlike the *Kupetz* case, data distribution agreements only have clauses indicating general risks. Therefore, one can analyse that such a clause can be construed as a sufficient defence against liability.

A typical data supplier's liability disclaimer attempts to waive any potential liability. For example:

"In no event shall Spot Image, nor anybody having contributed to development and/or production and/or delivery of the PRODUCT, be liable for any claim damage or loss incurred by the END-USER, including without limitation indirect, compensatory, consequential, incidental, special, incorporeal or exemplary damages arising out of the use of or inability to use the PRODUCT, and shall not be subject to legal action in this respect."[56]

Although the phrasing of liability disclaimers provided by different entities varies in detail, the central philosophy is the same: the data suppliers do not wish to be held liable for consequential damage suffered by data recipients, or by any other third party, that may arise in association with the data; however, the damage associated with the satellite images are usually consequential.

It is common for data providers to disclaim liability vis-à-vis a third party. For instance, terms and conditions for the DMCii data states that: "Under no circumstances can DMCii be held responsible for all indirect and or immaterial

[54] Article 5.1 of the Terms and Conditions for the sale of DMC Satellite Data or Products, DMCii, August 2007.

[55] See *Kupetz v. Deere & Co*, 644A 2d 1213(1994).

[56] Article 4.3 of the Standard License of Spot Image, November 2007.

losses (...) caused to the customer or to any third party."[57] These are unilateral waivers, as they are not agreed upon by the potential third-party claimants.

With respect to the liability amongst the supply chain – such as between data generator and their subcontractors (e.g., value added entities), currently little is covered.[58] Unlike launch services contracts, which deal more concretely with third party liability such as insurance against third party loss,[59] little is arranged by entities within the data supply chain to allocate responsibilities and share liability properly in case of damage to a third party in the field of EO.

6.3.1.5. *The Enforceability of Liability Disclaimers*

The question of whether liability disclaimers are actually enforceable in court should be considered in part by understanding under what circumstances such clauses are enforced through existing statutes and case law, with respect to both contracting parties and third parties.

Whether a particular disclaimer or exclusion clause is enforced by the courts depends upon the contract terms and the nature of the disclaimer.[60] A very broad liability disclaimer, including the defences such as *force majeure*, is not enforceable in all cases and appears to be conditional. Generally, a disclaimer is enforceable in a commercial contract setting where the parties have equal bargaining power, for claims of economic loss, and for claims that do not involve either the intentional breach of contract or gross negligence.

Within the available case law, decisions dealing with launch services contracts support the rules stated above on enforceability of a liability disclaimer. The inclusion of an exemption clause, which is commonplace in commercial contracts in the space field, is seen in inter-party waivers of liability, including the Space Shuttle launch agreement and the 1973 Spacelab agreement.[61] It is important to examine the enforceability of the liability disclaimer in order to analyse the legal trend.

There are certain trends with respect to the enforceability of liability disclaimer. In *Boucher v. Riner*[62], Boucher, a customer, sued the company,

---

[57] Article 12.4 of the Terms and Conditions for the sale of DMC Satellite Data or Products, DMCii, August 2007.

[58] According to personnel of SERTIT, the liability associated with the images is not covered sufficiently by the contract between CNES and SERTIT.

[59] Third party liability – 17.2.1. Martin Maritetta shall procure and maintain Third Party liability insurance to protect Martin Marietta and INTELSAT against injury, death and property damage caused to Third Parties, arising out of or resulting from Launch Services provided under this Contract at Cape Canaveral Air Force Station or Kennedy Space Centre. Such insurance shall be in an amount of US$ 100 million.

[60] See K. Henaku, The International Liability of the GNSS Space Segment Provider, XXI-I *Annals of Air and Space Law* (1996),155.

[61] Space Laboratory: Cooperative Program, 24 *UST* 2049, *TIAS* 7722.

[62] *Boucher v. Riner*, 68 Md. App. 539, 543, 514 A 2d 485, 487 (1986).

'Parachutes Are Fun, Inc.' after Boucher parachuted into power lines. Although Boucher signed the waiver, it was held that the waiver could not exempt 'Parachute Are Fun, Inc'. from its liability on the ground that contractual waivers cannot shift risks of a party's own wanton, reckless behaviour. In *State Highway Admin. v. Greiner Engineering Sciences*[63], the Court held that an unambiguous no-damage-for-delay clause was enforceable, but noted that the exemption clause is not enforceable if the damage is caused by intentional wrongdoing, gross negligence, fraud, or misrepresentation. These two cases clearly demonstrate that a disclaimer does not apply in case of intentional breach of contract or gross negligence. Enforcement of an exemption clause was barred with a finding of gross negligence in *Martin Marietta v. INTELSAT*[64] and also illustrates the language of the liability disclaimer affecting its enforceability. Here are the details of the case:

The dispute concerned the failure of the Titan III of Martin Marietta to launch the INTELSAT satellite into the planned orbit.[65] The failure resulted from a lack of communication between software and hardware engineers at Martin Marietta, a fact to which Martin Marietta stipulated.[66] Following a motion for declaratory judgement filed by Martin Marietta, the Federal District Court of Maryland dismissed the INTELSAT claims on the ground that the concerned parties possessed equal bargain power; thus, the liability disclaimer was enforceable. The gross negligence claim was also dismissed on public policy grounds.[67] INTELSAT appealed the decision of the Federal District Court to the US Court of Appeals in January 1992.[68] INTELSAT argued the ambiguity of the contract, relying on Article 6, which stated that "replacement launch is the sole remedy" whereas Article 17 suggested the possibility of other claims.[69]

---

[63] *State Highway Admin. v. Greiner Engineering Sciences*, 83 Md. App. 621, 577 A.2d 363(Md. Ct. Spec App.).

[64] Civil Action number MJG90-1840, US District Court, Maryland. In this case, ambiguity of the contract terms was indicated. The issue was that Article 6.7 states that "replacement launch is INTELSAT's sole and exclusive remedy" whilst the limitation of liability clause of Article 17 puts a damage cap on claims arising under contract in negligence, strict liability, or under any other theory of tort or liability.

[65] See I.H.Ph. Diederiks-Verschoor, Practical Consequences of Space Activities, *Recent Trends in International Space Law and Policy*, Lancers Books, New Delhi, 1997, 375.

[66] See I.H.Ph. Diederiks-Verschoor, Practical Consequences of Space Activities, *Recent Trends in International Space Law and Policy*, Lancers Books, New Delhi, 1997, 375.

[67] It was held that the Commercial Space Launch Act Amendments of 1988 established a federal policy that overrode Maryland's common law rule that parties ordinarily cannot waive liability for gross negligence.

[68] 991 F. 2d 94 (4th Cir. 1993).

[69] Article 17.6 stipulates that Martin Marietta's liability to INTELSAT and to persons claiming by or through INTELSAT, whether or not arising under contract, or in

The Court of Appeal sustained the claim[70] reversing the District Court's dismissal of INTELSAT's breach of contract claim and overturning the dismissal of the gross negligence claim.[71] Ultimately, on 6 October 1993, the claim was settled between the parties confidentially and dismissed.[72] This case has demonstrated (1) that the liability disclaimer may be invalidated in case of gross negligence, and (2) that the ambiguity of a contract significantly affects its validity.

The fact that enforceability hinges on the phrasing of the exemption clause was also illustrated in *Appalachian Insurance Co. v. McDonnell Douglas.*[73] This case concerned the failure to place the satellite for Western Union into the planned orbit because of the malfunction of the perigee kick motor – a rocket motor fired at a perigee of an orbit to establish more elliptical orbit in order to place a payload into a higher orbit. There was an inter-party waiver of liability clause incorporated into the standard McDonnell Douglas contract as follows: "McDonnell Douglas extends no warranty of any kind, express or implied, including any implied warranty of merchantability or suitability for purpose with respect to the PAM or with respect to services provided by McDonnell Douglas hereunder (…) Under no circumstances will McDonnell Douglas be liable to purchaser under or in connection with this Agreement, under any tort, negligence, strict liability contract or other legal or equitable theory, for incidental or consequential damages or for Purchaser's cost of effecting cover."[74] The court concluded that the liability disclaimer effectively exempted McDonnell Douglas from liability.[75]

Ambiguous clauses will be interpreted by the court favourably for the defendant, which is known as *contra proferentem,*[76] contractual interpretation

---

negligence, strict liability, or under any other theory of tort or liability, shall not include any loss or use or loss of profit or revenue or any other indirect, special, incidental or consequential damages.

[70] In light of these ambiguities, it does not appear beyond A REASONABLE doubt that Martin Marietta DID NOT USE ITS BEST EFFORTS, and Intelsat could not prove that Martin Marietta breached the contract.

[71] It was held that "the District Court erred in dismissing INTELSAT's gross negligence claim."

[72] See I.H.Ph. Diederiks-Verschoor, Practical Consequences of Space Activities, *Recent Trends in International Space Law and Policy*, Lancers Books, New Delhi, 1997, 375.

[73] *Appalachian Insurance Co. v. McDonnell Douglas,* 214 Cal ap 3d 1, 262 Cal Reporter 716 (Cal Ap Fourth District 1989).

[74] See I. Awford, Commercial Space Activities: Legal Liability Issues, *Recent Trends in International Space Law and Policy*, Lancers Books, New Delhi, 1997, 414.

[75] See K.H. Böckstiegel, Case on Space Activities, *Space law: development and scope*, N. Jasentuliana (Ed.), Prager Publisher, Wesport, 1992, 211.

[76] M. Whincup, *Contract Law and Practice*, Kluwer Law International, Hague, 2001, 194.

which provides that an ambiguous *term* will be construed against the party that imposed its inclusion in the contract. Awford emphasizes the importance of phrasing in enforcement of a liability disclaimer, pointing out that exemption clauses are interpreted strictly so that any ambiguity or doubt will be construed or resolved against the party who seeks to rely on the exclusion.[77] The clauses to disclaim any liability 'whatsoever' are not sufficient[78] and they need to refer specifically to the terms, such as negligence, from which the party intends to be exonerated.[79] It can be argued that the clearer the language of the liability disclaimer, the higher the possibility that it will be construed as valid.

Having seen the case law, one can draw a conclusion that, as far as agreements between parties of equal bargaining power are concerned, enforceability of liability disclaimers depends upon the phrasing of the contract, but the interpretation is open to the court unless there is intentional tort or gross negligence.

There is a question as to the enforceability of a liability disclaimer on a third party. Generally, the waiver of liability is agreed between the two concerned parties and hence, is not enforceable on third parties.[80] Usually, third parties are not aware of the existence of a liability disclaimer.

The invalidity of a disclaimer of liability was illustrated in the UK case *Ministry of Housing and Local Government v. Sharp*;[81] furthermore, it was demonstrated in the dispute over the loss of the Westar VI communication satellite. In contrast to *Appalachian Insurance Co. v. McDonnell Douglas*, where the parties signed the waiver of liability and the liability disclaimer was enforceable, in *Lexington v. McDonnell Douglas*[82] the insurers of the Palapa B-2 were able to sue McDonnell Douglas, Morton Thiokol, and Hitco. The reason was that a clause

[77] See I. Awford, Commercial Space Activities: Legal Liability Issues, *Recent Trends in International Space Law and Policy*, Lancers Books, New Delhi, 1997, 415.

[78] See I. Awford, Commercial Space Activities: Legal Liability Issues, *Recent Trends in International Space Law and Policy*, Lancers Books, New Delhi, 1997, 415.

[79] See I. Awford, Commercial Space Activities: Legal Liability Issues, *Recent Trends in International Space Law and Policy*, Lancers Books, New Delhi, 1997, 415.

[80] See K. Henaku, The International Liability of the GNSS Space Segment Provider, XXI-I *Annals of Air and Space Law* (1996), 159.

[81] *Ministry of Housing and Local Government v. Sharp*, [1970] 2 QB 223. The Ministry of Housing was entitled by statute to receive payments and the Ministry's entitlement, known as a compensation notice, was entered on the local land charges registry. An employee of the registry carried out the search carelessly and issued a certificate to the owner without the compensation notice. As a result, the Ministry lost its right to receive payments due to the carelessly conducted search and sued the registry; the Ministry could recover its loss despite the disclaimer contained in the registry's reply because the Ministry was not the recipient of the search.

[82] *Lexington v. McDonnell Douglas,* Filed by the California court of Appeal on 8 May 1992.

of inter-party waiver of liability incorporated in the standard McDonnell Douglas contract had only been valid between McDonnell Douglas and Hughes and the Indonesian government was not party to the waiver of liability.[83] The case, which ended up in lengthy six-month trial where the parties incurred enormous litigation costs,[84] illustrates how the inter-party waiver of liability clause in contract may not be enforceable on a third party. A liability disclaimer vis-à-vis third parties in a sense constitutes a one-sided approach, as such a disclaimer is not agreed upon by the potential claimant and hence the disclaimer may not be enforceable. The *Lexington* case demonstrates exposure of liability of data providers and other concerned parties in the supply chain of satellite data with respect to any third party outside contractual relations. This issue invokes the need for a clear regime concerning liability vis-à-vis a third party.

Case law highlights a certain trend regarding the enforceability of liability disclaimers. To summarize, liability disclaimers tend to be enforceable when (1) the two parties in dispute need to be in a contractual relation; (2) the two parties have equal bargaining power; (3) the language of the liability disclaimer is unambiguous; and (4) the damage does not result from deliberation or gross negligence.

The current practice amongst data generators regarding data use contracts by no means properly addresses the potential liabilities of the parties to those contracts. One can argue that the contractual provisions are generally written much more favourably for the side of the suppliers. Indeed, there are many layers of defence against liability available to the suppliers of remote sensing data. Contractually, there is little language in the distribution agreements upon which data recipients can rely to support their claims in case of damage. Considering the nature of the contractual provisions, it is more likely that the data suppliers can recover damages on the ground of breach of contract, than data recipients can recover damage for unsatisfactory performance of data providers. Data suppliers only need to demonstrate that data recipients have committed any of the long lists of prohibited activities written into the contract. In contrast, it is difficult for data recipients to demonstrate a breach of contract because little is guaranteed by the data providers under the contract; hence, the interests of data recipients are not safeguarded sufficiently. The role of the legal regime to provide a solid compensation channel to the user community is therefore essential.

On the other hand, the examination of cases has highlighted that the enforceability of a variety of defences as well as exoneration clauses is not guaranteed.

---

[83] See L.S. Kaplan, Recent Developments in Space Law Litigation, *Research and Invention in Outer Space, Liability and intellectual Property Rights,* Martinus Nijhoff Publishers, London, 1995, 123.

[84] See I. Awford, Commercial Space Activities: Legal Liability Issues, *Recent Trends in International Space Law and Policy,* Lancers Books, New Delhi, 1997, 414.

The extensive use of the liability disclaimer by suppliers may discourage potential claims, but once action is brought, the outcome is rather unpredictable. Mere inclusion of an exoneration clause does not eradicate all potential risks for liability. Indeed, gaps exist between the practice of satellite data suppliers and the general rules that enforce liability disclaimers, particularly as to: (1) a liability disclaimer without regard to the bargaining power of the data recipient, and (2) a third party disclaimer. The problem for data suppliers lies in the fact that a liability disclaimer is not likely to be enforceable in certain situations.

In disputes between suppliers and end-users, problems may arise with respect to certain end-users. The target for end-user agreements includes individuals, who are not considered to have equal bargaining power vis-à-vis suppliers, and in such cases a liability disclaimer is not likely to be enforceable.[85]

With respect to suppliers and subcontractors, liability disclaimers are likely to be enforceable because the parties are in equal bargaining positions in these commercial settings. With respect to liability disclaimers between contractors and subcontractors, it is possible to have an inter-party waiver of liability amongst the entities in the supply chain if the concerned parties so agree. Such inter-party liability precludes the defendant from laying the blame on the other relevant parties in the supply chain. Inter-party waiver of liability minimizes the litigation risks only amongst the supply chain, and does not properly resolve the issues of risk allocation and assignment of liability if the claim is brought by an external third party.

With regard to suppliers and third parties, despite the existence of an exoneration clause, any unilateral waiver is likely to be invalid, as discussed in Chapter 4, under the Disaster Charter's waiver of liability.

The enforceability of a liability disclaimer would impact both the claimant and defendant. These uncertainties are an impediment not only to data recipients, but also to suppliers. The data generators' attempt to exonerate themselves from any potential liability by relying on the contractual liability disclaimer may fail, resulting in unnecessary litigation costs when lawsuits are brought against the data supplier. In the end, uncertain liability leaves both parties unprepared for the risks of bearing unforeseen costs; therefore, it is necessary to clarify the rules and establish criteria articulating when liability disclaimers are enforceable or not, so that potential recovery for damage is predictable for both data recipients and data providers.

The legality of specific defences, as well as liability disclaimers and conditions for enforceability, are three essential issues that need to be addressed under

[85] Regardless of types of data recipients – such as professional users or knowledgeable users – licenses have a standard form. As mentioned in the previous Chapter, the only distinctions made in the licenses are1) whether recipients are end-users or value added resellers, and 2) whether recipients are single users or multiple users sharing the same data.

an improved regime. It is necessary to establish the fundamental rules of liability and clarify when a data supplier is and is not liable and to identify the legal defences available to data suppliers. Third-party loss, unaddressed under the branch of contractual liability, is examined in the next section discussing the general tort law framework.

6.3.2. *Tort Law*

In considering a potential remote sensing legal regime, third-party loss is an inevitable element. Before examining how tort liability should be introduced to the regime of remote sensing, it is necessary to understand the meaning of the term 'tort'. Tort is "a civil wrong, other than breach of contract, for which a remedy may be obtained, usually in the form of damages."[86] Unlike obligations of parties who chose to be bound through contracts, the duties imposed on the parties in a tort case are mandatory for all in that jurisdiction. In contrast to contractual liability, where a breach of any contractual provision gives rise to liability, tort liability is invoked if damage is caused to others, whether there is a contract or not.

6.3.2.1. *The Concept of Tort Liability*

Under a contract, a breach of the contract terms is the key element that gives rise to liability, whereas under tort, it is the harm actually incurred by the claimant that gives rise to liability. Tort law, like contract law, is governed by national frameworks, although there has been an effort to establish basic principles of tort in an international context such as the European Principles on Tort Law. Tort liability is defined in Black's Law Dictionary as "liability that arises from the breach of a duty that (1) is fixed primarily by the law, (2) is owed to persons generally, and (3) when breached, is redressable by an action for unliquidated damages."[87] As noted above, tort matters associated with remote sensing are addressed under the limited scope of the Liability Convention and the UN Remote Sensing Principles. This section undertakes an analysis by referring to examples of existing tort regimes in a national context, based largely on US and UK examples, with an examination of the scope of tort liability as it applies to contractual parties and third parties, including liability with and without negligence.

Tort law covers claims in any setting under all circumstances if damage is done: the parties in dispute include contractual parties and between third parties. In the British case of *Junior Books v. Veichi*,[88] concerning the claim

[86] Black's Law Dictionary, Ninth Edition, 2009, 1626.
[87] Black's Law Dictionary, Ninth Edition, 2009, 998.
[88] *Junior Books v. Veichi*, 1983 A.C.520.

between a contractor and sub-contractor, the finding of tort liability was based on the contractual reliance between the plaintiff and the subcontractor who laid a floor defectively in a factory built for the plaintiff. Tort law may be invoked despite the presence of a contract, as in the earlier discussed case of *INTELSAT v. Martin Marietta*, where INTELSAT claimed both gross negligence of Martin Marietta and its breach of contract.

Tort law deals with claims of third parties vis-à-vis the tort-feasor – one who commits a tort, a wrongdoer, who is outside contractual relations. The Liability Convention as well as the Rome Convention are good examples of international documents addressing third-party loss where there is no contractual relation between the victims and the parties responsible for causing damage.

The scope of tort liability includes damage resulting from mere action or omission, as well as damage resulting from fault or negligence.[89] The elements necessary for tort liability in negligence are: breach of duty of care; loss or damage to the interests of another person; and a reasonable close causal connection between the conduct and the resulting loss or damage.[90] In a negligence action, claimants must demonstrate that the defendant has been negligent in act or omission, and that such negligence caused damage.

The critical question concerns the boundary between when there is liability for damage without fault and when proof of negligence is required. This issue will be further explored later in this Chapter.

6.3.2.2. *Remedies*

As illustrated by most of the cases, the European Principles on Tort Law provide monetary compensation as follows: "Damages are a money payment to compensate the victim, that is to say, to restore him, so far as money can, to the position he would have been in if the wrong complained of had not been committed. Damages also serve the aim of preventing the harm".[91] The Principles also recognise restoration under Art. 10 104: "Instead of damage, restoration in kind can be claimed by the injured party as far as possible and not too burdensome to the other party". The Rome Convention provides remedies to the third parties as long as the damage is a direct result of the incident.[92]

---

[89] The French Civil Code's Article 1383 adds that everyone is liable for the harm which he has caused not only by his act, but also by his negligence or by his carelessness.

[90] See C.F. Prosser, and W.L. Keeton, *Prosser and Keeton on the law of torts*, 5th edition, 1991, 164.

[91] Art. 10 101 of the European Principles on Tort Law.

[92] Article 1 of the Rome Convention states that: "Any person who suffers damage on the surface shall, upon proof only that the damage was caused by an aircraft in flight or

6.3.2.3. *Defence Available to Tort Liability*

In a certain sense, although not precisely, defences available under tort overlap those available under contract, including Act of God/*force majeure*; contributory negligence; and the specific tort defence using the Good Samaritan Principle.

As in the defence for contractual liability, a defence based on Act of God/*force majeure* is available under tort. International conventions on third-party loss caused by aircraft such as the Convention on Damage caused by Foreign Aircraft to Third Parties on the Surface usually incorporate *force majeure* provisions. In the British case *Nicolas v. Marsland,*[93] the defendant who created some artificial ornamental lakes by damming up a natural stream was not held liable when a violent rainfall caused the embankments to collapse, destroying four nearby bridges. Incidents beyond the control of a responsible party are conceivable and should be available as a legitimate defence.

An example of contributory negligence in tort can be well illustrated in the case of *Stubley v. The London and North Western Railway Co,*[94] concerning the contributory negligence of pedestrians who were injured crossing the railway line. The court held that the railway was not negligent in failing to take precautions against possible injury to people crossing a railway line by a footpath on the ground where they had the opportunity to confirm safety and that the company did not bear any duty to warn people.[95] Under the Liability Convention, a launching state may be exonerated from liability upon a showing of contributory negligence on the part of the claimant, as provided under Article VI(1).[96] Contributory negligence is a critical element to limit or exclude the liability of the defendant if the claimant is responsible, in whole or in part, for the damage.

As previously examined in Chapter 4, the Good Samaritan Principle is codified in US and Canadian law, and may serve as a defense against tort liability. The Good Samaritan Principle is applied to waive liability of the rescuer who attempts to help but inadvertently causes damage. It applies only when a rescuer is a Good Samaritan, that is, one who has no obligation to help. The Good

by any person or thing falling therefrom, be entitled to compensation as provided by this convention. Nevertheless there shall be no right to compensation if the damage is not a direct consequence of the incident."

[93] *Nicolas v. Marsland,* [1876] 2 Ex. D 1.

[94] *Stubley v. The London and North Western Railway Co* [1865-66]LR 1 Exch 13, 18.

[95] *Stubley v. The London and North Western Railway Co* [1865-66]LR 1 Exch 13, 18.

[96] "Exoneration from absolute liability shall be granted to the extent that a launching state establishes that the damage has resulted either wholly or partially from gross negligence or act or omission done with the intent to cause damage on the part of a claimant State."

Samaritan Principle does not apply to parties in contractual relations or rescuers who owe an obligation. The application and interpretation of the Good Samaritan Principle varies by jurisdiction: in most civil law countries, obligations to help persons in peril were established, whilst in common law countries duty existed in case of a special relationship. As mentioned in Chapter 4, the United States has considered adopting the Good Samaritan Principle specifically when dealing with natural disasters. A draft Good Samaritan Act in the United States proposed that volunteers are exempted from liability unless they have acted with intentional and gross negligence. The conditions for applicability of the Good Samaritan Principle should be harmonized and codified in the context of satellite remote sensing.

6.3.2.4. *Consideration for Applying Tort Law to Satellite Images*

The analysis shows that tort law may be used to address complex issues of liabilities, which involve a variety of parties including a supplier and an end-user, a supplier and a sub-contractor, a supplier and a third party, and between the third parties under different circumstances. A tort law approach is essential to address conceivable claims associated with satellite remote sensing – such as third parties suffering damage from the use of inaccurate data reported by the media. The role expected of an improved regime with regard to tort matters is to provide potential claimants with a legal recourse for consequential damage.

Unlike contractual liabilities, which have certain bases in the language of the distribution contracts by the data suppliers, tort liability associated with remote sensing has no legal basis whatsoever. It is necessary, therefore, to analyse the existing rules in the broader framework of tort law and draw a conclusion about how it could be applied to cases in the area of remote sensing.

The defences available to tort liability are highly relevant and should be incorporated into a potential satellite remote sensing legal regime. *Force majeure* occurs irrespective of contract or tort, and thus needs to be incorporated in the regime; in addition, contributory negligence involving third parties is critical to permit waiver and/or apportion the liability of the concerned parties outside contractual relations.

The Good Samaritan Principle, a defence specifically available in tort under certain jurisdictions, is necessary to address the important liability aspects involved in both public and humanitarian services. The Good Samaritan Principle is, in theory, applied to the partners of the Disaster Charter; however its application raises important questions because the formal recognition for, or the interpretation of, the Good Samaritan Principle depends on jurisdiction. As discussed previously, the application of the Good Samaritan Principle to potential claims related to the Disaster Charter is uncertain, since the Charter partners make the prior arrangements to help afflicted states who request assistance; hence, it is necessary to formally establish the provision to exonerate service providers from liability associated with voluntary humanitarian assistance based on remote sensing so as to protect good-will rescuers against liability risks and create incentives to help the states affected by disasters.

In addition to the codification of certain available defences, the central question to consider is to determine on what basis liability arising from remote sensing activities should be addressed. To that end, certain issues need closer examination, such as elaboration of the concepts of intentional tort, gross negligence, and negligence apart from gross negligence.

6.4. The Analysis of Different Bases of Liability

Currently, the outcome of disputes concerning remote sensing activities that are not addressed by contract law remains largely unpredictable. To address the question of determining the basis upon which the defendant should be responsible, this section examines two different bases of liability, namely, strict and fault-based liability, and liability for product or service, including the legal trends in each category of liability. Strict liability places liability regardless of fault; fault-based liability is invoked only when parties causing damage are proven to be at fault. Products and services give rise to liability on different bases: product liability is imposed on a strict basis, whilst service liability is determined under a fault-based liability standard. In almost all international and national legal systems, there is a distinction between strict and fault-based liability.

6.4.1. *Strict Liability*

In strict liability, the claimant only needs to demonstrate the damage and the causation of the damage, without the necessity to prove any negligence on the part of the defendant.[97] The absolute/strict liability regime has been adopted internationally or nationally by many states, particularly for dangerous operations and/or activities accompanied by a high risk of accidents that result in direct damage, including death and personal injury. Such operations include those involving nuclear installations,[98] the launching of a space object,[99]and

[97] See F.P. Schubert, An International Convention on GNSS Liability: When does Desirable Become Necessary?, XXIV *Annals of Air and Space Law* (1999), 261.

[98] For instance, the 1963 Vienna Convention on Civil Liability for Nuclear Damage 2 *ILM* 727, and 1960 Paris Convention on Third Party Liability in the Field of Nuclear Energy, 956 *UNTS* 251.

[99] See Article II of the Liability Convention

international air transport,[100] where rules have been formulated into international conventions.

In the earlier examination of the Liability Convention, two distinct bases of liability for damage were introduced: 1) damage on the surface of the Earth or to aircraft in flight; and 2) damage caused elsewhere than to the Earth surface or to aircraft.[101] This section analyses closely what elements determine the application of those two bases. Adoption of a two-tiered liability regime (depending upon the geographical location where damage was caused) appears to reflect the difference in probability of damage as well as the helplessness of victims in the two settings.

The Liability Convention contains a totally victim-oriented position for damage on the Earth's surface, taking into consideration the helplessness of victims in the sense that launching space objects is extremely hazardous and, in the case of falling satellites, victims can do little to avoid devastating impact. It does not matter whether the launching state let the space object fall intentionally or accidently, or that negligence was involved in the operation. The responsible state parties must compensate if damage on the Earth's surface is actually sustained by the fall of the space object. As mentioned in Chapter 2, only one incident of such circumstances has taken place thus far.

In 1978, Cosmos 954, a Russian nuclear powered satellite, disintegrated in space and its debris was precipitated on Canadian territory. As a result, although Canada claimed that the incident was caused by the faulty motor of the satellite, the fault of Russia in letting Cosmos 954 fall on Canadian territory was not the central question. Rather, the point for negotiation was the extent of harm and the clean-up operations needed as well as the cost thereof.[102] After lengthy negotiations, in accordance with the diplomatic channels available under the Liability Convention and other international law, Canada requested compensation and the dispute was settled when Russia paid $3 million to Canada on an

---

[100] Warsaw Convention for the Unification of Certain Rules Relating to International Transportation,137 *LNTS* 11, 1933 International Convention for the Unification of Certain Rules Relating to Damage caused by Aircraft to third Parties on the Surface; 1952 Rome Convention on Damage caused by Foreign Aircraft to Third Parties on the Surface 310 *UNTS* 181.

[101] Article II of the Liability Convention states that: "Launching State shall be absolutely liable to pay compensation for the damage caused by its space object on the surface of the earth or to aircraft in flight."

[102] See A.F. Cohen, Cosmos 954 and the International Law of the Satellite Accidents, 10 *Yale Journal of International Law* (1984), 78–91. For instance, the Soviet Union declared that the remains of Cosmos 954 posed a minimal radiation hazard, whilst Canada argued that fragments found are lethal in terms of radioactivity. The Soviet Union believed that their duty to compensate was limited to the incremental costs for repair, whilst Canada claimed that the Soviet Union should bear the total costs of cleaning up the debris.

ex-gratia basis in 1981.[103] Although this case was settled without litigation, it created a precedent for the applicability of strict liability for damage caused to the Earth's surface by a space object. It is clearly the hazardous nature of activities accompanied by high likelihood of accidents – as opposed to helplessness of victims and their inability to avoid such damage – that make the launching of space objects an appropriate subject for an absolute/strict liability regime.

6.4.2. *Fault-Based Liability*

Fault-based liability holds defendants liable only if they are proven to be at fault. The Liability Convention has a less victim-oriented approach for damage caused in locations other than on the surface of the Earth. In contrast to damage on the Earth's surface, how the damage occurred and any blameworthiness of the concerned parties will be taken into account. On this basis, whilst a launching state would not be liable for an inevitable accident, it would be held liable for damage caused intentionally (e.g., changing the orbit of a satellite deliberately to cause a collision), or with fault (e.g., negligent loss of control of a satellite leading to crash). Cohen contends that it is reasonable to apply fault-based liability, since launching states have the ability to reduce the risk of collisions and owe a mutual responsibility to try to avoid injury amongst satellite operators, in contrast to strict liability applied to victims on the ground, who can do little to protect themselves against a fall of a satellite. [104]

Damage caused in outer space had long been a theoretical matter; however, it became reality when the US and Russian satellites collided in orbit on 10 February 2009.[105] The US Iridium 33, a commercial telecommunication satellite and Cosmos 2251, a defunct Russian military telecommunication satellite, collided at the altitude of 790 km and both of them were destroyed. It was the first incident of its kind and the probability of such collision is extremely low. Whilst Iridium 33 was part of the Iridium satellite phone constellation of 66 satellites, Cosmos 2251 had ceased its operation around 1996. In that respect, the damage, at least from the economic point of view, was incurred solely to Iridium – even if it did not result in the interruption of service to its customers.

[103] M. Whincup, *Contract Law and Practice*, Kluwer Law International, Hague, 2001, 206.

[104] See G.H. Reynolds & R.P. Merges, *Outer Space Problems of Law and Policy*, Second Edition, Westview Press, Colorado, 1997, 188.

[105] See generally, Iridium 33-Cosmos 2251 Collision, Fact Sheet, Secure World Foundation, http://www.secureworldfoundation.org/siteadmin/images/files/file_273.pdf. See also D. Wright, Colliding Satellites: Consequences and Implications, available on the Internet at: http://www.ucsusa.org/assets/documents/nwgs/SatelliteCollision-2-12-09.pdf.

It is rather difficult to establish fault. In absence of a definition, it is not easy to establish fault of both parties. Whilst it is not impossible to argue that Iridium 33 could have changed its orbit to avoid the collision, the prompt removal of a defunct Cosmos 2251 from orbit would have avoided the collision. This incident gave rise to the need to elaborate the term 'fault' under the Liability Convention.

As already mentioned in the discussion of tort law, it is critical to gain better understanding of the point at which fault is established. The different levels of fault are an indicator to delineate when the party is definitely held responsible. A line can be drawn between one category that consists of intentional tort and gross negligence, and the other category of ordinary negligence. The general rule is that defendants cannot be exonerated from liability when they act with intent or gross negligence, as seen in the contractual context – even with the presence of a liability disclaimer in the contract. This section attempts to clarify the concepts of intentional tort, gross negligence, and negligence, so as to set standards for establishing fault associated with satellite remote sensing activities.

The intentional tort and gross negligence are recognized in different conventions[106] and case law serving as a critical element to invalidate a defendant's exoneration from liability or to limit the liability of a claimant in contributory negligence cases. An intentional tort is a tort committed by someone acting with general or specific intent. Intentional tort can be found for example, in the Warsaw Convention, under the term 'wilful misconduct' as referenced under Article 25. Article 25 states that "The carrier shall not be entitled to avail himself of the provisions of this Convention which exclude or limit liability if the damage is caused by his wilful misconduct or by such default on his part as, in accordance with the law of the Court seized of the case, is considered to be equivalent to wilful misconduct."[107] This wilful misconduct is defined in *Horabin v. British Overseas Airways Corporation*[108] as: "to be guilty of wilful misconduct the person concerned must appreciate that he is acting wrongfully or omitting to act yet persist in so acting or omitting to act regardless of the consequences, or acts or omits to act with reckless indifference as to what the result may be."[109] Intentional tort is committed when one has the awareness of

---

[106] Article VI(1) of the Liability Convention and Article 25 of the Warsaw Convention.

[107] *Horabin v. British Overseas Airways Corporation,* [1952] 2 All E.R. 1016 (Q.B. 1952 Article 25 of the Warsaw Convention.

[108] *Horabin v. British Overseas Airways Corporation,* [1952] 2 All E.R. 1016 (Q.B. 1952).

[109] *Horabin v. British Overseas Airways Corporation,* [1952] 2 All E.R. 1016 (Q.B. 1952), 1022E.Wilful negligence was also established in a British case *Wilkinson v. Downton* [1897] 2 Q.B. 57.

potential harm as a consequence of particular action or omission and yet deliberately commits such action or omission. Under such circumstances, one cannot be exonerated from liability. One can recall the Kansas case concerning misrepresentation of an image as an intentional tort committed through deliberately misrepresenting the aerial photography provided to be a satellite image.

Moving on to the concept of gross negligence, generally speaking, reckless behaviour appears to constitute gross negligence. For instance, in the French case *Lamberth v. Guiron*,[110] an accident occurred in which a helicopter crashed while flying under the cables of an aerial railway, killing all on board. The court held that it was impossible for the pilot not to realize that he was near a very dangerous installation.[111] Likewise, in the case of *Diop v. Cie Air France*,[112] the pilot attempted to land under bad weather conditions and the plane crashed. Repetitive failure can also be construed as gross negligence. A French court found the City of Paris liable for its gross negligence in consistently failing to enforce a parking prohibition where the parked cars impeded access to a narrow road.[113] In these ways, one can draw a conclusion that damage resulting from reckless behaviour and/or repetitive failure could constitute gross negligence.

In the context of national legal frameworks, ordinary negligence leaves room for exoneration from liability. The elements to establish negligence are complex and have to be examined through a number of case law examples. Fault-based liability holds the defendant liable in case of negligence by taking into account the elements such as the reasonableness of the defendant's behaviour, the foreseeability of the harm, and the cost-benefit analysis.

The foreseeability of the harm is applied to the case in such way that the defendant is at fault only if the damage was foreseeable. For instance, in the British case of *Bolton and Stone*,[114] a ball hit by the defendant went out of the ball field and onto the highway, injuring the plaintiff; yet the defendant was not held liable because the harm was not foreseeable.

The reasonableness of the defendant's behaviour leading to the damage was considered in the British case *Latimer v. AEC Ltd*,[115] when the plaintiff, a workman, injured himself when he slipped on the slippery factory floor after the factory was flooded. The House of Lords held that it had not been established

[110] *Lamberth v. Guiron*, (1963) 26 R.G.A. 185.

[111] See G. Miller, *Liability in International Air Transport*, Kluwer Publisher, Deventer, 1977, 208.

[112] *Diop v. Cie Air France*, (1966) 29 R.G.A.E. 42 (T.G.I. Seine, 6 November 1964).

[113] See G. Monti, Osman v. UK-Transforming English Negligence Law into French Administrative Law? 48 *International and Comparative Law Quarterly* (1999), 766.

[114] *Bolton and Stone*, [1951] AC 850.

[115] *Latimer v. AEC Ltd*, [1953] AC 643, [1953] 2 All ER 449, HL.

that a reasonably careful employer ought to have taken the drastic step of closing the factory.[116]

The cost-benefit approach was developed by the judge as a measure to determine negligence in deciding *United States v. Caroll Towing Co.*[117]: duty exists if the cost of preventing the damage is smaller than the probability of its occurrence, multiplied by the expected amount of losses. Although it is difficult to quantify each required element, this method can provide a certain objective standard to determine liability.

6.4.3. *Consideration for Applying Fault-Based Liability to the Remote Sensing Activities*

Having examined the issues of strict liability and fault-based liability, the answer to the question of which basis is more suited to remote sensing appears to be obvious. Victims that incur damage arising from remote sensing are not as helpless as those on the Earth's surface who become victims as a result of the crash of space objects or airplanes. Indeed, it is likely that users of satellite data may be able to rely on alternative source of information.

From the standpoint of remote sensing activities – including data collection, generation, and supply, and use of the remote sensing data – these are not hazardous. The probability of damage associated with satellite remote sensing is much lower than the damage done on the Earth's surface. Even in case of damage, the nature of the damage is likely to be consequential rather than direct. From these reasons, one can conclude that fault-based liability is an appropriate basis for a remote sensing liability regime.

Elements considered under fault-based liability such as foreseeability of the harm, reasonableness of the defendant's behaviour, and the cost-benefit approach, are highly relevant to decisions in remote sensing legal disputes. A fault-based regime would allow courts to determine which cases concerning the misuse and misinterpretation of data involve intentional conduct, and which do not, and can further assist the court to determine whether unintentional misinterpretation or misuse of the data was within or beyond the reasonable limit under the particular circumstances. An approach considering a number of elements is essential to apply to the cases related to remote sensing activities, because uncertainties and complexities exist in the field. In addition, circumstances under which data is supplied are different for emergency humanitarian services than for commercial services, and these contrasts need to be taken into consideration.

---

[116] See R.A. Buckley, *The Modern Law of Negligence*, Butterworths, London, 1999, 42.

[117] *United States v. Caroll Towing Co.*, 350 U.S. 61(1955).

The examination of tort law in this section focused on the overall sphere of liability and did not provide details as to the circumstances under which data suppliers should be liable with respect to data recipients. Certain elements need to be taken into consideration: for instance, data recipients who have equal knowledge as satellite data suppliers and those who do not have such equal knowledge should not be treated the same.

There are certain categories of liability that touch upon these issues more comprehensively, warranting a discussion that considers a potential regime for remote sensing. These issues are explored in the next section with a different approach to the basis of liability: product liability – or service liability – focusing on the unequal bargaining power between the supplier and the consumer.

6.4.4. *Liability for Products*

An examination of the product liability regime is useful to clarify when and how data providers are liable, as they control duties and responsibilities owed by the entities that produce and provide those products. Product liability gives us certain standards to determine under what conditions the parties in the supply chain are liable. In some national systems, the laws specialising in product and service liability exist, but the main focus of this Chapter will be on European law.

The application of a strict basis for product liability is established in US case law including *La Rossa v. Scientific Design Co.*,[118] as well as stipulated in the EC Directive concerning Liability for Defective Product, amended by Council Directive 1999/34/EC[119] and the Draft Directive on the Liability of Service Providers, respectively. The analysis will refer also to certain parallels with GNSS that can be drawn with the application of law and nature of damage.

Product liability is the liability of producers of goods in the event those goods cause death or injury or damage to private property.[120] The Product Liability Directive adopted and binding in the Member States of the European Union serves as the significant guideline to address the issue of product liability. The fundamental purpose of the Product Liability Directive is to protect consumers from unsafe products.

---

[118] *La Rossa v. Scientific Design Co.*, 402 F. 2d 937 (1968).

[119] Council Directive 85/374/EEC on the approximation of the laws, regulations, and administrative provisions of the Member States concerning liability for defective products (OJ 1985 L 210 /29). Amended by Directive 1999/34/EC of the European Parliament and of the Council of 10 May 1999. [Hereafter: Product Liability Directive]

[120] See A. Geddes, *Product and Service Liability in the EEC,* Sweet and Maxwell, London, 1992, 1.

The concept of product liability was first established in *MacPherson v. Buick*,[121] where the plaintiff was seriously injured in an accident caused by the broken drive shaft of a car. The defendant was held responsible for damage caused by the dangerous product it provided to the plaintiff. In *Greenman v. Yuba Power Products*[122] the application of the strict liability regime to a product was established: the damage was caused by a specific defect and proof of negligence on the part of the producer was not required.[123] Product liability applies only between parties with unequal bargaining power, such as a producer and a consumer, thus excluding the commercial settings of parties with equal bargaining power.[124]

Article I of the Product Liability Directive stipulates that: "The producer shall be liable for damage caused by a defect in his product."[125] This means that there is no requirement of fault before a producer will be held responsible if the defects of its product caused the damage. 'Product' is defined as: "all movables even if incorporated into another movable or into an immovable" and "includes electricity."[126] In theory, therefore, a wide range of products including satellite images and derived products are considered to fall within the scope of the Product Liability Directive.

What constitutes a defect under the Product Liability Directive is provided by Article 6 as "A product is defective when it does not provide the safety which a person is entitled to expect, taking all circumstances into account, including: (a) the presentation of product, (b)the use to which it could reasonably be expected that the product would be put; (c)the time when the product was put into circulation."[127] Circumstances could be represented, for example, by "(…) a trailer connection that fractures after a year of use owing to metal fatigue" which will be considered as defective.[128]

Safety is not defined further under the Produce Liability Directive and case law must be consulted.[129] One can easily assume some products, including food

[121] *MacPherson v. Buick* 217 N.Y. 382, 111 N.E. 1050 (1916).

[122] *Greenman v. Yuba Power Products* 59 Cal. 2d 57 (1963).

[123] See H. Karasawa, *International Transaction*, Dobunkan Publisher, Tokyo, 1996, 177.

[124] See *Kaiser Steel Corp. v. Westinghouse Corp*, 55 Cal. App. 3d 737; 127 Cal. Reptr. 838(1976), See also Diederiks-Verschoor, I.H.Ph., Product Liability and Insurance on Aircraft and Satellite, *The utilization of the world's air space and free outer space in the 21st century: proceedings of the International Conference on Air and Space Policy, Law and Industry for the 21st Century*, 23-25 June 1997, Seoul, 198.

[125] Article 1 of the Product Liability Directive.

[126] Article 2 of the Product Liability Directive.

[127] Article 6 of the Product Liability Directive.

[128] A. Geddes, *Product and Service Liability in the EEC*, Sweet and Maxwell, London, 1992, 32.

[129] US legislation related to product liability provides a more concrete definition: The Second Restatement of the Law of Torts (There is no unified federal law on

products and vehicles, can be unsafe if they are made defectively. For instance, manufacturers of airplanes are sued on the basis of strict liability;[130] a restaurant was held liable for serving poached turbot which contained botulism even though the court accepted that the restaurant could not have detected or removed the offending organism.[131] The user's expectation of safety appears to be that using or consuming the product in a reasonable manner will not generate direct damage.

It is important to note that the EC Product Liability Directive invalidates liability disclaimers under Article 12.[132] In the setting where parties have unequal bargaining power, the mere inclusion of the exemption clauses, as examined in the previous section, is not likely to exonerate the data suppliers from liability vis-à-vis data recipients. There are certain defences available under the Product Liability Directive as stipulated in Article 7, for example: if the producer proves that damage was due to a defect that did not exist at the time of circulation of a product;[133] if the state of scientific and technical knowledge available did not allow discovery of the defect;[134] if damage was due to a design defect;[135] if the producer did not put the product into circulation; and if a

---

product liability in the US and the Restatement of Torts serves as a general guideline). 402A states: "One who sells any product in a defective condition unreasonably dangerous to the user or consumer o to his property is subject to liability for physical harm."

130 See for example, in the case of crash of Turkish Airlines DC-10 on 3 March1974 leading to the death of 346 passengers, McDonnell Douglas was held liable. Diederiks-Verschoor, I.H.Ph., Product Liability and Insurance on Aircraft and Satellite, *The utilization of the world's air space and free outer space in the 21st century: proceedings of the International Conference on Air and Space Policy, Law and Industry for the 21st Century, 23-25 June 1997,* Seoul, 198.

131 TGI Poitiers 7, 7 January 1969.

132 Article12 of the Product Liability Directive states that "The liability of the producer arising from this Directive may not, in relation to the injured person, be limited or excluded by a provision limiting his liability or exempting him from liability."

133 Article 7(b) of the Product Liability Directive states:

"The producer shall not be liable as a result of this Directive if he proves: (b) that, having regard to the circumstances, it is probable that the defect which caused the damage did not exist at the time when the product was put into circulation by him or that this defect came into being afterwards."

134 Article 7(e) of the Product Liability Directive states: "The producer shall not be liable as a result of this Directive if he proves: (e) that the state of scientific and technical knowledge at the time when the producer put the product into circulation was not such as to enable the existence of the defect to be discovered."

135 Article 7(f) of the Product Liability Directive states: "The producer shall not be liable as a result of this Directive if he proves: (f) that in the case of a manufacturer of a component that the defect is attributable to the design of the product in which the component has been fitted or to the instructions given by the manufacturer of the product." The example of a case where plaintiff producer was not responsible for the alleged design defect leading to an accident was demonstrated in *Lexington v.*

producer did not manufacture the product for sale or for business use.[136] The state of scientific and technical knowledge is particularly relevant to remote sensing activities and shall be examined further in detail later in this Chapter.

How a satellite image can be considered to be defective under the Product Liability Directive warrants a discussion. The liability associated with the satellite images can be roughly divided into 1) the damage arising from the unsuitable use of products and, 2) the use of products of an insufficient quality, including misinterpreted images.

It is conceivable that one set of satellite data may be fit for several different purposes, such as map making, environmental studies, assessment of impact to the state of the environment, and the detection of changes. Furthermore, the image can be fit for a particular purpose yet may not bring the desired results to the users. The position of the data suppliers is that it is the responsibility of customers to determine whether the particular data are fit for their intended purpose.[137] One can easily assume that the intention of the data supplier is that the quality of the product is not expected to meet the needs of all potential uses within the reasonable sphere. Consequently, even if users incur damage from the reasonable use of products, particularly outside the declared purpose for their purchase, suppliers will not be liable for the results that may be undesirable to users.

The Product Liability Directive refers to the safety expected of a product,[138] but does not set concrete standards to determine the defect in other terms,

---

McDonnell Douglas, where negligent design of the perigee kick motor, an identical motor, which was used and burned in the earlier incident, resulted the failure placement of Indonesian telecommunication satellite Palapa B-2.

[136] Article 7(c) of the Product Liability Directive states: "The producer shall not be liable as a result of this Directive if he proves: (c) that that the product was neither manufactured by him for sale or any form of distribution for economic purpose nor manufactured or distributed by him in the course of his business."

[137] For example, Article 4.3 of Spot Image standard license stipulates that "it does not warrant that the product shall meet the end user's requirements or expectations, or shall be fit for user's intended purposes. There are no express and implied warranties of fitness or merchantability given in connection with the sale or use of the product." DMCii is more explicit in exonerating their liability in this regard stating in Article 12.1 of terms and conditions that "Customer is fit to define the characteristics of his needs and suitability of the data or Products to their satisfaction, as DMC shall provide no advice or recommendation whatsoever in this respect. It shall under no circumstances be held responsible as regards the suitability of these data or products concerning any use whatsoever." See, Terms and Conditions for the sale of DMC satellite Data or products, DMCii, August 2007.

[138] Article 6(1) of the Product Liability Directive states that: "A product is defective when it does not provide the safety that a person is entitled to expect, when taking all circumstances into account, including: (a) the presentation of the product; (b) the use to which it could reasonably be expected that the product would be put; (c) the time when the product was put into circulation."

particularly as to the quality of a product. The only provision referring to quality is Article 6.2 of the Directive, which states that "a product shall not be considered defective for the sole reason that a better product is subsequently put into circulation."[139] This provision excludes a defect based on its relative quality from the scope of the Directive; hence, even if customers could find products with a better quality elsewhere, it will not be considered a defect on such grounds. Indeed, no entities warrant that products will meet customer's needs or expectations.[140]

In terms of a warranty, suppliers provide minimal warranties to accompany satellite data. Indeed, the warranty provided by the suppliers is generally limited in terms of data period of validity and data content.[141] The provision of a warranty by no means guarantees the quality of the information content.

It may be interpreted that unreadable information is defective; for example, the satellite image with a 90 % cloud cover or extremely unfocused image. Although it can be argued that misinterpreted satellite images may be consequentially unsafe if, by someone relying on them, they could lead to death or personal injury (for example, during an evacuation instruction for a population affected by a disaster), it is difficult to see how an inaccurate image could be classified as defective under the strict liability regime. In fact, data suppliers consider a defect to be valid only if it is apparent.

The stance of data suppliers is that their product is defective if the data transfer medium is broken, which basically means if the information is unreadable.[142] A defective satellite image is more associated with direct damage to a person and/or property: for example the crash of a computer after putting in the CD-Rom containing the satellite image.[143] Furthermore, if the product is not in compliance with the specification – such as the maximum cloud cover, the location of the image, and the angle of the sensor agreed between suppliers and recipients in advance – such a product would be replaced. The data supplier would not be responsible, however, for the consequential damage incurred to the customers as a result of using of their products.

---

[139] Article 6(2) of the Product Liability Directive.

[140] See for instance, Euromap and Spot Image.

[141] An example of a warranty provided by MacDonald Dettwiler (MDA) Geospatial Services Inc: "MDA Geospatial Services Inc. warrants that the storage media on which the Data is supplied to Licensee is free from defects in materials and workmanship under normal use and service for a period of thirty (30) calendar days from the Licensee's receipt of product." See Article 3 of the Radarsat-2 End User License Agreement, LI-11525-6 V1.6 01OCT2009.

[142] Article 4.2 provides that "in case of medium on which the product is supplied by Spot Image to the end-user is deficient, as demonstrated by the end user and accepted by the Spot Image, Spot Image shall replace the said medium."

[143] Interview conducted with Stéphane Dupont of Spot Image on 12 January 2006.

One can argue that it is unsuitable to apply the standards expected of products under the strict liability regime to a satellite image. A fundamental reason lies in the nature of a satellite image as a product.

6.4.4.1. *The Analysis in Association with Liability for Intellectual Products*

A satellite image is not the type of product typically considered under the strict liability regime such as food, tools, machines, and vehicles. Considering the nature of satellite images, it is useful to introduce a special product category, known as "information products"[144] or "intellectual products."[145] These products convey information, such as books, maps and computer software, and are unlikely to cause direct damage; but reliance on the false information contained therein may lead to consequential damage.[146] A distinction should be drawn between the intellectual contents of a product and the product itself.[147]

For information products, it is evident that the risk of damage associated with such products does not rest with a product itself, but with the intellectual content of the product. Much of the conceivable damage would result from user/consumer reliance on the content of satellite data and derived products, but no suppliers cover such defects in terms of the information content. It is worthwhile to note that the nature of satellite imagery and the damage it might cause tie in with satellite navigation, where reliance on a defective signal will give rise to consequential damage.

Bearing in mind the fundamental difference between information products and other products, the Product Liability Directive is unclear as to whether the strict liability regime should be applied to intellectual products. By definition, satellite images constitute a 'product' under the scope of the Product Liability Directive,[148] as well as under US law. Specifically, the Restatement of Torts 3rd stresses that a product has to be tangible,[149] but excludes information by itself in the definition of a 'product'.

[144] D. Nolan, Product Liability, *The Law of Tort*, Grubb, A. (Ed.), Leed Elsevier, 2002, 823.

[145] S. Whittaker, European Product Liability and Intellectual Products, 105 *Law Quarterly Review* (1989), 127.

[146] See S. Whittaker, European Product Liability and Intellectual Products, 105 *Law Quarterly Review* (1989), 823.

[147] See C.J. Miller & R.S. Goldberg, *Product Liability*, Oxford University Press, New York, 2004, 321.

[148] Article 2 of the Product Liability Directive states: "For the purpose of this Directive, 'product' means all movables even if incorporated into another movable or into an immovable."

[149] Product is defined as: "tangible personal property distributed commercially for use or consumption."

As far as the applicability of the strict liability regime is concerned, the view is that the Consumer Protection Act – the UK law implementing the Product Liability Directive – does not apply to damage resulting from the reliance on the inaccurate information provided through such products.[150] If there is no human intervention and the information causes the damage directly, then the law applies.[151]

Cases dealing with intellectual products generally appear to support the view that they do not fall under the strict liability regime. In the US case of *Wilhelm Winter v. G.P. Putnam's Sons*,[152] mushroom enthusiasts who became severely ill from picking and eating mushrooms after relying on information in a book could not recover against the book's publisher under a product liability theory on the grounds that the information contained in a book is not a product for the purposes of strict liability. It can be argued that subjective judgement of the reader contributed to the damage.

In the European Court of Justice case *Levob Verzekeringen OV Bank v. Staatssecretaris van Financiën*,[153] it was held that software is considered a service for tax purposes. Furthermore, there is a trend that French courts are considering that some 'products' sold through the Internet become, from the tax standpoint, 'services'.[154] The ECJ case and the trend in French court decisions, taken together, strongly support the view that satellite imagery constitutes a service.

Although a product liability regime may be applicable to defects inherent in a satellite, satellite imagery is an intellectual product and most sources of damage lie in the content of the information. It is more appropriate to consider satellite imagery just like satellite navigation: as an 'information provision' service accompanied by a physical product, rather than a product to which the strict liability regime is applicable.

6.4.4.2. *Defences Available to Product Liability*

As previously mentioned, defences that are of high relevance to remote sensing activities are provided under the Product Liability Directive, and this section analyses how they may affect the claims associated with satellite images.

---

[150] See S. Whittaker, European Product Liability and Intellectual Products, 105 *Law Quarterly Review* (1989), 823. This view is also supported by the US Restatement of Torts 3rd: Product Liability 19, states that in the case where the plaintiff's grievance is with the information and not with the tangible medium, most courts have appropriately refused to impose strict liability.

[151] See S. Whittaker, European Product Liability and Intellectual Products, 105 *Law Quarterly Review* (1989), 823.

[152] *Wilhelm Winter v. G.P. Putnam's Sons*, 938 F 2d 1033(1991).

[153] *Levob Verzekeringen OV Bank v. Staatssecretaris van Financiën*, CJCE, 27 October 2005 (affaire C-41/04).

[154] Interview conducted with Stéphane Dupont of Spot Image on 12 January 2006.

As has been discussed, the Product Liability Directive provides several defences available to producers under Article 7. Of particular relevance to remote sensing data is the defence on the grounds of the state of scientific knowledge, often referred to as the 'development risk defence'. Article 7(e) of the Product Liability Directive provides: "(...) the state of scientific and technical knowledge at the relevant time was not such that a producer of products of the same description as the product in question might be expected to have discovered the defect if it had existed in his products whilst they were under his control."[155] The provision offers a useful defence for manufacturers in case of risks developed over time with respect to technologically innovative products. Manufacturers of product may escape liability by relying on this defence if the state of knowledge available to the manufacturers at the time of the product's circulation did not enable the manufacturers to foresee the risk of the product, despite every precaution they took. With satellite remote sensing being a fast-evolving field, the development of, and changes in, scientific and technical knowledge is a key factor to take into account for damage associated with satellite data.

The state of scientific and technical knowledge associated with remote sensing and data handling operations is evolving in accordance with continuous research and development. It is more justifiable, for instance, to argue over uncertainties involved in the interpretation of a new sensor than a commonly used sensor. *Bolam v. Friern Hospital*[156] exemplifies this argument, as the court held that it will not expect the defendant to have anticipated future developments in knowledge or practice: he will be judged by the state of knowledge of the normal professional at the time of the alleged tort.[157]

Another critical issue in determining the applicability of the risk development defence is whether the knowledge was, indeed, accessible. It was established in *Commission v. United Kingdom*[158] that before the scientific and technical knowledge could be relevant for the purposes of defence, such knowledge must have been accessible at the time when the product in question was put into circulation.[159] In this case, the Advocate General approached the case by determining the accessibility of the knowledge in terms of objective factors – such as source of the knowledge, and the language in which it was given. It was considered unreasonable to expect a European product manufacturer to know of a study published in a local scientific journal in Chinese, and the

---

155 Article 7(e) of the Product Liability Directive.

156 *Bolam v. Friern Hospital*, 1957 1 WLR 582, 586.

157 See B. Markesinis & S. Deakin, *Markesinis and Deakin's Tort Law*, Oxford University Press, Oxford, 2003, 175.

158 *Commission v. United Kingdom*, Case C –300/95 [1997] ECR I—2649, I-2670.

159 *Commission v. United Kingdom*, Case C –300/95 [1997] ECR I—2649, I-2670, paragraph 29.

European Court of Justice rejected the view that a producer would be liable on the ground of one such person's discovery of a defect, irrespective of the country and language of the study.[160] This view can be applicable, for instance, to such specialized knowledge as information concerning a new sensor using an uncommon spectral band, or data handling-activities conducted in the closed facility of a particular country. Under such circumstances, the knowledge would not be considered as available to the general remote sensing communities. Establishing a reasonable limit of liability in terms of the risk development defence is a necessary element to be incorporated into an improved regime on satellite remote sensing.

6.4.5. *Liability for Services*

Since the analysis in the previous section contends that satellite imagery is closer to service, this section evaluates how the damage arising from remote sensing activities or products fits under the regime of service liability. It provides an overview of service liability, and thorough analysis using an analogy with the liability regime governing the services in the space field – in particular, GNSS.

6.4.5.1. *The Concept of Service Liability*

Liability for service is much less recognized and established than product liability. In order to understand how liability is applied to services, however, it is worthwhile to look at the Council Draft Directive on liability of suppliers of services.[161] Unlike the Product Liability Directive, the Draft Service Liability Directive has not been adopted thus far, although it could serve as a major reference to liability as applied to services.[162] Article 1(1) of the Draft Service Liability Directive provides that "the supplier of a service shall be liable for damage to the health and physical integrity of persons or physical integrity of movable or immovable property, including the persons or property which were the object of the service, caused by a fault committed by him in the performance of his service."[163] As it is clear from this provision, service liability is a

[160] *Commission v. United Kingdom*, Case C –300/95 [1997] ECR I—2649, I-2670, paragraph 24.

[161] Proposal for a Council Directive on the liability of suppliers of services COM (90) 482 final- SYN 308 OJ 1991 C12/8. [Hereafter: Draft Service Liability Directive]

[162] Service if defined under Article 2 of the Draft Service Liability Directive as "any transaction carried out on a commercial basis or by way of a public service and in an independent manner, whether or not in return for payment, which does not have as its direct and exclusive object the manufacture of movable property or the transfer of rights or intellectual property rights."

[163] Article 1(1) of the Draft Service Liability Directive.

fault-based liability for physical injury, as well as for material damage to property, with the burden of proof on the service provider. Article 1(3) of the Draft Service Liability Directive further elaborates that "In assessing the fault, account shall be taken of the behaviour of the supplier of the service, who, in normal and reasonably foreseeable conditions, shall ensure the safety which may reasonably be expected."[164]

Since draft legislation has no binding effect, one should consider approaching the issue of service liability from the standpoint of case law. The case law under different jurisdictions shows that service providers are held responsible only when they are proven to be at fault. For instance, in one French case a hairdresser had treated the customer with a bleaching agent that resulted in his suffering a skin complaint. The lower court's decision, which held the hair dresser liable, was reversed on appeal on the grounds that the hairdresser had used the bleaching agent in accordance with its instructions.[165] In a US case, the New Jersey courts held that strict liability was not to be imposed on a dentist for breaking a hypodermic needle in the plaintiff's jaw.[166]

From these cases, it is evident that the duty of care standard expected of the providers of services is lower than the standard for producers of products, which are more likely to directly affect the safety of users. In contrast to the strict liability applied for products, fault has to be established in order for a service provider to be held liable. The key element here is that a service provider is not presumed to be at fault, and a service provider can escape liability in the absence of fault even if the causal relationship between the performance of service and the damage is evident. Service liability considerations with regard to a finding of negligence involve not only whether duty of care existed but whether it was exercised to the reasonable extent expected under the particular circumstances.

6.4.5.2. *The Analysis in Association with Observation of Liability of GNSS Providers*

As has been mentioned previously, satellite remote sensing and GNSS share several common features. Both of them are global applications to provide communities worldwide with critical information based on space technology. Indeed, satellite remote sensing and GNSS can be used for the same applications, sometimes combined together to provide the useful value-added services to aid emergency rescue operations and to verify activities that are adverse to the Earth's environment. The consequential damage arising from the information generated by the operation of both systems is not a speculative matter.

---

164 Article 1(3) of the Draft Service Liability Directive.

165 Civ (1) 4 Oct, 1967, D 1967, 652.

166 See C.J. Miller & R.S. Goldberg, *Product Liability*, Oxford University Press, New York, 2004, 95.

Yet, complexities inherent in both operations and service provision channels involving affected communities over multiple jurisdictions have resulted in difficulties with establishing a satisfactory legal solution.

Since GNSS is ahead of satellite remote sensing in terms of both the number of studies that have been done and the degree of its legal development achieved thus far, it is useful to analyze the approach that is taken for a present regime surrounding GNSS as a reference for a potential legal regime on satellite remote sensing. These issues will be examined first through the overall liability regime applicable to GNSS, followed by that established for the specific programmes within GNSS.

First of all, a brief background of GNSS-related damage and liability should be given. GNSS application is based on a provision of signals from a constellation of satellites that help users to identify their location. As has been discussed, applying product liability to a defective service related to GNSS is not appropriate, as it would impose liability for the manufacture and or sale of a product by which, in the course of using it, damage has been caused.[167] Instead, a central concept for GNSS-related liability stems out of the operation itself, and the damage that arises as a result of such activity. The primary focus of this section will be on the liability arising from the operations and service provisions based on GNSS. For this analysis it is necessary to consider the characteristics of defective service that may result in damage to end users or other affected third parties.

The scope of defective services and the magnitude of damage is considerably large. Causes of damage include signal failure, signal delay, discontinuity of services, and transmission of false signals – all of which could result in economic loss, damage to property, and personal injury or death in the worst case.[168] All of these damages could result from three types of signal/services provided under GNSS: basic signal, augmented signal, and value-added services – each of which are provided by different categories of entities.[169] The complexity arises in the fact that the different causes of damage would give rise to distinct legal relations and varying consequential liabilities.

An overview of the liability regime surrounding GNSS is necessary to an analysis of how the different type of liability – product, contractual and tort liability – can be invoked amongst the different categories of supply chain, service providers, recipients, and third-party victims. Product liability is fairly

---

[167] See F.G. von der Dunk, Liability Analysis for GPS and Galileo, 30 *Journal of Space Law* (2005), 135.

[168] See L.J. Smith, Where is Paradise? The EU's Navigation System Galileo – Some Comments on Inherent Risks (or Paradise Lost), *Proceedings of the Fiftieth Colloquium on the Law of Outer Space*, 2008, 356.

[169] F.P. Schubert, An International Convention on GNSS Liability: When does Desirable become Necessary? XXIV *Annals of Air and Space Law* (1999), 150–151.

straightforward, as product liability can only be invoked against manufacturers. As to contractual liability, it can certainly be invoked by basic signal providers against augmented signal and value-added service providers; however, whether contractual liability can be invoked by service recipients depends on the programme and type of services. In principle, liability arising from all types of signals can be invoked by service recipients and third-party victims against any service providers on the basis of fault. Indeed, the position of the GNSS operator has been that the potential liability arising from GNSS is not placed on a strict liability basis.[170] This complex legal environment surrounding GNSS becomes even more complicated because the detailed application of liability differs by programmes and jurisdictions, most notably, the US GPS and European Galileo.

The United States makes available GPS signals free of charge to the global community; however, the US authorities deny the presence of a contractual relationship of any form with respect to a basic signal provision.[171] The United States has disclaimed any type of contractual liability for such open access. It would place a severe burden on the United States if it were to be liable for every single incident that may arise from its free service to the global community. On the other hand, the US does not deny the possibility for liability claims under US tort law.[172]

Tort liabilities were inferred from a letter sent to the President of the ICAO in 1991: "The United States shall take all necessary measures to maintain the integrity and reliability of the service."[173] Likewise, in a 1994 unilateral agreement, the US government agreed to accept liability at a national level, as the GPS provider, for damages caused by deficiencies in the provision of GNSS signals under the general liability regime governing Air Navigation Systems – that is, fault-based and unlimited liability.[174] Therefore, the United States may be liable only when it is at fault; however, there are limitations to the application of liability in terms of a geographical scope. Liability does not apply to

---

[170] For example, during the 28th Session of the ICAO Legal Committee many delegates thought that the United States and Russia were indirectly denying worldwide liability of damage caused by their primary navigation signals.

[171] See F.G. von der Dunk, Liability Analysis for GPS and Galileo, 30 *Journal of Space Law* (2005), 140.

[172] See F.G. von der Dunk, Liability Analysis for GPS and Galileo, 30 *Journal of Space Law* (2005), 140.

[173] Letter of D. Hinson to A. Kotaite December 1994 State Letter LE 4/49.1- 94/89. Cf. K. Henaku, The International Liability of the GNSS Space Segment Provider, XXI-I *Annals of Air and Space Law* (1996), 154.

[174] State Letter LE4/49.1-94/89(27 October 1994), Exchange of letters between ICAO and the USA. Cf. F.P. Schubert, An International Convention on GNSS Liability: When does Desirable Become Necessary? XXIV *Annals of Air and Space Law* (1999), 253.

claims arising outside the United States;[175] hence, it does not apply to users that suffered damage as a result of receiving and/or using the signal in a foreign country. Furthermore, it is often burdensome for non-US citizens to present claims in the US courts.[176] Given the global nature of the GPS service provision and its use, the existing national regimes may not function as a sufficient mechanism of protection for all the communities affected by it.

In contrast to the US GPS, the contractual liability for certain Galileo services is recognized. Five types of services are envisaged for Galileo: Open Service, Commercial Service, Safety of Life Service, Public Regulated Services, and Search and Rescue Service.[177] The Open Service consists of a combination of open signals that provide positioning, navigation, and timing. It is suited for mass-market applications such as car navigation and positioning applications, targeted for an unlimited number of users. The other four services are designed for more specified needs and users. The Commercial Service is targeted for business applications using two extra signals that are encrypted. The Safety of Life service is used to provide improved Open Service performance with timely alerts to users, targeted mostly for government users. The Search and Rescue Service is Galileo's future contribution to the COSPAS-SARSAT effort on humanitarian search and rescue activities where Galileo will detect the signals and improve position accuracy.

There are fundamental differences between the Open Service and the other three services: the Open Service is free of charge and the other three are fee-generating services. The intention behind the stance taken by Europe for Galileo appears to be equivalent to that of the United States on GPS: they do not wish to accept contractual liability for free service. Hence, for the Open Service, the contractual relationship does not exist between the Galileo signal providers and value added service providers or end users.[178] On the other hand, such contractual relationships are recognized as to the aforementioned fee-generating services. Under such contractual relations, liability aspects may be determined within the terms agreed between/among the contractual parties.

---

[175] See Federal Tort Claims Act, 28 USC 1346(b), 2671–2680 (1988).

[176] See F.G. von der Dunk, Liability Analysis for GPS and Galileo, 30 *Journal of Space Law* (2005), 142.

[177] Communication from the Commission to the European Parliament and the Council State of progress of the Galileo programme, COM(2002) 518 Final, Official Journal 248, 2–22.

J. Benedicto and D. Ludwig, Galileo System Architecture and Services, Seventh International Workshop on Digital Signal Processing Techniques for Space Communications, 1-3 October 2001, Seismbra, Portugal, 6-8. See also P. Manzini, and A. Masutti, International Civil Liability Regime for the Galileo Services: A proposal.

[178] See F.G. von der Dunk, Liability Analysis for GPS and Galileo, 30 *Journal of Space Law* (2005), 148.

A proper contractual service would guarantee services up to certain standards of accuracy, continuity, and availability.[179]

Von der Dunk contends that "Galileo would also refuse to accept such contractual liability in jurisdictions other than those of the European states constituting the Authority." [180] Even if entities in Africa receive signals from Galileo, and suffer damage as a result of defective service, they do not have the right of recourse on the basis of contract as their European counterparts would have.

A detailed regulation for Galileo is yet to be developed, but a noteworthy step has been taken by the Italian government in developing a Draft Liability Regulation for Galileo,[181] which was submitted to the European Commission in June 2007. The draft regulation proposes to distribute liability between registered qualified providers and the GNSS operator on proof of GNSS malfunctioning[182] as well as granting the jurisdiction to hear the case to states where the damage or incident occurs.[183] This initiative highlights key issues that constitute a gap in the current regime surrounding GNSS, and emphasizes the need for clarification in actual operations.

Looking at the overall picture of GNSS operation and legal framework as well as the regimes applicable to GPS and GNSS, the following findings can be made:

- The analysis of GNSS made clear the intention of service providers to deny liability with an unlimited number of beneficiaries; denial of contractual liability under the regime governing GPS, as well as the limitations on liability for fee-involved services from Galileo within European jurisdiction leave a large population of beneficiaries unprotected;
- No GNSS-specific regime exists as to protection of third parties outside contractual relations;
- It is questionable whether the existing regime at a national level provides adequate solutions for all the conceivable scenarios (e.g., uncertainties as to the applicability of the US Federal Tort Claims Act to GNSS-related damage that took place outside the United States); and

---

[179] See F.G. von der Dunk, Liability Analysis for GPS and Galileo, 30 *Journal of Space Law* (2005), 149.

[180] See F.G. von der Dunk, Liability Analysis for GPS and Galileo, 30 *Journal of Space Law* (2005), 153.

[181] Proposed Regulation on Civil Liability and Compensation for damage resulting from the Malfunctioning of European GNSS Services and Equipment, 2007. [Hereafter: Draft Liability Regulation for Galileo]

[182] See L.J. Smith, Where is Paradise? The EU's Navigation System Galileo – Some Comments on Inherent Risks (or Paradise Lost), *Proceedings of Fiftieth Colloquium on the Law of Outer Space*, AIAA, 2008, 356.

[183] See Article 14 of the Draft Liability Regulation for Galileo.

- For foreign claimants, bringing action in the jurisdiction of a service provider is often burdensome and not feasible in reality.

The above legal problems and uncertainties experienced by GNSS operators are common to satellite remote sensing. There are two key issues presented by GNSS that need to be clarified for a regime encompassing satellite remote sensing:

1) Absence of a specific tort regime opens an opportunity for creating solutions to approach the issue by applying varying general liability regimes that exist at a national level, which give rise to the problem of jurisdictions;
2) The Italian Draft Liability Regulation on GNSS points to the need for improvements in satellite remote sensing, particularly as to the need for distribution of liability within the supplier chain, and granting jurisdictions to courts of the states where the damage occurs.

It is particularly essential to determine how beneficiaries and affected communities not covered under the contractual relations may recover damages in a cross-border legal situation.

6.4.5.3. *Consideration when Applying GNSS Liability Regime to Satellite Remote Sensing*

In the light of the existing GNSS liability regime, and consolidating the analysis undertaken thus far, one can show a circle of liability encompassing different stakeholders of satellite remote sensing in generation, processing, value adding, and use of data/derived products. (See Figure 6.1: Liability Model surrounding Stakeholders of Satellite Remote Sensing).

Although certain overlaps may be seen, there are roughly five varying legal relations between and/or among different stakeholders:

1) contractual liability between/amongst the entities in the supply chain;
2) contractual liability between data distributors and end users in accordance with the distribution agreements;
3) tort liability, which could be invoked by end-users against data generators, value added entities, image processing wholesalers, and data distributors;
4) tort liability which could be invoked by third-party victims; and
5) product liability, which could be invoked against manufacturers of a remote sensing satellite.

Whilst product liability claims against manufacturers of satellites should be recognized under both the regimes of GNSS and satellite remote sensing, for the latter, special attention should be paid to product liability claims against the data manufactures. Following the previous analysis on an intellectual or information product, if product liability is applicable to satellite imagery or derived

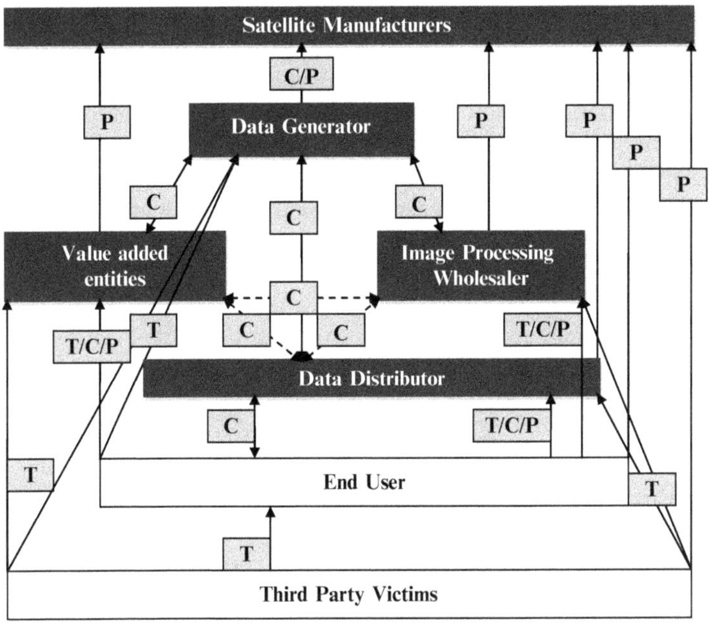

T Tort Liability
C Contractual Liability
P Product Liability
* ◄---► means whether a contractual relation exists between these entities is decided case by case as value added entities and processing wholesaler could act as data distributor
■ means that these entities constitute a supply chain of satellite data and derived products.

Figure 6.1: Liability Model Surrounding Stakeholders of Satellite Remote Sensing

products, it should be applicable only to the physical exterior of the product itself. Whilst such possibilities do exist, it is more likely that a non-product liability type of damage would occur rather, than product liability damage, if the damage is to arise associated with satellite products.

A key aspect that is not depicted in the diagram is the basis of tort liability. A review of product and service liability as well as the analysis of GNSS liability supports the earlier conclusion that a fault-based regime is most suitable for addressing consequential damage arising from satellite remote sensing activities.

Service liability, adopting fault-based liability, should be the primary basis of liability applicable for the intellectual content of the product. Whilst fault based liability is suited to govern the consequential damage and liability arising from the misuse or use of the data, defences available in product liability should be applied for data providers and other concerned parties.

The GNSS analysis illustrates both the problems with identification of the responsible party/parties in complex situations, and the difficulties in establishing consistent legal solutions over multiple jurisdictions. Such issues would be common in a remote sensing regime, and must be clarified if that regime is to be effective.

6.5. Further Examination of Relevant Issues Towards a Recommended Regime

This final section details the issues that are essential elements to fill the final gaps in the current legal framework toward an effective legal regime governing the supply and the use of remote sensing data. Three important elements will be considered. The first one covers the varying degrees of application of fault-based liability to different circumstances in terms of bargaining power, taking into account the balance of knowledge/expertise of data providers and data recipients. The second element addresses the distribution of liability when more than one party shares responsibility for damage. The last deals with how the appropriate jurisdiction is determined and how claims presented in parallel can be minimized.

6.5.1. *Higher Duty of Care Required of Professionals*

Under both product and service liability, a higher duty of care is required of professionals when selling a product or supplying a service. For those who claim a special skill, a standard of professional liability is applied that requires those individuals to use that skill in a proper manner. Article 7 of the EC Product Liability Directive imposes a higher duty of care for a professional seller/distributor: "if a product was not made for sale or in the course of business, the producer can be exonerated from liability."[184] Under both common law and civil law, professional negligence is recognized.

Under the French legal system, the specialty of a claimant vis-à-vis the defendant is also taken into consideration. If a business buyer has the same speciality as a business seller, he may not be able to claim compensation in case

[184] Article 7 of the EC Product Liability Directive.

of a hidden defect.[185] Just like the civil law category of professional seller – *'vendeur professionnel'* – for product liability, the common law concept of professional negligence exists in liability for service. From case law, it can be shown that when a third party suffers loss or damage because of the defective performance of a service, that party can also recover damages from the service providers.

In the United Kingdom, the higher duty of care expected of professionals is illustrated in several cases. The leading case regarding professional negligence is *Hedley Bryne v. Hellers Partners,*[186] which established that professionals could be held responsible for negligent misstatements leading to the defendant's financial loss. In this case, Bryne incurred financial loss as a result of a statement by a bank that negligently gave favourable reference on the creditworthiness of the company, which soon went bankrupt; however, due to the disclaimer, the defendant bank was not held liable. In deciding this case, the court introduced the concept of special relationship, which exists when the plaintiff relies upon the defendant, with the latter being aware that such reliance existed.[187]

Later, establishing professional negligence was further clarified in *Smith v. Eric Bush,*[188] which set forth the elements that must be considered in determining professional negligence: 1) equality of bargaining power; 2) the practicability of obtaining advice from an alternative source; 3) the difficulty of the task being undertaken; and 4) the consequences to the parties of upholding the disclaimer.[189]

It should be pointed out here that advice, which eventually turns out to have been wrong, does not necessarily constitute negligence. In *Stafford v. Conti Commodity Services Ltd.,*[190] an investor relied upon advice from his broker and subsequently incurred large losses on the commodities market; he sued his broker for damages claiming that the broker's advice was negligent. The court held that a broker cannot always be right in the advice that he gives, taking into

185 See S. Whittaker, European Product Liability and Intellectual Products, 105 *Law Quarterly Review* (1989), 94.

186 [1964] AC 465, [1963] 2 All ER 575, HL. See also BGE 111 II 471, *Spring v. Guardian Assurance* [1995] 2 AC 296 [1994] 3 All ER 129, HL; *White v. Jones* [1995] 2 AC 207, [1995] 1 All ER 691, HL; *Henderson v. Merrett Syndicates* [1995] 2 AC 145, [1994] 3 All ER 506, HL.

187 See further *Mutual Life and Citizens' Insurance Co. Ltd. v. Evatt* 1971 AC 793 at 805.

188 *Smith v. Eric Bush,* [1990] 1 AC 831. In this case, Lord Griffins expressed his view that "it is unreasonable for professional men in all circumstances to seek to exclude or limit their liability in negligence."

189 See R. A. Buckley, *The Modern Law of Negligence,* Butterworths, London, 1999, 128.

190 *Stafford v. Conti Commodity Services Ltd,* [1981]1 All ER 691.

account the rapidly changing market. Where advice takes the form of prediction, much will therefore depend in practice upon the degree of accuracy generally to be expected in the field.[191] This holding is particularly relevant to space-based services involving prediction and forecasting such as weather forecasting, and systems designed to provide early warnings of hurricanes or tsunamis. As previously mentioned, not all earthquakes cause tsunamis; therefore a warning that turns out to be unnecessary is not considered to give rise to liability.

The provider of a free service cannot automatically assume an exemption from responsibility and liability. It is stipulated under Article 2 of the Draft Service Directive that responsibility and liability attach, "whether or not in return for payment."[192] This is illustrated by the Swiss case of the Emile Gallé lamp.[193] In this case, the owners of the Emile Gallé lamp suffered loss after selling the unique lamp of considerable value at a low price following the assessment given by an art auction company that erroneously assessed the lamp to be of a low value because of an internal communication mistake. Despite the fact that the firm had intended to render the information as a favour, the court held that the parties had entered into a binding contract to provide the service.[194]

One can draw a conclusion here that the provision of a free service does not always exclude the responsibility or lower the degree of duty of care expected of the service provider, but the degree of reliance by the plaintiff and whether a special relationship existed between the disputing parties determines the extent of negligence. To that end, it is useful to recall the aforementioned Good Samaritan Principle that invokes the special relationship and reliance, supported by several cases such as *Indian Towing Co. v. United States*[195], and *Engham Air v. Eastern Airlines.*[196] These decisions may have an impact on the potential regime of remote sensing so that even when the data providers are providing satellite images for free – such as in the case of the Disaster Charter and the free supply of data to developing countries – they may be still held responsible, and hence, liable. The next section examines the circumstances under which data suppliers can actually be exonerated from their liability.

---

[191] See R.A. Buckley, *The Modern Law of Negligence*, Butterworths, London, 1999, 103.

[192] Article 2 of the Draft Service Directive.

[193] BGE 112 II 347. See Roberto, V., and Widemer, C., Tort Liability for Services, *Swiss Reports Presented at the XVI International Congress of Comparative* Law, Zurich – Bale – Geneva, Schulthess, 2002.

[194] BGE 112 II 347, 350 et seq.

[195] *Indian Towing Co. v. United States*, 350 U.S. 61(1955).

[196] *Engham Air v. Eastern Airlines*, 373 F. 2d 227 at 236 (2nd Cir.).

6.5.2. *Joint and Several Liability Shared amongst the Supply Chain*

As briefly mentioned in the section dealing with contracts, the clarification of joint and several liabilities is crucial for an unambiguous liability regime. Where there is more than one party causing the same damage, liability may be joint or several.[197] In such instances, a claimant can sue any or all of the concerned parties who are potentially responsible for his loss.[198] Joint and several liability[199] is significant to the liability associated with the supply and use of remote sensing data, as a number of entities are involved in the course of operations before the delivery of the final products. Potential defendants may be data distributors, information extractors, processing wholesalers, satellite operators, and satellite manufacturers.

Liability may also extend to the manufacturers of a camera on board the satellite, or to the manufacturer of a computer which conducts automatic processing. In the absence of a clear contractual agreement determining liability issues, potential defendants can remain exposed to liability risks.

Remote sensing application programmes, such as the Disaster Charter, have several different types of actors involved. The Disaster Charter brings together divergent bodies, such as governmental entities (e.g., member space agencies like ESA) and relief agencies (e.g., Canadian Civil Protection); cooperating bodies, such as the European Union and United Nations Office of Outer Space Affairs (UN-OOSA); value-added entities (e.g., Service Régional de Traitement d'Image et de Télédétection (SERTIT)); and private entities (e.g., Digital Globe) for its complementary data provision, and often there are even different individuals as Project Managers.

In case of claims related to satellite remote sensing data, the final data distributors are likely to be the first to be sued because they are the interface with the data recipients; however, the entities in the supply chain all have liability exposure. Indeed, damage associated with data may be attributed to different stages of remote sensing activities where different entities/individuals are responsible. Damage may be attributable to: defects in the design and/or a manufacturing a satellite; errors during the tasking, such as inappropriate command given by the satellite operators; and processing errors, as well as misinterpretation of data performed by the processing wholesaler and/or information extractor.

---

[197] See V. Harpwood, Joint and Several Liability in Tort, *The Law of Tort*, Grubb, A. (Ed.), Leed Elsevier, London, 2002, 131.

[198] See C.J. Miller & R.S. Goldberg, *Product Liability*, Oxford University Press, New York, 2004, 268.

[199] Article 5 of the Product Liability Directive and Article V of the Liability Convention refer to joint and several liabilities.

An ambiguity exists in the current practice as to where liabilities rest amongst the supply chain, as they are not dealt with properly in the contracts. Consequently, it is unclear as to how the parties in the supply chain share their liability in cases where several parties contribute to the damage.

It is necessary to look further at the apportionment of liabilities because increasing numbers of entities are becoming involved in producing satellite data, and multiple sources of information – including GIS, statistical data, and GNSS – are combined to produce the final products. Two issues to be addressed are the risk of passing on the loss to another entity in the supply chain, and the apportionment of liability.

6.5.2.1. *The Risk of Shifting the Responsibility for the Loss*

Since contractual agreements do not cover sufficiently the matter of sharing liabilities, the possibility of further legal proceedings within the supply chain is highly conceivable.[200] In an attempt to shift the responsibility for the loss to others, defendants may sue the other parties in the chain. Endless proceedings amongst concerned parties, in attempts by the parties themselves to be exonerated from liability, are likely to generate unnecessary litigation costs. The inter-party waiver of liability, if applied, would solve the problem only partially, as it precludes proceedings amongst the entities in the supply chain, and therefore generates burdens upon the parties who are held responsible unjustly.

If claims are brought amongst the parties in the supply chain, the question may arise as to the identification of the applicable jurisdiction. It is possible that legal or juridical persons of a supply chain are all located in different jurisdictions. There may be a conflict of laws amongst multiple jurisdictions. Depending on the choice of forum, the decision of the court may differ. Claimants may take advantage of this fact and may choose the most favourable jurisdiction or court in which a claim might be heard – what is known as 'forum shopping'. Therefore, the parties in the supply chain should come to an agreement beforehand as to the applicable law and potential court to hear the case.

6.5.2.2. *The Problem of Apportionment of Liability*

If the concerned parties decide that they will actually share the liability and any compensation, then the question arises as to how their liability is to be apportioned. The problem is that concrete guidelines or method to assess the apportionment of liability are often absent. For example, as discussed in Chapter 2, the Liability Convention approaches the issue by stating that "the burden of compensation for the damage shall be apportioned between the first two States

[200] Similar consideration exists in the field of air transport. See F.P. Schubert, An International Convention on GNSS Liability: When does Desirable Become Necessary? XXIV *Annals of Air and Space Law* (1999), 260.

in accordance with the extent to which they were at fault; if the extent of the fault of each of these States cannot be established, the burden of compensation shall be apportioned equally between them."[201] This approach would not adequately address the issue, as it is quite conceivable that more than two entities in the supply chain are responsible for damage. Also, dividing the liability in half does not always reflect the proper apportionment. Domestic legislation also fails to establish a concrete rule. United Kingdom law – for instance, Section 2(1) of the Civil Liability Act 1978 – states that "the amount of the contribution recoverable from any person shall be such as may be found by the court to be just and equitable having regard to the extent of that person's responsibility for the damage in question."[202] The definition of what constitutes "just and equitable" is not provided.

The issue of apportionment of liability proportionate to the degree of contribution to the damage was elaborated in case law: for example *in Miraflores v. Owners of the George Livanos,*[203] "the investigation is concerned with fault, which includes blameworthiness as well as causation and no true apportionment can be reached unless both those factors are borne in mind."[204] The case of *Stennett v. Hancok and Peters*[205] is an illustration of the application of joint and several liability and apportionment of liability reflecting the degree of contribution to the damage. In this case, the first defendant, the owner of a lorry, was not held liable, but rather the contractor, a motor repairer, was found to be responsible for the negligent repairs.[206] This judgement may be applied to remote sensing cases in a way that if the data generator subcontracts to the value-added entity for the interpretation of data, and if the data is misinterpreted and then distributed by the supplier, the data supplier may be the first defendant, but the value-added entity will be the only party to be held liable.

In some cases, the apportionment of liability is concrete in accordance with the contribution to the damage determined by the court. *Prendergast v. Sam*

---

[201] Article IV(2) of the Liability Convention.

[202] Section 2(1) of the UK Civil Liability Act 1978.

[203] [1967]1 AC 826, 845, [1967] 1 All ER 672, 677-8.

[204] See C.J. Miller & R.S. Goldberg, *Product Liability*, Oxford University Press, New York, 2004, 272.

[205] [1939] 2 All ER 578. "He knew that if it was not repaired with due care, with this wheel so assembled as to make it keep together and not fly apart upon the road, in all probability somebody would be injured as the result of his not having done that which he should have done."

No liability was incurred by the first defendant who had exercised reasonable care in relying upon an independent contractor to repair the vehicle.

[206] See C.J. Miller & R.S. Goldberg, *Product Liability*, Oxford University Press, New York, 2004, 757.

Dee Ltd.[207], is a case concerning a doctor's illegible handwriting being misread by a pharmacist leading him to prescribe 50 times more than usual dose of medicine to the claimant patient, who consequently suffered from unconsciousness and irreparable brain damage. Liability was apportioned 25 percent to the doctor and 75 percent to the pharmacist.[208] Like the concrete approach taken for this case to apportion the liability in question, there is a need to systematically assess the degree of blameworthiness for the damage associated with satellite remote sensing.

6.5.3. *Issues of Identification of Applicable Jurisdiction*

There are several issues to be considered when it comes to actual proceedings. Establishing guidelines as to the applicable jurisdiction and minimizing parallel proceedings would contribute to effective and prompt dispute settlement. In many legal systems, the place in which the incident occurs is the criterion for establishing applicable jurisdiction.[209] Indeed, as introduced in the analysis of GNSS, the Draft Liability Regulation for Galileo proposed criteria based on the location where the damage or incident occurred. This criterion is not completely clear, however, as it can be interpreted in two ways: either the place in which the damage occurs or the place in which the action that produced the damage occurs.[210]

When applied to the field of satellite remote sensing, such criterion does not clarify the current uncertain situations. In fact, multiple jurisdictions may exist with respect to the interpretation of a location of the damage. For example, an error involved in data collection could be considered as having taken place in a location in outer space. Furthermore, the place where value-added entities misinterpreted the satellite image could be in one jurisdiction whilst the place of use of data that leads to the damage is in another.

Adoption of jurisdiction on the basis of the location of damage is not an appropriate solution to the problem of jurisdictions. Instead, it is best to adopt jurisdiction on the basis of the claimant's residence or principal place of business. In order to minimize the burden of a potential claimant to present claims in foreign jurisdiction (e.g., if he or she is an individual claiming against the end user or parties in the supply chain that belong to foreign jurisdictions),

---

207 *Prendergast v. Sam Dee Ltd,* [1989] 1 MLR 36.

208 See C.J. Miller & R.S. Goldberg, *Product Liability*, Oxford University Press, New York, 2004, 737.

209 See P. Manzini & A. Masutti, An International Civil Liability Regime for the Galileo Services: A proposal, 116.

210 See P. Manzini & A. Masutti, An International Civil Liability Regime for the Galileo Services: A proposal, 116.

principal domicile of a claimant should be approved as the primary jurisdiction, regardless of where the damage occurred.

Given the global nature of satellite remote sensing that involves data generation and supply of a cross-border nature, parallel proceedings may not actually be eradicated; however, certain rules should be established to minimize them. For example, presentation of claims for the same damage to more than one court in parallel should be avoided. Whilst contractual terms often indicate the applicable jurisdiction in case of disputes, multiple jurisdictions may be applicable for damage outside contractual relations. Such circumstances generate an opportunity for forum shopping and promote continued inefficiency and confusion in dispute settlement, obviating the need for a common approach.

6.6. Conclusion

In recent years, satellite remote sensing has become a significant part of the services related to disaster management and environmental protection. Information supply using remote sensing from space has become a service of public and commercial significance. At the same time, it has increased the possibility for damage as well as exposure to liabilities. A remote sensing legal framework is critical both to maximize remote sensing utility and to address anticipated claims. An unregulated liability regime would be of potential harm to all the parties concerned in the generation, supply, and use of remote sensing data, and this situation urgently calls for improvement. Different sets of conventions such as those in air law are not necessary, but a single regime should serve as a regulatory framework, and guidelines are needed to implement the rules covering both contractual liability and tort liability. First of all the regime should allow for recovery of consequential damage arising from the supply, use, and misuse of data; furthermore, fault-based liability should be adopted for all situations.

The regime should set guidelines so that the key elements determining fault will be the foreseeability of the harm, the reasonableness of the alleged party causing the damage, and the equal bargaining power of the concerned parties. The regime should also specifically incorporate several elements that serve as defences against liability. Amongst others, the scope and definition of *force majeure* should be set to exonerate defendants from liability for insuperable circumstances. Reasonable limits regarding the state-of-the-art scientific and technical knowledge should be established to distinguish avoidable errors from negligent errors involved in the entire course of remote sensing operations. The Good Samaritan Principle should be incorporated for the benefit of public service providers, and it should codify specifically the exoneration of liability arising from voluntary assistance of a humanitarian character using remote sensing, unless damage was caused wilfully or through gross negligence.

Concerning the contractual liability, existing contractual agreements should be left as they are but a certain harmonization should be achieved. It is recommended that the regime address the legality of liability disclaimers and establish the conditions under which such disclaimers will apply: 1) the alleged parties did not cause the damage intentionally or by gross negligence; 2) the disputing parties do not have equal bargaining power; and 3) a liability disclaimer is only enforceable on the contracting parties. Varying degrees of duty of care are expected of alleged parties vis-à-vis the claimant, depending upon the bargaining power of the parties, as well as the expertise of a claimant.

Clarification of joint and several liability is the essential element for the harmonization of rules under an improved regime and for the implementation of rules with regard to contractual liability. Clear and coherent responsibilities and the division of liability need to be established amongst the parties comprising the supply chain. The clear demarcation between competencies of different actors and coherent divisions of responsibilities and apportionment of liabilities are critical. The agreements between the parties should clearly indicate the scope of responsibility and division of liability: which parties are liable for which operations, potential allocation of compensation, and procedures for indemnification. Error detection, identification of the cause of the problem, and apportionment of liability should be facilitated by documenting all the operations using a suitable electronic authenticity tool.

Finally, specific guidelines should be established as to the identification of applicable jurisdictions for actual proceedings. The rule of adopting the forum where the damage/incident occurs is not suited for satellite remote sensing cases, as the determination of such locations is more complex, may lead to multiple jurisdictions and generate confusion. Rather, the principal domicile of a claimant should be determined as the primary jurisdiction, particularly when the claimant is an individual third-party victim claiming in tort against foreign persons in the supply chain. Even if the claimant is bound by a contract that indicates specifically the applicable jurisdiction, if filing claims in that jurisdiction seems not feasible, the claimant should be able to choose the jurisdiction of his principal domicile. On the other hand, the applicable jurisdictions for proceedings amongst the supply chain should be pre-determined through contract. Apart from the circumstances where the claims in parallel may be justifiable, once jurisdiction is chosen, other options should be abandoned.

Chapter Seven

Conclusion

This book takes a novel approach to actual problems inherent in satellite remote sensing, which is a relatively new and rapidly progressing field with topical issues addressed on a worldwide basis. The originality vests in setting problem solving as the research goal through pursuing a practice-oriented and multi-dimensional approach to seek realistic solutions. Legal documents and literature were extensively researched and examined closely to identify where the problems lie, and state-of-the-art practice was surveyed through communication with experts to understand precisely the gaps existing between the regulations and actual practice. The book then seeks solutions in legal disciplines outside space law, where rules are more concretely established and applied more substantially for longer periods of time. The book scrutinizes rules in relevant disciplines, extracts their key essence and tailors them to match the practice of satellite remote sensing. The result of this research is a set of proposals that can be directly applied in real situations.

In summary, this book provides the first complete overview of the legal aspects of remote sensing activities by:

- Conducting a comprehensive review of the current legal framework surrounding remote sensing;
- Identifying the legal and policy shortcomings of the current regime of space law, environmental law, and disaster management law;
- Stating the urgent need for review and improvement of the current legal regime on satellite remote sensing;
- Recommending that there should be a common data policy applicable for public benefit services;
- Proposing guidelines to establish an overall framework for data policy;

- Recommending that there should be uniform rules for data authentication and accuracy validation;
- Proposing a concrete mechanism for data traceability and authentication; and
- Proposing a potential regime for consequential liability arising from remote sensing activities.

7.1. Review of Key Issues Examined

7.1.1. *Clarification of the Current Regime Achieved in Part I*

Part I clarifies the existing legal framework surrounding the use of satellite remote sensing for environmental protection and disaster management applications. Chapter 2 addresses the question of the adequacy of a current regime on satellite remote sensing. Through a review of the legal norm of space activities, the underlying substance of rules established by international and national space law are elucidated. An overview identifies the problems of the present regulations by indicating the inadequacies of existing rules with the resulting uncertainties and their potential impact. Chapter 2 further defines the inherent ambiguities and missing elements of current rules that hamper widespread use of data that impede effective dispute settlement for damage associated with the use of satellite remote sensing.

Whilst regulations at a domestic level are more up-to-date with ongoing practices, the analysis of those regulations confirms that their nature and objectives are primarily to protect national interests, and to facilitate an individual government's control over non-governmental activities under its jurisdiction. For that reason, domestic regulations would not suffice as a complementary source of law that could serve as a solution to the problems not addressed under the international space law.

Chapters 3 and 4 further narrow down and study areas in need of rectification and demonstrate the rationales for such rectifications. The discussion focuses on how inadequacies inherent in space law would serve as impediment for each application of disaster management and environmental protection where both regimes do not function effectively enough. The Chapter discussions lead to the observation that, whilst the two regimes are ineffective for distinct reasons, the levels of achievement in both applications can be improved if the inadequacies of remote sensing are overcome.

Chapter 3 demonstrates the need for clear data policies and guidelines for the assessment of data reliability. The initial analysis identifies the source of problems in effective environmental protection that lie in the insufficient implementation of the multitude of existing environmental laws. The effect of environmental laws, as analysed, can be enhanced through strengthened monitoring and assessment as well as using remote sensing to improve compliance and enforcement. The discussion leads to the observation that the first issue is the hurdle in data access and use caused by uncertain data policies; consequently, user communities are restricted in the opportunity to use available resources. The second problem identified is the uncertain value of remote sensing data as reliable information. A variety of case examples demonstrates an overwhelming trend of cautious application of remote sensing data for verification purposes, and how such data is used as a source of supporting information. A major barrier to the reliance on data by users lies in the absence of objective guidelines to assess the information value.

Chapter 4 serves as the key to demonstrate the need for clarification of liability. The analysis highlights the characteristics of disaster management, governed by the principle of state sovereignty that led inevitably to the present model of international disaster response, which is predominantly voluntary and without the assumption of liability by the donor of assistance. Chapter 4 then presents the example of space-based disaster response and discusses its success based on the voluntary model, indicating the exposure of liability faced by service providers as documented in court cases following disasters. Following a close look at the problem of the hesitation of states to provide assistance because of such liability risk, the conclusion is drawn that the clarification of responsibility and liability of a goodwill service provider associated with voluntary provision of data is necessary to facilitate widespread contributions of data for public emergencies. The analysis identifies the essential need for a clear definition of liability associated with data use and distribution. Particular emphasis should be placed on the necessity for a mechanism to limit and/or waive liability of data providers in cases of insufficient or unsuccessful assistance.

Overall, Chapters 3 and 4 present three specific hurdles which need to be overcome;

1) restrictive data policies that hamper access and use of data;
2) lack of clear guidelines to establish the suitability and reliability of data; and
3) uncertainty of liability for damage arising from the supply and use of the data.

7.1.2. *Improvements to the Current Legal Framework Recommended in Part II*

Part II of this book is devoted to seeking and providing answers to the questions raised in Part I and proposing solutions to the three hurdles as identified. Possible solutions are contemplated through the analysis of current practices and policies within the remote sensing supplier communities and scrutiny of the relevant international and national frameworks. The examination of the three issues has highlighted a number of gaps that need to be closed as proposed for the new regime described in Part II.

7.1.2.1. *Data Policy*

The first part of Chapter 5 provides answers to the issues raised with regard to uncertain data policies, including intellectual property rights. Thorough investigation was undertaken on the polices set for different satellite data to show the type of anarchy currently imposed by individual suppliers. Whilst achieving a single data policy for all satellites/ data products is not a feasible solution, attention is given to the dominant trend of suppliers to distinguish between public and commercial use, and to the potential for further regulations.

Studies of distribution agreements explored the key substance of restrictive data policies – specifically, users are only given the license to use data whilst the data provider retains the associated intellectual property rights. The closer analysis clarifies where copyright is applicable and where it is questionable. A conclusion is drawn that copyright protection can only be ensured with manually processed data and value-added products.

Issues of protection on the basis of *sui generis* rights applicable to a database is then discussed as another approach to address the protection of satellite data products. The studies of regulations present the suitability, level of security, and extent of protection that a right of *sui generis* would bring to data generators in the scenarios of actual claims. It was observed that protection on the basis of *sui generis* should be used in combination with copyright wherever applicable, rather than as an alternative to copyright.

Other key issues that must be clarified are then examined in consideration for sufficient application of rules in situations specific to satellite remote sensing. The issue of protection of value-added products in terms of joint authorship, and the question of a juridical person being a copyright holder are then discussed with the focus of how the existing gaps identified in the practice can be adjusted to existing intellectual property regulations.

7.1.2.2. *Data Authenticity and Accuracy*

In response to the question raised regarding the reliability of satellite data products, the remainder of Chapter 5 is devoted to the discussion of the current practice, problems, and the need to rectify the mechanisms for both authentication and accuracy validation of satellite data products. The section initiates the discussion of the inadequacy of the present basis for users to know the quality of products, and reveals the reality in which only limited measures are available to users for assessing data accuracy. Particular attention is given to the fledging nature of image processing technology and the resulting impact of inevitable inaccuracy, and subjective judgments by individual data analysts in their interpretation. The discussion of issues then identifies that the problems lie in the 1) digital nature of satellite data products susceptible to modification and manipulation, and 2) absence of definitive mechanisms and common principles to verify authenticity and accuracy of all the processes before product delivery.

The analysis suggests the need to establish a common mechanism to assess the reliability of data. In order to achieve certain standard of quality, the following needs have been identified: 1) disclosure of metadata of different data sets; 2) registering data history in an unalterable format; and 3) documenting the degree of certainty in the respective data handling operations. The discussion then turns to the concept of traceability already applied in food products, and evaluates its relevance and potential application to satellite remote sensing products. Finally, the issues of an electronic authenticity tool as a mean for implementing traceability are addressed.

7.1.2.3. *Liability*

Chapter 6 serves to provide solutions to the problem of uncertain liability associated with satellite remote sensing. First, the key rules are examined in a general liability context of both contractual and tort liability; then, a trend in the existing case law for actual recovery of damage is analysed, examining the liability situations specific to satellite remote sensing in reference to distribution agreements between a data supplier and an end user, as well as to various cases laws on breach of contract. The evaluation of contracts leads to the analysis of agreements that are commonly written to favour the data suppliers and limit the rights of data recipients. The overall analysis reveals that gaps exist between the practice of satellite data suppliers who intend to waive their liability in all circumstances, and the general conditions on enforceability of liability disclaimers.

The examination of tort liability focuses on the analysis of, and determination of different bases of liability and leads to the suggestion that fault-based liability is a better-suited basis to address damage associated with satellite remote sensing activities. This conclusion was further exemplified in the relevant legislation addressing liability with a particular emphasis on product liability.

An examination of product liability was then undertaken to highlight key elements of fault and clarify the standard expected of manufacturers of products. The analysis leads to observations that whilst adopting product liability based on strict liability is not suitable, the defences available in the product liability regime can be extracted for a new regime as proposed in Part II. The suitability of fault-based liability was further supported by analogies drawn with the regime relating to GNSS, which is primarily fault-based liability. The Chapter then further refers to consideration of the liability circle of GNSS as a practical liability model that could be applied to satellite remote sensing.

The analysis is complemented by the discussion of issues such as professional negligence bridging remaining gaps toward a comprehensive liability regime. Chapter 6 generally closes the gap of practice and rules, extracts the essential rules of the general context of liability and adapts it to the regime on remote sensing, and proposes a feasible legal model to address potential claims.

7.2. Recommendations

In order to realize the full benefits of remote sensing from space, the result of research shows the incontestable need for improvements to the current remote sensing legal regime, and provides recommendations to achieve this imperative. An unambiguous and comprehensive regime is required so that both access to, and utility of, data is maximized; data use is to be widespread. An improved regime is reiterated, with concrete recommendations establishing certain rules for the first time, as well as supplanting current inadequate remote sensing regulations.

7.2.1. *Goal of the Regime*

The goal of the proposed regime is to cover the multiple dimensions of remote sensing activities, and to properly regulate the generation and supply of data, as well as application and use of data. The overall objective is to strike a balance between the current widespread practice of remote sensing, and the protection of the user community and other concerned parties. Three key goals are to: 1) regulate and facilitate the supply of accurate data and proper use of data, 2) promote data use without discouraging incentive for data generation and knowledge creation, and 3) adequately regulate the settlement of potential claims.

In attaining specific goals to overcome the identified hurdles, three pillars of the goal are reaffirmed herewith. The first pillar of the goal is to establish a clear regime with a set of rules to determine data accessibility, and exploitability to promote data use for public benefit services. The second pillar of the goal is to provide an objective standard and guidelines for users to assess the reliability of data. The third pillar of the goal is to clarify liability involved, and establish an adequate legal mechanism to provide remedies, or at least a channel for recourse in case of damage.

7.2.2. *Recommendation on Data Policy*

The recommendation is to establish a common access and pricing policy only for public benefit services, where data should be made available: under effective coordination; free-of-charge or at marginal cost; and for unlimited use and number of users without a requirement for case-by-case authorization. Both public benefit services and commercial service should have a common intellectual property regime although some intellectual rights of data generators need to be compromised for the public services. It is proposed that copyright can only be securely claimed for products with certain intellectual input. It is recommended that the right of *sui generis* or mere credit should be adopted for the data that does not involve such input. Recommendations for certain harmonization of protection for value added products of different types are also provided, in a manner that rights are coherently shared between data generators and value added entities.

7.2.3. *Recommendation on Data Verification*

It is recommended that certain standards of quality be achieved for authenticated and accuracy-verified products where they are provided 1) in an unalterable format, registering each operation involved with the names of persons who conducted the operations, 2) with complete metadata, and the degree of certainty in the processing and analysis of images, 3) with a common label that indicates proof of the authentication and accuracy verification. The recommendation to use an electronic authenticity tool as a means to trace data history can be applied either through use of a two dimensional bar code or through digital watermarking, depending on the priority of the intended purposes. A further step would be that in the long term, a common standard should be developed to assess the reliability of data authentication as well as to validate accuracy of processing and analysing data. For use in court, reliability of data should be examined by an organization independent of any data generator and distributor. Such an organization should also coordinate the common standards for traceability adopted for EO information.

7.2.4. *Recommendation on Liability*

The overall recommendation is to establish a single regulatory framework addressing liability with 1) harmonization of rules in the contractual context 2) fault based liability for consequential damage associated with satellite remote sensing, and 3) guidelines in identifying an applicable jurisdiction.

With regard to contractual liability, harmonization should be achieved as to 1) a legitimate liability disclaimer, and conditions on its application, and 2) division of liability amongst the parties in the supply chain. It is recommended that agreements be pre-arranged between/amongst the actors concerned to clarify demarcation between competencies, the scope of responsibility, and division of liability. The allocation of cost may be quantified objectively, taking into account the cost for processing and information extraction as well as the price paid by users for the final products.

Clarification of how fault-based liability is applied to damage associated with remote sensing should include determination of fault, with a special emphasis on varying degrees of duty of care expected of parties in the supply chain vis-à-vis the claimant. It is recommended that the regime should: specifically establish defences against liability: set a clearer scope and definition of *force majeure*, incorporate Good Samaritan Principles for the benefit of public service providers; and establish reasonable limits on liability in relation to state-of-the-art scientific and technical knowledge.

The final points to be addressed are what measures are needed for cross-border claims, and the rules to identify the applicable jurisdiction. For contractual claims amongst the supply chain, the applicable jurisdiction indicated in the contract should be sufficient, whereas for a contractual claim between the data supplier and end users – as well as for tort claims – the jurisdiction of the claimant's principal domicile (for natural person) or the principal place of business (for juridical person) should be the primary options. Whilst this recommendation is generally for liability-related claims, it can also be applied to claims associated with infringement of intellectual property rights on satellite products.

7.3. How the Proposed Recommendations are Applied in Practice

A recommendation for common access and pricing policy would be an impetus for creating a user-friendly environment for access to and use of satellite remote sensing products. Establishing common data policies is the first step, but it should be applied further to facilitate the realization of a single framework that integrates all EO programmes dedicated for public benefit services as a next step. Just as the data of different satellites from the Disaster Charter is available under a single point of contact, a functional mechanism should be established combining the data available from all EO programmes dedicated to disaster management and environmental applications under a single gateway – including the Disaster Charter, DMC, and GMES. A single UN agency should be responsible for coordinating and organizing the information, to allow much easier data access by states, entities, and individuals.

A harmonized data policy for public benefit service should have two tiers: the first category would provide data free of charge for disaster response and urgent environmental applications that affect mankind; in the second category, data would be made available for non-urgent environmental services at only marginal costs. The existing framework providing the services to disaster response, such as the Disaster Charter, partial capacity of DMC, and (pre-operational) crisis service of GMES, will be included in the first category. Other non-urgent environmental services should be available at only marginal costs, so pricing schemes set differently according to application or purpose are achieved and implemented beyond individual programmes. In this way, there will be increased opportunities for entities to participate within their allowed capacity without dedicating themselves to a particular programme, and the INSPIRE Directive discussed in Chapter 5 would be implemented within a category of the satellite-based information as a step for further regulation of all spatial information targeted under the Directive.

The percentage of data devoted to contribution to services for the public benefit over the total capacity of each programme or entity should be determined. Allocation for such purposes is usually less than 10% of the total capacity, but should be set individually by the different data generators at a rate that can be covered by revenues from other services, so as to make the data supply for public benefit services financially sustainable. If commercial service is to be left unregulated, the current data policies for the commercial portions of the specific programmes that involve public service – such as commercial distribution of DMC data through DMCii and commercial service of GMES – do not require modification. This way, contribution to public benefit services would not be burdensome to the data generator and, in the meantime, it would be beneficial to end users because of easier access. Such practice is likely to stimulate data use without discouraging the incentive for commercial operations of data generators.

With the proposed recommendations, common intellectual property rules applicable to all products under existing and future programmes will be established, uncertainties with respect to the joint authorship of products will be clarified, and consistencies amongst different entities will be achieved. A uniform system of notation for both joint copyright and credit for information extraction and map creation are to be achieved by the Disaster Charter, DMC, and GMES. Inter-programme data and knowledge would then be facilitated.

If clear guidelines for, and a suitable mechanism of, data authentication and accuracy validation are achieved, satellite data could be used more extensively for rigorous enforcement of environmental laws. Data would then be used with much higher confidence, for instance, in fulfilling reporting obligations relating to the level of emissions of greenhouse gases under the Kyoto Protocol, and in enforcing environmental obligations (e.g., to identify a vessel responsible for an oil spill). With improved authentication and objective assessment of data accuracy, decision makers have a much more reliable source of information to make suitable decisions. When presented by an expert who can objectively demonstrate its authenticity and accuracy, data from remote sensing can contribute to the clarification of facts sought by a court.

The codification of the Good Samaritan Principle in the recommended regime would further support the implementation of the Disaster Charter, and improve the situation where the current uncertainties with regard to liabilities discourage both the provision and acceptance of aid. Clearly defining the applicability of the Good Samaritan Principle to EO data and incorporating it into the legal regime of remote sensing will encourage data suppliers to provide assistance without fear of being liable for operations undertaken in good faith. With clarification of liability, more entities would be willing to join coordinated frameworks to provide humanitarian services based on remote sensing from space.

The proposed recommendations should be applied to actual claims in providing clearer guideline for courts to decide the case, such as the lawsuit associated with the 2004 Indian Ocean Tsunami. The approach to the claim would be primarily based on fault; in this case, the fault of NOAA's Pacific Tsunami Warning Center (PTWC). The point to clarify would be whether PTWC was at fault by not warning the affected countries of the impending tsunami. The reasonableness of the PTWC's attempt to contact the afflicted states, combined with the fact that the PTWC only covered the Pacific region – as well as the lack of systems of dissemination for warning the affected nations – would all be considered. With such a scenario (of determination of fault), it can be easily envisaged that PTWC would not be liable.

7.4. Contribution to the State-of-the Art Knowledge and Practice

If the recommendations proposed in this book are adopted, then the procedures for the supply and use of remote sensing data will be much improved. Uniform data access and pricing policies applied through an integrated information distribution system for public service would result in easier access and more effective data sharing at different levels. If an integrated distribution system is to be achieved, data/products would be offered to wider communities under a common database. With the proposed common intellectual property regime, coherence in the security of rights associated with the satellite data products will be achieved and infringement on those rights of a cross-border nature will be addressed more sufficiently – including in the jurisdictions where only juridical persons can be the copyright holder. On the other hand, such a regime does not significantly damage the remote sensing business because data generators retain the discretionary freedom to set the access and pricing policies; they can still exploit the intellectual property rights associated with the data for commercial distribution and will continue to benefit from such activities.

The misinterpretation of data will be minimized and misuse discouraged, as the clarified liability regime creates disincentives for any parties intending to engage in such practices. The identification of errors will be improved extensively with the adoption and real implementation of traceability. If the data handling activities from the raw data to final products is to be registered in an irreversible file, the location and source of errors can be more easily identified through comparing the original data with the final files of distinct operations by different actors.

The greater number of suppliers that adopt a traceability mechanism, the greater the achievement of objective standards for the classification of data/products with differing qualities. With a uniform mechanism for data authentication and validation, users would be able to distinguish between authenticated and unauthenticated data, and could also make a more objective assessment of the relative accuracy of data. Users then would be in a much better position to choose the products suitable for their needs and expectations. This type of mechanism enhances the possibilities that satellite data will be treated more often as primary evidence rather than as secondary evidence in court cases. Overall, having a means to trace the data history through the proposed mediums would largely resolve questions on the reliability of data.

Implementing clearly established guidelines using suitable electronic authenticity tools will make it easier to process information from a combination of different sources to produce value-added products that could serve as useful knowledge for immediate use. Such improved access to information facilitates effective management for satellite-derived data and particularly achieves their improved integration with the other information databases.

Liability issues arising from remote sensing will be addressed with more clarity. The proposed regime would spell out the expectation and responsibility of different stakeholders to the benefit of the end users. Should a claim arise, it will be addressed more objectively and substantially in accordance with the newly established rules, which provide potential claimants with remedies and potential defendants with a better means of quantifying the risk of damages.

Wherever the action is brought, the proposed liability regime should serve as a legal basis upon which potential claimants can rely. Right of recourse can be invoked based on the regime, even if the contract between the data supplier and users barred such right. In addition, the recommendation on the applicable jurisdictions, if adopted, would no longer discourage claimant from bringing action because of impracticality.

Clarification of the responsibilities of the concerned parties in the supply chain will benefit them individually. Concerned parties would be able to minimize the risk that they will be held liable for damage they did not cause. They would better prepare for liability risks, and unexpected costs related to claims are likely to be reduced. The supplier community could take much more suitable measures in preparing for liability risks (e.g., insuring against conceivable damages) knowing types /circumstances of damage for which they might be potentially held liable.

It is anticipated that the proposed recommendations would create predictable scenarios for the outcome of disputes, thereby facilitating the settlement of disputes either out of court, or more promptly and coherently in court.

A concise liability regime on satellite remote sensing would facilitate a further liability regime surrounding operational services with the combination of different sources of information – such as different satellites, GPS data, and GIS information. Overall, the proposed regime would provide for the first time the clarity and mechanisms urgently needed for the full beneficial exploitation of EO data in support of disaster management and environmental protection.

7.5. Further Issues and Future Work to be Done by Others

The implementation of a more comprehensive regime will improve the legal environment surrounding the supply and use of data and problems identified will be better addressed; however, several issues remain to be studied further. These issues are mainly in three areas: 1) dual use for remote sensing data; 2) coordination with the existing regulations that apply to other sources of information apart from satellites (e.g., GIS); and 3) the establishment of a framework to address the issues of privacy.

In recent years, the delineation between military and civil remote sensing activities has become ambiguous: the same systems are often used for both applications. Issues concerning international military and dual use programmes, such as GMES, should be further examined in terms of data policies and responsibilities. The complexities arise because of the sensitivity of data collection, dissemination, and use of information related to national security. Military applications are usually conducted outside the scope of both domestic and international legal regimes. It is critical to address the question of whether to establish a common regime or completely different regimes for civil and military activities within one system over the long term. How control is placed on highly sensitive data beyond the national framework requires such technological considerations as encryption of data and applications, as well as considerations for establishing international regulations on information.

Although the issues of privacy are not addressed in this book, they will be of increasing concern as the spatial resolution of remote sensing satellites continues to become higher, coupled with the emerging trend for entities engaged in the business of non-space/EO operations such as the media, toward possessing their own remote sensing capabilities. The risks range from the exposure of private life to the confidentiality of operations breached through publication of information based on data gathered by satellites. It is possible that lawsuits will be brought against data suppliers by both natural and juridical persons for the infringement of privacy as a result of the disclosure of information that may be detrimental to their interests. It may be particularly serious if the disclosed information is false. Indeed, defamation cases brought by individuals against the media are conceivable. The current legal framework protecting privacy is in the context of human rights, and is not sufficient to safeguard the rights of parties whose privacy is infringed. Further research is required and ultimately certain rules need to be established as a part of the satellite remote sensing legal framework.

Finally, further research needs to be done for achieving legal coherence amongst the regulations associated with different sources of information that may be combined with remote sensing data. These sources include aerial photography, meteorological information, geographic and other ground information systems, and GNSS. Particularly, coordination with GNSS is important as it is highly likely that satellite images will be combined with guaranteed GNSS signals for verification purposes and/or to be used as evidence. Authenticated images, together with authenticated positioning/time signals, would be useful to verify – for instance, the precise location for environmental harm disputed in a potential case. Legal coherence is needed so that the results of the overall service that combine satellite images and guaranteed GNSS signal will be reliable. Divisions of responsibilities and liabilities, as well as an authentication mechanism as proposed for the different steps in the remote sensing supply chain, need to be established. It is hoped that this book will serve as a stepping stone for the further development of a legal framework addressing issues in the generation and use of space-based information.

Annex 1

Principles Relating to Remote Sensing of the Earth from Outer Space

Principle I

For the purposes of these principles with respect to remote sensing activities:

(a) The term "remote sensing" means the sensing of the Earth's surface from space by making use of the properties of electromagnetic waves emitted, reflected or diffracted by the sensed objects, for the purpose of improving natural resources management, land use and the protection of the environment;

(b) The term "primary data" means the raw data that are acquired by remote sensors borne by a space object and that are transmitted or delivered to the ground from space by telemetry in the form of electromagnetic signals, by photographic film, magnetic tape or any other means;

(c) The term "processed data" means the products resulting from the processing of the primary data, needed to make such data usable;

(d) The term "analysed information" means the information resulting from the interpretation of processed data, inputs of data and knowledge from other sources;

(e) The term "remote sensing activities" means the operation of remote sensing space systems, primary data collection and storage stations, and activities in processing, interpreting and disseminating the processed data.

Principle II

Remote sensing activities shall be carried out for the benefit and in the interests of all countries, irrespective of their degree of economic, social or scientific and technological development, and taking into particular consideration the needs of the developing countries.

Principle III

Remote sensing activities shall be conducted in accordance with international law, including the Charter of the United Nations, the Treaty on Principles Governing the Activities of States in the Exploration and Use of Outer Space, including the Moon and Other Celestial Bodies, and the relevant instruments of the International Telecommunication Union.

Principle IV

Remote sensing activities shall be conducted in accordance with the principles contained in article I of the Treaty on Principles Governing the Activities of States in the Exploration and Use of Outer Space, including the Moon and Other Celestial Bodies, which, in particular provides that the exploration and use of outer space shall be carried out for the benefit and in the interests of all countries, irrespective of their degree of economic or scientific development, and stipulates the principle of freedom of exploration and use of outer space on the basis of equality. These activities shall be conducted on the basis of respect for the principle of full and permanent sovereignty of all States and peoples over their own wealth and natural resources, with due regard to the rights and interests, in accordance with international law, of other States and entities under their jursidiction. Such activities shall not be conducted in a manner detrimental to the legitimate rights and interests of the sensed State.

Principle V

States carrying out remote sensing activities shall promote international cooperation in these activities. To this end, they shall make available to other States opportunities for participation therein. Such participation shall be based in each case on equitable and mutually acceptable terms.

Principle VI

In order to maximize the availability of benefits from remote sensing activities, States are encouraged, through agreements or other arrangements, to provide for the establishment and operation of data collecting and storage stations and processing and interpretation facilities, in particular within the framework of regional agreements or arrangements wherever feasible.

Principle VII

States participating in remote sensing activities shall make available technical assistance to other interested States on mutually agreed terms.

Principle VIII

The United Nations and the relevant agencies within the United Nations system shall promote international co-operation, including technical assistance and co-ordination in the area of remote sensing.

Principle IX

In accordance with article IV of the Convention on Registration of Objects Launched into Outer Space and article XI of the Treaty on Principles Governing the Activities of States in the Exploration and Use of Outer Space, including the Moon and other Celestial Bodies, a State carrying out a programme of remote sensing shall inform the Secretary-General of the United Nations. It shall, moreover, make available any other relevant information to the greatest extent feasible and practicable to any other State, particularly any developing country that is affected by the programme, at its request.

Principle X

Remote sensing shall promote the protection of the Earth's natural environment. To this end, States participating in remote sensing activities that have identified information in their possession that is capable of averting any phenomenon harmful to the Earth's natural environment shall disclose such information to States concerned.

Principle XI

Remote sensing shall promote the protection of mankind from natural disasters. To this end, States participating in remote sensing activities that have identified processed data and analysed information in their possession that may be useful to States affected by natural disasters, or likely to be affected by impending natural disasters, shall transmit such data and information to States concerned as promptly as possible.

Principle XII

As soon as the primary data and the processed data concerning the territory under its jurisdiction are produced, the sensed State shall have access to them on a non-discriminatory basis and on reasonable cost terms. The sensed State shall also have access to the available analysed information concerning the territory under its jurisdiction in the possession of any State participating in remote sensing activities on the same basis and terms, taking particularly into account the needs and interests of the developing countries.

Principle XIII

To promote and intensify international co-operation, especially with regard to the needs of developing countries, a State carrying out remote sensing of the Earth from space shall, upon request, enter into consultations with a State whose territory is sensed in order to make available opportunities for participation and enhance the mutual benefits to be derived therefrom.

Principle XIV

In compliance with article VI of the Treaty on Principles Governing the Activities of States in the Exploration and Use of Outer Space, including the Moon and Other Celestial Bodies, States operating remote sensing satellites shall bear international responsibility for their activities and assure that such activities are conducted in accordance with these principles and the norms of international law, irrespective of whether such activities are carried out by governmental or non-governmental entities or through international organizations to which such States are parties. This principle is without prejudice to the applicability of the norms of international law on State responsibility for remote sensing activities.

Principle XV

Any dispute resulting from the application of these principles shall be resolved through the established procedures for the peaceful settlement of disputes.

Annex 2

Charter on Cooperation to Achieve the Coordinated Use of Space Facilities in the Event of Natural or Technological Disasters

Article I – Definitions

For the purposes of this Charter:

The term "natural or technological disaster" means a situation of great distress involving loss of human life or large-scale damage to property, caused by a natural phenomenon, such as a cyclone, tornado, earthquake, volcanic eruption, flood or forest fire, or by a technological accident, such as pollution by hydrocarbons, toxic or radioactive substances;

The term "Charter" means this text;

The term "crisis" means the period immediately before, during or immediately after a natural or technological disaster, in the course of which warning, emergency or rescue operations take place;

The term "space data" means raw data gathered by a space system controlled by one of the parties, or to which that party has access, and transmitted or conveyed to a ground receiving station;

The term "information" means data that have been corrected and processed by the parties using an analysis program, in preparation for use in crisis management by one or more associated bodies in aid of the beneficiaries; it forms the basis for the extraction of specific products for use on location;

The term "space facilities" means space systems for observation, meteorology, positioning, telecommunications and TV broadcasting or elements thereof such as on-board instruments, terminals, beacons, receivers, VSATs and archives;

The term "parties" means the agencies and space system operators that are signatories to the Charter;

The term "associated bodies" means the rescue and civil protection, defence and security bodies or other services referred to in Articles 5.2 and 5.3;

The term "cooperating bodies" refers collectively to the various bodies and institutions, referred to in Article 3.5 of the Charter, with which the parties cooperate;

The term "crisis victims" means any State or community for whose benefit the intervention of the parties is sought by the associated bodies.

The term "beneficiary bodies" means all the bodies benefiting from information intended for crisis management; for example, the authorities and bodies concerned in countries affected by a disaster. Certain associated bodies may also be beneficiaries at the time of a disaster.

Article II – Purpose of the Charter

In promoting cooperation between space agencies and space system operators in the use of space facilities as a contribution to the management of crises arising from natural or technological disasters, the Charter seeks to pursue the following objectives:

- supply during periods of crisis, to States or communities whose population, activities or property are exposed to an imminent risk, or are already victims, of natural or technological disasters, data providing a basis for critical information for the anticipation and management of potential crises;
- participation, by means of this data and of the information and services resulting from the exploitation of space facilities, in the organisation of emergency assistance or reconstruction and subsequent operations.

Article III – Overall Organisation of Cooperation

3.1 The parties shall develop their cooperation on a voluntary basis, no funds being exchanged between them.

3.2 The Charter shall be open, in accordance with the provisions of Article VI below, to space agencies and national or international space system operators wishing to cooperate in it.

3.3 The administrative, operational and technical coordination needed to achieve this cooperation shall be provided by a Board on which each party is represented and an executive Secretariat for implementation of the Charter.

3.4 The authorities and bodies concerned in a country affected by a disaster (beneficiary bodies) should request the intervention of the parties either directly through the rescue and civil protection, defence and security bodies of the country to which one of the parties belongs or of a State belonging to international organisations that are parties to the Charter (associated bodies) or where appropriate via a cooperating body acting in partnership with an associated body.

The country affected by a disaster may also make a direct approach to the parties' Secretariat but, for the purposes of the intervention itself, the bodies concerned in that country must engage a partnership with one or more associated bodies.

The above provisions in no circumstances prevent parties intervening on their own initiative.

3.5 The European Union, the UN Bureau for the Coordination of Humanitarian Affairs and other recognised national or international.6 organisations, whether governmental or non-governmental, are bodies with which the parties may have cause to cooperate in pursuance of the Charter (cooperating bodies). The Board shall maintain a regularly updated list of cooperating bodies.

Article IV – Contributions by the Parties

The parties shall use their best endeavours in the conduct of this cooperation, which shall proceed on the following basis:

4.1 Space facilities available for use The parties shall undertake to maintain an up-to-date list of the available space facilities under their management and, as far as possible, of such space facilities under the management of private or public operators as may be called upon to supplement the parties' own facilities. In particular, the list shall specify for each space system the following details:

- mission characteristics
- orbital characteristics
- operational condition
- programming procedure
- products and services provided by ground systems.

4.2 Scenario-writing

The parties shall together analyse recent crises for which space facilities could have provided or did provide effective assistance to the authorities and rescue services concerned. A report, structured according to the crises identified and the types of situation encountered, and highlighting possible contributions by existing facilities, shall be prepared by the Secretariat in consultation with the associated bodies described in Article V below and where appropriate with cooperating bodies.

Moreover, the parties shall keep abreast of new methods being developed in applied research for warning of, anticipating and managing disasters. Once these new methods (or technologies) have been identified and validated by the design authorities and associated bodies, they may, with the Board's approval, be subjected to pre-operational implementation testing. A test report and an assessment of the areas of application of the method would then be prepared by the Secretariat.

Lastly the Secretariat shall be responsible for designing and proposing, on the basis set out above, scenarios for each type of crisis. Each scenario shall state the conditions under which the parties.7 would coordinate, in the event of a crisis being identified, their action in supplying appropriate information and services, access to the available space facilities being planned accordingly. These scenarios, approved by the Board and regularly updated, shall constitute the basis for action in the event of identification of a crisis.

4.3 Identification of a crisis situation

A crisis situation exists primarily where so identified by a country affected by a disaster and at least one associated body seeking the intervention of the parties under the terms of the Charter, in accordance with the provisions of Article 3.4 above.

The Secretariat shall handle all associated body requests and shall thus have the authority, once it has identified a crisis situation, to arrange for the appropriate action to be taken.

4.4 Planning of space facility availability in the event of a crisis

In the event of a crisis, the parties shall use their best endeavours to plan the availability of space facilities or arrange for it to be so planned.

Such planning shall reflect the provisions described in the corresponding scenarios defined in Article 4.2 above. In the event of an alert or potential crisis, the parties may, in anticipation, plan the availability of the satellite systems under their control.

4.5 Organisation and assistance on completion of planning arrangements

The parties shall use their best endeavours, in accordance with the identified crisis scenarios, to supply associated bodies and, where appropriate, beneficiary bodies with data, and if necessary associated information and services, gathered by the space facilities. Implementation of the procedures described in the scenarios implies coordination of tasks between the parties, possibly leading to combining of the available resources:

- access to data archives
- merging of the data to aid understanding of pre-crisis situations
- access to data acquired at the time of the crisis
- merging of those data to report on the crisis
- routing of information to the user
- access to all the technological resources available -telecommunications, data collection, navigation.

The procedures for accessing and integrating data or other services (telecommunications, data collection, navigation) to obtain specific products shall, as far as possible, be stipulated in the scenario descriptions.

Article V – Associated bodies

5.1 The role of associated bodies in intervention by the parties is defined in Article 3.4.

5.2 An associated body shall, for the purposes of this Charter, be an institution or service responsible for rescue and civil protection, defence and security under the authority of a State whose jurisdiction covers an agency or operator that is a party to the Charter, or of a Member State of ESA or of an international organisation that is a party to the Charter.

5.3 Any entity or service authorised to this effect by the Board may also be considered an associated body.

5.4 The parties shall ensure that associated bodies which, at the request of the country or countries affected by a disaster, call on the assistance of the parties undertake to:

- alert the Secretariat as soon as possible in the event of a crisis and designate their points of contact;
- promptly provide the Secretariat with the necessary details;
- use the supplied information only for the purposes defined with the Secretariat;

- take part as necessary in the relevant meetings organised by the Secretariat;
- report on the use made of the data, information and services supplied and prepare an assessment of each case for which intervention took place;
- confirm that no legal action will be taken against the parties in the event of bodily injury, damage or financial loss arising from the execution or non-execution of activities, services or supplies arising out of the Charter;
- meet any other condition agreed with the Secretariat or Board.

Article VI – Accession

6.1 It is the intent of the parties to encourage the widest possible accession to the Charter by agencies and national or international space system operators.

Requests to adhere to the Charter may be made by any space system operator or space agency with access to space facilities which agrees to contribute to the commitments made by the parties under Article IV above and is willing to assume the responsibilities of a party under the terms of the Charter.

6.2 The Board shall examine accession requests and formulate its recommendations to the parties to the Charter within 180 days of their submission. In doing so, it shall consider that any new accession must, in particular:

- bring a significant contribution by the acceding party to the intervention capacity required for the purposes of the Charter and a commitment to bear its share of the common costs;
- help to achieve the objectives of the parties;
- be such as not to compromise normal deployment of the systems already in place.

On the basis of such recommendations by the Board, any accession shall require the unanimous approval of the parties to the Charter.

Article VII – Entry into Force, Expiry and Withdrawal

7.1 The Charter shall enter into force on the day of its signature by at least two parties. It may be terminated at any time by mutual consent of the parties. Any party may withdraw from the Charter after notifying, with 180 days' notice, the other party or parties in writing of its intention to do so. The possibility of pursuing the mission in a modified form shall be examined by the parties. The party intending to withdraw shall endeavour to maintain continuity of its current contribution.

7.2 Subject to the provisions of Article 7.1 above, the Charter shall remain in force for a period of five years from the date of its entry into force, and shall be automatically extended for subsequent periods of five years.

Article VIII – Implementation

The implementation arrangements for this Charter shall be defined by the parties meeting in the Board.

IN WITNESS WHEREOF, the undersigned have signed the Charter in two originals, one in the French and one in the English language.

List of Documents

Treaties and Conventions

- 1886 Berne Convention: Berne Convention for the Protection of Literary and Artistic Works of 1886 UNTS 11850 vol.828.
- 1909 Canada-United States of America: Washington Treaty Relating to Boundary Waters and Questions Arising Along the Boundary Between the United States and Canada, 11 January 1909, 12 Bevans 319, 36 Stat. 2488, TS 548, 102 BFSP 137.
- 1929 Warsaw Convention: Warsaw Convention for the Unification of Certain Rules Relating to International Transportation 137 LNTS 11, 1952.
- 1933 London Convention relative to the Preservation of Fauna and Flora in Their Natural State, 8 November 1933, UKTS No. 27 (1930), 172 LNTS 241.
- 1944 Chicago Convention: Chicago Convention on International Civil Aviation 7 December. 1944, Article 44, 15 UNTS 295.
- 1948 European Convention on Human Rights 10 Dec.1948.
- 1949 Geneva Convention I for the Amelioration of the Condition of the Wounded and Sick in Armed Forces in the Field Aug. 12, 1949, 6 U.S.T. 3114, 75 UNTS 31.
- 1949 Geneva Convention II for the Amelioration of the Condition of Wounded, Sick and Shipwrecked Members of Armed Forces at Sea, Aug. 12, 1949, 75 UNTS 85.
- 1949 Geneva Convention III Relative to the Treatment of Prisoners of War, Aug. 12, 1949, 75 UNTS 135.

- 1949 Geneva Convention IV Relative to the Protection of Civilian Persons in Time of War, Aug. 12, 1949, 75 UNTS 287.
- 1952 Rome Convention: Rome Convention on Damage caused by Foreign Aircraft to Third Parties on the Surface 310 UNTS 181.
- 1952 Universal Copyright Convention 943 UNTS 178.
- 1954 Convention relating to the Status of Refugees 189 UNTS 150.
- 1963 Vienna Convention on Civil Liability for Nuclear Damage 2 ILM 727.
- 1967 Outer Space Treaty: 1967 Treaty on Principles Governing the Activities of States in the Exploration and Use of Outer Space, including the Moon and Other Celestial Bodies, 27 Oct. 1967, 610 UNTS 205.
- 1967 Protocol to the Convention relating to the Status of Refugees 606 UNTS 267.
- 1968 Rescue Agreement: Agreement on the rescue of astronauts, the return of astronauts and the return of objects launched into outer space(1968) 672 UNTS 199.
- 1971 Ramsar Convention on Wetlands of International Importance especially as Waterfowl Habitat 2 February 1971, 996 UNTS 245.
- 1972 Liability Convention: Convention on International Liability for Damage Caused by Space Objects, 29 Nov. 1971 24 UST 2389.
- 1972 Convention concerning the Protection of World Cultural and Natural Heritage 1972, 1037 UNTS 151.
- 1973 Kyoto Convention on the Simplification and Harmonization of Customs Procedure TIAS 6633.
- 1973 International Convention for the Prevention of Pollution from Ships November 1973, 1313UNTS 3.
- 1975 Registration Convention: Convention on Registration of Objects Launched Into Outer Space, 12 Nov. 1974 28 UST 695.
- 1977 Bonn Convention for the Protection of the Rhine against Chemical Pollution, 3 December 1976, 16 ILM 242 (1977).
- 1978 Moscow Convention on the Transfer and Use of Data of Remote Sensing of the Earth from Space UN Doc. A/33/162 of 29 June 1978.
- 1979 Moon Agreement Governing the Activities of States on the Moon and Other Celestial Bodies, (1979) 18 ILM 1434.
- 1982 United Nations Convention on the Law of the Sea, Dec. 10, 1982, 21 ILM 1261.
- 1986 Vienna Convention on Early Notification in the case of Nuclear Accident or Radiological Emergency IAEA INRIRC 335, 25 IM 1370(1986).
- 1991 Convention on Environmental Impact Assessment in Transboundary Context, 30 ILM 802.
- 1992 Convention on Biological Diversity 5 June 1992, 31 ILM 822.
- 1992 United Nations Framework Convention on Climate Change 31 ILM 849.

- 1992 Convention on the Protection and Use of Transboundary Water Courses and International Lakes Mar. 17, 1992, 31 ILM 1312.
- 1994 Trade Related Aspects of Intellectual Property Rights 33 ILM 1140-1272.
- 1996 World Intellectual Property Organisation Copyright Treaty 1 36 ILM 65.
- 1998 Tampere Convention on the Provision of Telecommunication Resources for Disaster Mitigation and Relief 1439 UNTS 275.
- 1997 Kyoto Protocol to the Framework Convention on Climate Change 16 March 1998, 37 ILM 22.
- 1999 Convention on Access to Information, Public Participation, and Decision Making and Access to Justice in Environmental Matters 38 ILM 517.

Declarations, Principles and Documents

- Universal Declaration of Human Rights, UN Doc. A/810 at 71 (1948).
- Statute of the Office of the United Nations High Commissioner for Refugees, UN General Assembly Resolution, UN Doc. A/RES/428 (V), 14, 14 December 1950.
- UNGA Resolution 1962(XVIII) on Declaration of Legal Principles governing the Activities of States in Exploration and Use of Outer Space, adopted on 13 December 1963.
- UNGA Resolution 2625 (XXV) 24 October 1970 on Declaration on Principles of International Law Concerning Friendly Relations and Co-operation Amongst States in Accordance with the Charter of the United Nations.
- UN Doc. A/AC.105/C.2/L.73, 26 (1970).
- Stockholm Declaration of the United Nations on the Human Environment, UN Doc. A/Conf/48/14(1972).
- UN Doc. A/AC/105/111 (1973).
- UN Doc. A/AC.105/C.2/L.99(1974).
- UN Doc. A/AC/1/1047, Oct 15, 1974.
- UN Doc. A/AC.105/C.2/L.103 (1975).
- UN Doc. A/AC.105/288 (1981).
- UNGA Resolution 2816 on Assistance in cases of natural disaster and other disaster situations, 1981.
- UNGA Resolution 36/225 on Strengthening the capacity of the United Nations system to respond to natural disasters and other disaster situations.
- Direct Broadcasting Principles: The Principles Governing the Use by States of Artificial Earth Satellites for International Direct Television Broadcasting, UNGA Resolution 37/92 of 10 December 1982.

- 1984 UNDRO Draft Convention on Expediting the Delivery of Emergency Assistance of the Disaster Relief, UN Doc. A/39/267/Add .2-E/1984/96/Add.2, 18 June, 1984.
- 1986 Remote Sensing Principles: UNGA Resolution 41/65 on The Principles Relating to Remote Sensing of the Earth from Outer Space adopted on 3 December 1986, UN Doc. A/AC.105/572.
- 1988 Humanitarian Assistance to victims of natural disasters and similar emergency situations, UN Doc. A/RES/43/131(1988).
- 1991 UNGA Resolution, 46/182 Strengthening of the coordination of humanitarian emergency assistance of the United Nations adopted on 19 December 1991, UN Doc. A/RES/46/182.
- 1992 Nuclear Power Principles: UNGA Resolution 47/68 on the Principles Relevant to the Use of Nuclear Power Sources in Outer Space adopted on 14 December 1992.
- 1992 Rio Declaration on Environment and Development, 31 ILM 874(1992).
- 1996 UNGA Resolution 51/122 on Declaration on International Cooperation in the Exploration and Use of Outer Space for the benefit and in the interests of All States, Taking into Particular Account the Needs of Developing Countries, UN Doc. A/AC.105/C.2/L.182/Rev.2(1995).
- 2006 Study on the possibility of creating an international entity to provide for coordination and the means of realistically optimizing the effectiveness of space-based services for use in disaster management, UN Doc. A/AC.105/C.1/L.285(2006).
- 2006 United Nations Platform for Space-based Information for Disaster Management and Emergency Response, UN Doc. A/RES/61/110, 14 December, 2006.

Europe

- Directive 85/374/EEC on the approximation of the laws, regulations and administrative provisions of the Member States concerning liability for defective products; OJ 1985 L 210 /29 Amended by Directive 1999/34/EC of the European Parliament and of the Council of 10 May 1999.
- Directive 96/9/EC of European Parliament and of Council on the legal protection of databases of March 1996;OJ L77/20(1996).
- Directive 2003/4/EC of the European Parliament and of the Council of 28 January 2003 on public access to environmental information and repealing Council Directive 90/313/EEC.
- Directive 2007/2/EC of the European Parliament and the council of 14 March 2007 establishing an infrastructure for spatial information in the European Community.

National Documents

Australia

- Space Activities Act, December 21, 1998, No. 123.

Canada

- Emergency Medical Aid Act of Alberta R.S.A. 2000, c. E.
- Good Samaritan Act of Ontario S.O.2001, c.2.
- Agreement between the Government of Canada and the Government of the United States of America concerning the Operation of Commercial Remote Sensing Satellite Systems, U.S.- Can., June 16, 2000, 2000 Can. T.S. No. 2000/14.
- Act Governing the Operation of Remote Sensing Space Systems, S.C. 2005, c. 45.
- Remote Sensing Space Systems Regulations, SOR/2007-66.

France

- French Intellectual Property Code, 1 July 1992, no. 92-597.
- Declaration by certain European Governments on the Launchers Exploitations Phase of Ariane, Vega, and Soyuz from the Guiana Space Centre, Paris 30 2007.
- French Law on Space Operations, Act no. 2008-518, 3 June 2008.
- Decree of 6 June 2009 supporting the applications of measures provided by Title VII of the French Space Operations Act.

Germany

- Law Dealing with Copyright and Related Rights (Copyright Law) of September 9, 1965, as last amended by the Law of June 23, 1995.
- Satellite Data Security Act Federal Gazette BGBI.2007 I. No.58, 28.11.2007.

India

- Remote Sensing Data Policy, ISRO:EOS:POLICY-01:2001.

Japan

- Disaster Countermeasure Basic Act, document no. 223, adopted on 15 November 1961.
- Japan's Fundamental Act of Outer Space, Law No. 43, 2008.

Korea

- Korean Space Exploration Promotion Act, December 1, 2005.

Netherlands

- Rules Concerning Space Activities and the Establishment of a Registry of Space Objects 13.06.2006.

Philippines

- Memorandum order no. 369, of 26 June 1991.
- Republic Act 7637 of 24 September 1992, and Presidential Proclamation 1201 of 19 March 1998.

Russia

- Law of Russian Federation on Space Activity, August 20, 1993, Presidential Edict No. 185, About structure of management of space activity in Russian Federation (25 February 1992).

South Africa

- Space Affairs Act of South Africa, September 6, 1993.

Sweden

- Act on Space Activities 1982:963, 18 November 1982, Decree on Space Activities (1982:1069).

Ukraine

- Ordinance of The Supreme Soviet of Ukraine, On Space Activity Law of Ukraine, 15 November 1996.

UK

- UK Outer Space Act, July 18, 1986, 1986 Chapter 38.
- Copyright, Designs and Patents Act 1988, 1988 Chapter 48.

US

- An Act for the general revision of the Copyright Law, title 17 of the United States Code, and for other purposes, January 1978, 90 Stat. 2541 (1976).
- Commercial Space Launch Act, Public Law 98-575, 98th Congress, H.R. 3942, 30 October 1984.
- Land Remote Sensing Commercialization Act, Public Law 98–365, 98th Congress, H.R. 5155, 17 July 1984.
- Land Remote Sensing Policy Act 1992, Public Law 102–555, 102nd Congress, H.R. 6133, 28 October 1992.
- Presidential Decision Directive 23, US Policy on Foreign Access to Remote Sensing Capability, 9 March 1994.
- 1996 Bill Emerson Good Samaritan Food Donation Act P.L. 104-210.
- 1997 Volunteer Protection Act 111 Stat. 218.

- 2003 Fact sheet: US Commercial Remote Sensing Policy of April 25, 2003.
- 2005 The Good Samaritan Liability Improvement and Volunteer Encouragement Act of 2005, S. 1747, 109th Congress.
- Federal Register Vol. 71, No. 79, April 25, 2006, 15 CFR Part 960, Licensing of Private Land Remote Sensing Systems; Final Rule.

Case Law

International

- Trail Smelter Arbitration (United States v. Canada) 16 April 1938 3 RIAA 190781941.
- *Nuclear Test Case*(Australia v. France)(New Zealand v. France) 1973 ICJ Reports 99.
- *Mali v. Burkina Faso* 1986 ICJ 554. ICJ Report.
- Gabcikovo-Nagymaros (Hungary v. Slovakia) ICJ 1997 Reports 7.
- *Commission v. United Kingdom*, Case C –300/95 [1997] ECR I—2649, I-2670.
- *Levob Verzekeringen OV Bank v. Staatssecretaris van Financiën 2005*, CJCE, 27 October 2005 (affaire C-41/04).
- *DIC of Delaware, Inc., et al. v. Tehran Redevelopment Corporation et al*, 8 Iran-US CTR 144 (1985).

National

France

- *Lamberth v. Guiron,* (1963) 26 R.G.A. 185.
- *Diop v. Cie Air France,* (1966) 29 R.G.A.E. 42 (T.G.I. Seine, 6 November 1964).
- *S.A. Electricité de Strasborg c. SNCF,* bull. I, No. 27 (Cass. Civ. Ire, 26 January 1971).

- *Compagnie des Courtiers v. Cellier des Halles,* CA Paris, 4 Ch., March 26, 1991, (1991) 150 R.I.D.A. 148.
- *Duchier v. Gueroult,* CA Paris, 4 Ch.; (1995) 166, R.I.D.A. 240.
- *Euro Vacancies v. Froment* Cass. I civ., Oct. 24, 1995(1996) 168 R.I.D.A. 276.

U.K.

- *Stubley v. The London and North Western Railway Co* [1865-66]LR 1 Exch 13, 18.
- *Nicolas v. Marsland,* [1876] 2 Ex. D 1.
- *G.A. Cramp & Sons Ltd. V. Frank Smythoson Ltd.,* [1944] A.C. 329, H.L.
- *Bolton and Stones* [1951]AC 850.
- *Horabin v. British Overseas Airways Corporation,* [1952] 2 All E.R. 1016 (Q.B. 1952).
- *Latimer v. AEC Ltd* [1953]AC 643.
- *Bolam v. Friern Hospital,* [1957] 1 WLR 582.
- *Hedley Bryne v. Hellers Partners* [1964] AC 465.
- *La Rossa v. Scientific Design co,* 402 F. 2d 937 (1968).
- *Ministry of Housing and Local Government v. Sharp,* [1970]2 QB 223.
- *Mutual Life and Citizens' Insurance Co Ltd v. Evatt,* 1971 AC 793.
- *Kaiser Steel Corp. v. Westinghouse Corp,* 55 Cal. App. 3d 737; 127 Cal. Reptr. 838(1976).
- *Stafford v. Conti Commodity Services Ltd,* [1981]1 All ER 691.
- *Interlego Ag v. Tyco Industries,* 1989 1 A.C. 217.
- *Smith v. Eric Bush,* [1990] 1 AC 831.
- *Henderson v. Merrett Syndicates* [1995] 2 AC 145, [1994] 3 All ER 506, HL.
- *Spring v. Guardian Assurance* [1995] 2 AC 296 [1994] 3 All ER 129, HL.
- *White v. Jones* [1995] 2 AC 207, [1995] 1 All ER 691.
- *British Horseracing Board v. William Hill,* HC 2000 1335

U.S.

- *MacPherson v. Buick,* 217 NY 382, 111 N.E. 1050.
- *United States v. Caroll Towing Co.* 350 U.S. 61 (1955).
- *Greenman v. Yuba Power Products* 59 Cal. 2d 57 (1963).
- *Ingham v. Eastern Airlines,* 373 F. 2d 227 at 236 (2nd Cir., 1967).
- *United States vs. Reserve Mining,* 380 F. Supp. 11 (D. Minn. 1974).
- *Batlin and Son Inc. v. Snyder,* 516 F. 2d (2nd Cir. 1976).
- *Chevron USA, Inc. v. US EPA,* 658 F. 2d 271 (5th Cir., 1981).
- *Oklahoma Water Resource Board v. Texas County Irrigation and Water Resource Association,* 711 P.2d 38 (OK 1984).

- *Boucher v. Riner*, 68 Md. App. 539, 543, 514 A 2d 485, 487 (1986).
- *Nickel Plate Mining Co. v. Office of Surface Mining Reclamation and Enforcement*, NX5-122-R (Dep't Interior 1987).
- *Gasser v. United States*,14 Cl. Ct. 476(1988).
- *Appalachian Insurance Co. v. McDonnell Douglas*, 262 Cal. Rptr. 716 (Ct. App. 1989).
- *Highway Admin. v. Greiner Engineering Sciences, 83 Md.App. 621, 577 A.2d 363* (*Md.*Ct.Spec.*App.*), cert. denied, 321 *Md.* 163, 582 *A.2d* 499 (1990).
- *Kern River Gas Trans. Co.v. Coastal Corp*, 899 F.2d 1458 (5th Cir. 1990).
- *Feist Publications Inc. v. Rural Telephone Service*, 111 S. Ct 1282 (1991).
- *Martin Martietta v. INTELSAT*, 763 F Supp. 1327 (D.Md.1991).
- *Mason v. Montgomery Data Inc* 967 F. 2d 135 (5th Cir. 1992).
- *Lexignton v. McDonnell Douglas*, 214 Cal ap 3d 1, 262 Cal Reporter 716 (Cal Ap Fourth District 1989)
- *Daubert v. Merrell Dow Pharmaceuticals*, 509 U.S. 579 (1993).
- *Kupetz v. Deere & Co*, 644A 2d 1213(1994).
- *Westel-Oviatt Lumber Co. v. United States*, 40 Fed. Cl. 557(1998)
- *Bridgeman Art Library Ltd v. Corel Corporation*, 36 F. Supp. 2d 191(S.D.N.Y. 1999).
- *Willingham v. Hudson*, 274 Ga App. 200, 617 S.E. 2d 192 Ga. App. 2005.

Bibliography

I. General Books

Aoki, S., *Japanese Space Strategy*, Keio University Publisher, Tokyo, 2006.
Benko, M., Garaff, W. & Reijnen, G.C.M., *Space Law in the United Nations*, Martinus Nijhoff Publishers, Dordrecht, 1985.
Birnie, P.W. & Boyle, A.E., *International Law and the Environment*, Oxford University Press, New York, 2002.
Buckley, R.A., *Modern Law of Negligence*, Butterworths, London, 1999.
Cheng, B., *Studies in International Space Law*, Clarendon Press, Oxford, 1997.
Christol, C., *Past, Present and Future of Space Law*, Kluwer Law and Taxation Publishers, Deventer, 1991.
Dailler, P. & Pellet, A., *Droit International Public*, 7th edition, LGDJ, Paris, 2002.
Gordon, A., *Contract Law*, Second Edition, Thomson W. Green Publisher, Edinburgh, 2003.
Gorove, S., *Cases on space law: texts, comment and references*, Mississippi State University Publisher, Starkeville, 1996.
Gorove, S., *Developments in Space Law*, Kluwer Academic Publishers, Dordrecht, 1991.
Gorove, S., *Studies in Space law: its challenges and prospects*, Sijthoff Publisher, Leyden, 1977.
Hays, P., (Ed.), *Spacepower for a New Millennium*, McGraw Hill Companies, 2000.
Haanappel, P.P.C., *The Law and Policy of Air Space and Outer Space*, Kluwer Publisher, Hague, 2003.
Ilardi, A. & Blakeney, M., *International Encyclopaedia of Intellectual Property Treaties*, Oxford University Press, New York, 2004.
Karasawa, H., *International Transaction*, Dobunkan Publisher, Tokyo, 1996.
Jenks, C.W., *Space Law*, Stevens Publisher, London, 1965.

Jenks, E., *Digest of English Civil Law*, Butterworth, London, 1947.

Kiss, A., *International Environmental Law*, Transnational Publishers, New York, 2000.

Lachs, M., *The Law of Outer Space: an Experience in contemporary law making*, Stijhoff Publisher, Leiden, 1972.

Louise Van Traa-Engelman, H., *Commercial Utilization of Outer Space*, SI Publisher, Utrecht, 1989.

Louka, E., *International Environmental Law: fairness, effectiveness, and world order*, Cambridge University Press, New York, 2006.

Malanczuk, P., *Akehurst's Modern Introduction to International Law*, Seventh Revised Edition, Routledge, London, 1997.

Markesinis, B., *Markesinis and Deakin's Tort Law*, Oxford University Press, Oxford, 2003.

Miyatake, H., *Intellectual Property and Creativity*, Misuzu Publisher, Tokyo, 2007.

Nieuwenhuis, G.J.A., Vaughan, R.A. & Molenaar, M., *Operational remote sensing for sustainable development*, Balkema, Rotterdam, 1999.

Munir, A.B., *Cyber Law Policies and Challenges*, Butterworth Asia, Kuala Lumpur, 1999.

Piradov, A.S., Belitsky. (trans.) *International Space Law*, Hawaii, University Press of Pacific, 2000.

Reijnen, G., *Utilization of Outer Space and International Law*, Elsevier Scientific Publishing Company, Amsterdam, 1981.

Reynolds, G.H. & Mergers, R.P., *Outer Space Problems of Law and Policy*, Second Edition, Westview Press, Colorado, 1997.

Roth, A., *La prohibition de l'appropriation et les regimes d'acces aux espaces extra-terrestres*, Presses Universitaires de France, Paris, 1992.

Sands, P., *Principles of International Environmental Law*, Cambridge University Press, Cambridge, 2003.

Shaw, M., *International Law*, Fifth Edition, Cambridge University Press, Cambridge, 2003.

Sterling, J.A.L., *World Copyright Law*, Sweet & Maxwell, London, 2003.

Stokke, O.S. & Thommessen, O.B., *Yearbook of International Cooperation on Environment and Development 2002/2003*, Earthscan Publications Ltd, London, 2002.

Tatsuzawa, K., *Space Law System – Legal System for Space Development*, Maruzen Planet, Tokyo, 2000.

Terrill, D.R., *Air Force Role in Developing International Outer Space Law*, Air University Press, Alabama, 1999.

Von der Dunk, F.G., *Private Enterprise and Public Interest in the European 'Spacesacpe' towards harmonized national space legislation for private space activities in Europe*, International Institute of Air and Space Law, Leiden, 1998.

Whincup, M., *Contract Law and Practice*, Kluwer Law International, Hague, 2001.

Whittaker, S., *Liability for Products*, Oxford University Press, Oxford, 2005.

II. Specialized Books

Amerasinghe, C., *Evidence in International Litigation*, Martinus Nijhoff Publishers, Leiden, 2005.

Beigbeder, Y., *The role and status of international humanitarian volunteers and Organizations*, Martinus Nijhoff Publishers, Dordrecht, 1991.

Bogsch, A., *The law of copyright under the Universal Convention*, Sijthoff publisher, Leiden, 1968.
Courteix, S., (Eds.), *Droit télédétection et environnement: actes du colloque international: le droit face aux techniques de télédétections par satellite au service de l'environnement*, 2-3-4 juin 1993, Sides, Antony, 1994.
Churchill, R.R., and Freestone, D.A.C., *International law and global climate change*, Graham & Trotman, London, 1991.
Day, D.A., and Logsdon, J.M., (Eds.), *Eye in the Sky: Story of the Corona Spy Satellites*, Smithsonian Inst Press, 1999.
Freestone, D., and Streck, C., *Legal Aspects of Implementing the Kyoto Protocol mechanisms: making Kyoto work*, Oxford University Press, Oxford, 2005.
Gahtan, A., *Electronic Evidence*, Carswell Publishing, Tronto, 1999.
Geddes, A. A., *Product and Service Liability in the EEC*, Sweet and Maxwell, London, 1992.
Grubb, M., Vrolijk, C., Brack, D., and Forsyth, T., *The Kyoto Protocol: a guide and assessment*, Royal Institute of Internal Affairs, London, 1999.
Harris, R., (Eds.), *Earth Observation Data Policy and Europe*, Balkema Publishers, Lisse, 2002.
Hettling, J.K., *Satellite for verification and enforcement of public international law*, Heymann, Koln, 2008.
Macalister-Smith, P., *International Humanitarian Assistance Disaster Relief Actions in International Law and Organization*, Martinus Nijhoff Publishers, Geneva, 1985.
Matsui, K., *Basics of Digital Watermarking*, Morikita Publisher, Tokyo, 1998.
Miller, G., *Liability in International Air Transport*, Kluwer Publishers, Deventer, 1977.
Miller, C.J. & Goldberg, R.S., *Product Liability*, Oxford University Press, New York, 2004.
Nawa, K., *Digital Copyright*, Misuzu Publisher, Tokyo, 2004.
Niyama, Y., *Traceability in Food Chain*, Showado, Tokyo, 2005.
Norman, S., *Practical Copyright for Information Professionals*, Facet Publishing, London, 2004.
Oberthur, S., and Ott, H.E., *Kyoto Protocol International Climate Policy for the 21st Century*, Springer, Berlin, 1999.
Okolie, C.C., *International Law of Satellite Remote Sensing and Outer Space*, Kendall/ Hunt Publishing Co., Dubnque, 1989.
Ricketson, S., Berne *Convention for the Protection of literary and artistic works*, Kluwer Publisher, Deventer, 1987.
Sagawara, K., *Global Warming and Responsibility of America*, Fujiwara Press, Tokyo, 2002.
Schauer, F., and Sinnott-Armstrong, W., *The Philosophy of Law: Classic And Contemporary Readings with Commentary*, Harcourt Brace College Publishers, Oxford, 1996.
Seitz, J., *Digital Watermarking for Digital Media*, Information Science Publishing, Hershey, 2005.
Wassenbergh, H.A., *Outer Space in Hindsight*, Kluwer Academic Publishers, Dordrecht, 1991.
Yamamoto, K., *Agricultural Products and Traceability*, Seibyundo Publisher, Tokyo, 2003.

Yamashita, T., *Intelligence Strategy for Risk Corresponding Society*, Nikkei BP Publisher, Tokyo, 2006.

Yamin, F., and Depledge, J., *International Climate Change regime: a guide to rules, institutions and procedures*, Cambridge University Press, Cambridge, 2004.

III. Academic Articles, Papers, Report

Aamoth, R.J., Use of Remote Sensing Imagery by the News Media, *Heaven and Earth: Civilian Uses of Near-Earth Space*, Martinus Nijhoff Publishers, Hague, 1997, 139.

Abeyratne, R., ICAO's Involvement in Outer Space Affairs – a Need for closer scrutiny? 30 *Journal of Space Law* (2004), 185.

Abeyratne, R., The Legal Effect of ICAO Decisions and Empowerment of ICAO by Contracting States, 32 *Journal of Space Law* (2007), 525.

Achache, J., From GMES to NOE: a European network for the management of the environment, 17 *Space Policy* (2001), 97.

Achilleas, P., French Remote Sensing Law, 34-1 *Journal of Space Law* (2008), 1.

Awford, I., Commercial Space Activities: Legal Liability Issues, *Recent Trends in International Space Law and Policy*, Lancers Books, New Delhi (1997), 383.

Arzt, C., Use of Satellite Imagery in Legal Proceedings, 24-4/5 *Air and Space Law* (1999), 195.

Back-Impalomeni, E., Legal Protection through agreements, contracts of scientific information and data *Proceedings of the International Conference: Satellite Remote Sensing in aid of development: legal considerations*, Tunis 26–27 September 2002, European Centre for Space Law, 2003, 57.

Bannon, V., and Fisher, D., 10-6 Legal Lessons in Disaster Relief from the Tsunami, the Pakistan Earthquake and Hurricane Katrina, *ASIL Insight*, 2006, 1.

Becker, R., 73/78: an overview in international environmental enforcement, 10-2 The *Georgetown International Environmental Law Review* (1998), 625.

Berkowitz, M.J., National Space Policy and National Defense, *Space Power for a New Millennium - Space and US National Security*, McGraw Hill Companies, 2000, 37.

Benkö, M. & Schrogl, K., Article I of the Outer Space Treaty reconsidered after 30 years "Free Use of Outer Space" vs. "Space Benefits", *Outlook on Space Law over the next 30 years – Essays Published for the 30th Anniversary of the Outer Space Treaty*, Lafferanderie (Eds.), Kluwer Law International, Hague-London-Boston, 1997, 67.

Bockstiegel, K-H., Case on Space Activities, *Space law: development and scope*, N. Jasentuliyana, N. (Eds.), Prager Publisher, Wesport, 1992, 211.

Brachet, G., From Initial ideas to a European plan: GMES as an exemplar of European space strategy, 20 *Space Policy* (2004), 7.

Brigham, A., Natural resources the view from above, 55-2 *Journal of Air Law and Commerce* (1989), 505.

Catalano Sgorosso, G., Prevention and Management of Natural Disasters, *Outlook on Space Law over the next 30 years – Essays Published for the 30th Anniversary of the Outer Space Treaty*, Kluwer Law International, Hague-London-Boston, 1997, 319.

Catalano Sgorosso, G., Sharing of Remote Sensing Data Concerning Environmental Protection for Public Benefit, *Proceedings of Thirty Ninth Colloquium of Law of the Outer Space*, IISL, AIAA, 1995, 94.

Cheng, B., The 1967 Outer Space Treaty: Thirtieth Anniversary, 23-4/5, *Air and Space Law*, (1998), 156.

Cheng, C.J., New Sources of International Space Law, The Use of Air and Outer Space Cooperation and Competition: Proceedings of the International Conference on Air and Outer Space at the Service of World Peace and Prosperity, Held in Beijing from 21–23 August 1995, Kluwer Law International, The Hague-Boston-London, 1998, 207.

Christol, C., Article 2 of 1967 Principles Treaty Revisited, 9 *Annals of Air and Space Law* (1984), 217.

Christol, C., Common interest in the exploration, use, exploitation of Outer Space for Peaceful Purposes: Soviet –America Dilemma, 8 *Annauire de Droit Maritime and Aérien*, 1985, 235.

Christol, C.Q., Remote Sensing and International Law, 5 *Annals of Air and Space Law* (1980), 375.

Clarke, D., Access Control of Remote Sensing Satellites, *Commercial Satellite Imagery and UN Peacekeeping*, Ashgate, Aldershot, 2003, 171.

Cocoa, A.A., Legal Principles on Remote Sensing From 1970 to 1986: a brief comparative survey of the first and the final text IISL, *Proceedings of Thirtieth Colloquium of Law of the Outer Space*, IISL, AIAA, Washington, 1986, 276.

Cohen, A.F., Cosmos 954 and the International Law of the Satellite Accidents, 10 *Yale Journal of International Law* (1984), 78.

Cooper, J.C., High Altitude Flight and National Sovereignty, *International Law Quarterly Review*, (4 July 1951), 411.

Courteix, S., Toward the Legal Recognition of a new method of proof for the defence of the environment: satellite images, *Proceedings of the Thirty-Seventh Colloquium on the Law of Outer Space*, IISL, Washington, AIAA, 1994, 228.

Danilenko, G.M., Principles relating to Remote Sensing of the Earth from Space: Territorial Sphere of Application, *Proceedings of the Thirtieth Colloquium on the Law of Outer Space*, IISL, AIAA, Washington, 1987, 289.

Dalbello, R., and Martinez, L., The Legal and Political Implications of Media newsgathering from Space, *Proceedings of the Thirtieth Colloquium on the Law of Outer Space*, IISL, AIAA, Washington, 1987, 279.

Deekshatulu, B. L., Venkataraman,V. R., and Chandrasekhar., Overview of Remote Sensing, *Recent Trends in International Space Law and Policy*, Lancers Books, New Delhi, 1997, 221.

Desaussure, H., Remote Sensing by Satellite: What future for an international regime, 71-4 *American Journal of International Law* (1977), 720.

Desaussure, H., Remote Sensing Satellite Regulation by National and International Law, 15 *Rutgers Computer & Technology Law Journal* (1989), 351.

De Santis, H., Satellites, Alliance, Relations and the Developing World, *Commercial Observation Satellites and International Security*, Hampshire, the Macmillan Press, 1990, 78.

Diederiks-Verschoor, I.H.Ph., Product liability and insurance on aircraft and satellite, The utilization of the world's air space and free outer space in the 21st century: proceedings of International Conference on Air and Space Policy, Law and Industry for the 21st Century, held in Seoul from 23–25 June 1997, 197.

Draetta, U., Les clauses de force majeure et de hardship dans les contrats internationaux, 3–4 *Revue de Droit des Affaires Internationales* (2002), 347.

Dufresne, L., Protection of space data products under the European Directive on the legal protection of databases, *Proceedings of the International Conference: satellite remote sensing in aid of development: legal considerations*, Tunis 26–27 September 2002, European Centre for Space Law, 2003, 39.

Evans, R., The Humanitarian Challenge: A Foreign Policy Perspective, 6-2 *African Security Review* (1997), 1.

Feder, H., The Sky's Limit? Evaluating the International Law of Remote Sensing, 23-2 *NYU Journal of International Law and Politics* (1991), 611.

Ferrazzani, M., International Agreements and Contractual Practice for Collecting and Distributing Remote Sensing Data, *Droit télédétection et environnement: actes du colloque international: le droit face aux techniques de télédétections par satellite au service de l'environnement*, Strasbourg, 2-3-4 juin 1993, Sides, Antony, 1994, 181.

Florini, A., and Dehqanzada, Y.A., No More Secrets?: Policy Implications of Commercial Remote Sensing Satellites, Governance, Transparency and Civil Society Global Policy Program, Project on Transparency, Working Paper no. 1, Carnegie Endowment for International Peace, Washington, July 1999, 1.

Florini, A., and Dehqanzada, Y.A., Commercial Satellite Imagery Comes of Age, *Issues in Science and Technology* (fall 1999), 1.

Florini, A., The Opening Skies: Third Party Imaging Satellites and US Security, 13-2 *International Security* (1998), 91.

Fomtchenko, M.M., and Movlyav, A.S., High Resolution Remote Sensing: New Aspects and Problems, *Proceedings of the Thirty-Ninth Colloquium on the Law of Outer Space*, IISL, AIAA, 1996, 85.

Foresman, T., and Williams, D., Remote Sensing: An Environmental Enforcement Tool, *Earth observation systems: legal considerations for the '90s*, American Society for Photogrammetry and Remote Sensing, Maryland, 1990, 30.

Gál, G., Legal principles of remote sensing in the Moscow Convention of 1978, International space law miscellaned: liber amicorum honouring Professor Dr Andrzej Gorbiel in his 65th anniversary, Palyga, E. (Eds.), Andrej Frycz Modrzewski Foundation, Warsaw, 1995, 91.

Galloway, E., Remote Sensing Satellites and Legality, *Legal Implication of Remote Sensing From Outer Space*, Matte, N., and Dessaure, H. (Eds.), Leyden, Sijthoff, 1976, 91.

Gabrynowicz, J.I., Defining Data Availability for Commercial Remote Sensing, 23 *Annals of Air and Space Law*, 1998, 93.

Gabrynowicz, J.I., Expanding Global Remote Sensing Services, *Proceedings of Workshop on Space Law in the Twenty-first Century*, United Nations, 2000, New York, 97.

George, H., Remote sensing of Earth resources: emerging opportunities for developing countries, 14 *Space Policy* (1998), 27.

Gorin, P.A., Zenit - The First Soviet Photo-Reconnaissance Satellite, 50-11 *Journal of the British Interplanetary Society*, 441.

Hampe, W., and Mueller, R., The International Legal Framework for Conducting Remote Sensing Activities, *Proceedings of the Thirtieth Colloquium on the Law of Outer Space*, 1987, 318.

Hampton, S., Kelly, P.R., and Carter, H.R., 31 Tank vessel operations, seabirds and chronic oil pollution in California, *Marine Ornithology*, 2003, 29.

Harpwood, V., Joint and Several Liability in Tort, *The Law of Tort*, Grubb, A., (Ed.), Reed Elsevier, London, 2002, 129.

Hedley, S., Professional Liability, *The Law of Tort*, Grubb, A. (Eds.), Reed Elsevier, London, 2002, 685.

Hayward, C., Remote Sensing: Terrestrial Laws for celestial Activities, 8 *Boston University International Law Journal,* (May 5), 1989, 157.

Hardcastle, R.J., and Chua, A.T.L., 325 *International Review of the Red Cross* (1998), 589.

Henaku, K., The International Liability of the GNSS Space Segment Provider, 21-1 *Annals of Air and Space Law* (1996), 143.

Harris, R., Comparison of regime: ESA, NASA and NOAA, *Proceedings of the International Conference: Satellite Remote sensing in aid of development: legal considerations, Tunis* 26–27 September 2002, 37.

Harris, R., Browning, R., Global Monitoring for Environment and Security: data policy considerations, 19 *Space Policy* (2003), 265.

Harris, R., and Browning, R., Data Policy Assessment for GMES Final Reports, EVK2-CT-2002-80012-DPAG, University of London, London, 2004, 1.

Holdaway, R., Is space global disaster waning and monitoring now nearing reality?, 17 *Space Policy* (2001), 127.

Ito, A., Legal Aspects of Implementing World Heritage Convention Using Remote Sensing Data, *Proceedings of the 48th Colloquium on the Law of Outer Space,* IISL, AIAA, Washington, 2006, 70.

Ito, A., Legal Aspect of International Charter on Space and Major Disasters, *Proceedings of the Forty seventh Colloquium on the Law of Outer Space,* IISL, AIAA, Washington, 2005, 233.

Ito, A., Issues in the implementation of the International Charter on Space and Major Disasters, 21-2 *Space Policy* (May 2005), 141.

Ito, A., Improvements to the Legal Regime for the Effective use of Satellite Remote Sensing Data for Disaster Management and Protection of the Environment, 34-1 *Journal of Space Law* (2008), 45.

Jakovljevic, B., International Disaster Relief Law, 34 *Israel Yearbook on Human Rights* (2004), 251.

Jasani, B., Security – A New Role for Civil Remote Sensing Satellites, *Air and Space Law in the 21st Century,* Heymann Publisher, Köln, 2001, 339.

Joyner, C.C. & Miller D.R., Selling satellites: the commercialization of LANDSAT, *26 Harvard International Law Journal* (1985), 63.

Kaplan, L.S., Recent Developments in Space Law Litigation, *Research and Invention in Outer Space,* S. Mosteshar (Eds.), Nijhoff Publisher, Dordrecht, 1995, 113.

Karjala, D., Copyright in Electronic maps, 35 *Jurimetrics Journal* (1995), 395.

Kerrest, A., Liability for Damage Caused by Space Activities, *Space Law: Current Problems and Perspectives for Future Regulation.* Marietta Benkö and Kai-Uwe Schrogl (Eds.), Eleven International Publishing, 2005, 91.

Kerrest, A., Remarks on the responsibility and liability for damages caused by private activity in outer space, *Proceedings of the Fortieth Colloquium on the Law of Outer Space,* 1997, 134.

Kopal, V., The Role of United Nations Declarations of Principles in the Progressive Development of Space Law, 16-1 *Journal of Space Law,* (1988), 85.

Kotz, H., Contract Remedies in Civil Law and Common Law, De tous horizons: melanges Xavier Blanc-Jouvan, Société de Législation comparée, Paris, 2005, 573.

Kunich, J.C., World Heritage in danger in the hotspots, 78-2 *Indiana Law Journal* (2003), 619.

Lee, Y., Registration of space objects: ESA member states' practice, 22 *Space Policy* (2006), 44.

Lessig, L., Creative Commons Intellectual Property in the Digital Era, NTT Publisher, Tokyo, 2005.

Longhorn, R., & Blakremore M., Re-visiting the Valuing and Pricing of Digital Geographic Information, 4-2 *Journal of Digital Information* (2004), 1.

Marchisio, S., Remote Sensing for Sustainable Development in international law, *Outlook on Space Law over the next 30 years – Essays Published for the 30th Anniversary of the Outer Space Treaty*, Kluwer Law International, Hague-London-Boston, 1997, 335.

Markowitz, K.J., Legal Challenge and Market Rewards to the use and Acceptance of Remote Sensing and Digital Information as Evidence, 12 Duke Environmental Law and Policy Forum (September 2002), 230.

Mejia-Kaiser, M., Propriety Rights in Remote Sensing Image, *Proceedings of the 38th Colloquium on the Law of Outer Space* (1995), 30.

Mejia-Kaiser, M., Satellite Remote Sensing Data in Databases Copyright or sui generis Protection in Europe?, 22-1 *Annals of Air and Space Law* (1997), 496.

Monti, G., Osman v. UK-Transforming English Negligence Law into French Administrative Law?, 48 *International and Comparative Law Quarterly* (1999), 757.

Myers, D.S., Remote Sensing and National Sovereignty over Natural Resources: Assessment of Mexican View, 14-1 *California Western International Law Journal* (1984), 22.

Nabet, E., Legal Aspects of the Use and Applications of Remote Sensing in South East Asia, 5 *Singapore Journal of International and Comparative Law*, 2001, 156.

Nolan, D., Product Liability, *The Law of Tort*, Grubb, A. (Eds.), Leed Elsevier, 2002, 795.

Ohlemacher, R., 19 The Earth Observation Summit: hearing a new era for international cooperation, *Space Policy* (2003), 277.

Onsurd, H.J., Liability in the Use of GIS and Geographical Datasets, Geographical Information Systems: Vol. 2 Management Issues and Applications, Inc, Longley, M., Goodchild, D., et al, John Wiley and Sons (Eds.), 1999, 643.

Özacayit, O., MARPOL and its Amendments, 10-5 *The Journal of International Maritime Law* (2004), 444.

Patrick, A., Propriety Aspects of Remote Sensing Imagery, 13 *Northwestern Journal of International Law and Business* (1992), 349.

Rajski, J., Les clauses limitatives exon ératatoires de responsabilité, 3-4 *Revue de Droit des Affaires Internationales* (2002), 321.

Rao, U.R., Space for Sustainable Development, *Recent Trends in International Space Law and Policy*, Lancers Books, New Delhi, 1997, 27.

Redgwell, C., The World Heritage Convention and Other Conventions Relating to the Protection of the Natural Heritage, The World Heritage Convention: a commentary, Oxford University Press, Oxford, 2008, 377.

Reichman, J.H., and Samuelson, P., Intellectual Property Rights in Data?, 50 *Vanderbilt Law Review* (1997), 51.

Roberto, V., and Widemer, C., Tort Liability for Services, Swiss Reports Presented at the XVI International Congress of Comparative Law, Schulthess, Zurich – Bale – Geneva, 2002, 203.

Roberts, M.A., US. Remote Sensing Data from Earth Observation – Law Policy and Practice, *Proceedings of the Thirty ninth Colloquium on the Law of Outer Space*, IISL, AIAA, Washington, 1996, 105.
Salin, P.A., Landsat Contracts Signed by US Agencies with Foreign Ground Stations: Commercial Remote Sensing from NASA Scientific Experiments to EOSAT Private Endeavours, 41 *German Journal of Air and Space Law* (1992), 165.
Schubert, F.P., An International Convention on GNSS Liability: When does Desirable Become Necessary?, XXIV *Annals of Air and Space Law* (1999), 245.
Schreier, G., Data Policy Implications on Archive Design, Earth Observation Policy and Europe, A.A. Balkema Publishers, Lisse, 2002, 175.
Serra, P.R.M., and Barbosa, M.N., The point of view of and user country: A) Brazil, *Droit télédétection et environnement: actes du colloque international: le droit face aux techniques de télédétections par satellite au service de l'environnement*, Courteix, S. (eds), Strasbourg 2-3-4 juin 1993, Sides, Antony, 1994, 209.
Shaffer, L.R., The American Distribution System for Satellite Remote Sensing Data, *Droit télédétection et environnement: actes du colloque international: le droit face aux techniques de télédétections par satellite au service de l'environnement*, Straousbourg, 2-3-4 juin 1993, Sides, Antony, 1994, 123.
Sloup, G.P., Mediasat, Gray Reconnaissance, and the new United Nations Principles, *Proceedings of Thirtieth Colloquium on the Law of Outer Space*, IISL, AIAA, Washington, 1987, 385.
Sneifer, Y., The Implications of National Security Safeguards on the Commercialization of Remote Sensing Imagery, 19 *Seattle University Law Review* (1996), 539.
Spector, L.S., The Not-so-Open Skies, *Commercial Observation Satellites and International Security*, Hampshire, the Macmillan Press, 1990, 164.
Staelin, D.H., and Kerekes, J., Remote Sensing Capabilities, *Heaven and Earth: Civilian Uses of New Earth Space*, Martinus Nijhoff Publishers, Hague, 1997, p. 161.
Spradling, K.K. International Liability Ramifications of the US NAVSTAR Global Positioning System, *Proceedings of the Thirty Sixth Colloquium on the Law of Outer Space* (1990), 98.
Sweeting, M. et al. Evaluation of the Utility of the Disaster Monitoring Constellation in support of Earth Observation Applications, *Selected Proceedings of the 5th International Symposium of the International Academy of Astronautics*, Berlin, April 4–8, 2005, 436.
Tanner, K.L., Eyes in the Sky: The Technology, Politics, and Law of Remote Sensing, Unpublished thesis presented to Princeton University, 1987, 96.
Terekhov, A.D. UN General Assembly Resolutions and Outer Space Law, *Proceedings of the Fortieth Colloquium on the Law of Outer Space*, IISL, AIAA, 1997, 14.
Terekhov, A.D., Passage of Space Objects through Foreign Airspace: International Custom?, 25-1 *Journal of Space Law* (1997), 4.
Trinder, R.B., Remote Sensing Via Satellite: The Need for a New International Regime, 22-1 *Annals of Air and Space Law* (1997), 519.
Vaver, D., The copyright Mixture in a Mixed Legal System: fit for Human Consumption?, 5.2 *Electric Journal of Comparative Law* (May 2001), 1.
Vereschetin, V.S., Custom as a source of International Law of Outer Space, 13-1 *Journal of Space Law* (1985), 119.

Von der Dunk, F.G., Non-discriminatory data dissemination in practice, *Earth Observation Policy and Europe*, A.A. Balkema Publishers, Lisse, 2002, 41.

Von der Dunk, F.G., United Nations Principles on Remote Sensing and the User, Earth Observation Data Policy and Europe, Harris, R. (Eds.), Lisse, A.A. Balkema Publishers, 2002, 33.

Von der Dunk, F.G., The Moon Agreement and the Prospect of Commercial Exploitation of Lunar Resources, 32 *Annals of Air and Space Law* (2007), 97.

Von Kries, W., The US Commercial Remote Sensing Policy of April 28, 2003: Some Comments, 52-4 *German Journal of Air and Space Law* (2003), 554.

Walker, P., Victims of natural disaster and the right to humanitarian assistance: a practitioner's view, 325 *International Review of the Red Cross* (1998), 611.

Werro, F., Strict liability in European tort law: an introduction, *The boundaries of strict liability in European tort law*, Carolina Academic Press, Durham, 2004, 3.

Whittaker, S., European Product Liability and Intellectual Products, 105 *Law Quarterly Review* (1989), 125.

Zimmerman, P.D., Superpower Relations, and Public Diplomacy, *Commercial Observation Satellites and International Security*, the Macmillan Press, Hampshire, 1990, 33.

Zhukov, G., Une expérience historique: la convention de moscou de 1978 sur le transfert de l'utilisation des donn ées de téléobservation de la terre a partir de l'espace, *Droit télédétection et environnement: actes du colloque international: le droit face aux techniques de télédétections par satellite au service de l'environnement*, Straousbourg, 2-3-4 join 1993, Sides, Antony, 1994, 189.

IV. Non-Academic Articles

Brehon, N. J., Le Satellite au Service de l'Environnement, Air & Cosmos, May 25 2001, 52.

Brugioni, D.A., The Art and Science of Photoreconnaissance, 274-3 Scientific American, 1996, 78.

Clarke, A., "Copying maps costs AA £20m", The Guardian, Tuesday March 6, 2001.

De Selding, P.B., "Courts Limit Use of Satellite Imagery for Prosecuting Suspected Polluters", Space News, 2 March 2005.

Envisat Radar Protecting Patagonian Toothfish From Pirate, Space News, November 18, 2005.

Ferester, W., "Firm Suspected of Misrepresenting Imagery", Space News, 16 January 1995.

« Foret Rwanda: des satellites au secours des gorilles » GEO, no. 319, September 2005, 102–105.

Frederick, F., "Arizona uses IKONOS Imagery in Dispute with Developer", Space News, September 5, 2005.

"Greenland Ice Loss Doubles in Past Decade, Raising Sea Level Faster", Science Daily, February 16, 2006.

Jayaraman, K.S., India Us Firm agree to Sale of 1 meter Imagery, Space News, 17 July 2000.

Lepp, N., Disaster Relief Politics Complicate South Asia Effort, The Dominion, January 3 2005
"Politics of the Afghan earthquake", The Economist, 14 March 1998.

V. References Without Authors and Internet Based Resources

Background paper on International Disaster Response Laws, Rules and Principles for the FEMA Legal Round table on Foreign Assistance in the United States, Washington DC, November 28, 2005.
Crowsey Incorporated, *Legal Assistant Guide to Legal Applications for geospatial information*, June 10, 2002.
Data User element DUE_DUP Directory 2002-2004, ESA, Paris, 2004.
Ghosh. Remote Sensing Policies. Available on the Internet at http://www.gisdevelopment.net/magazine/global/2007/july/54_1.htm.
ESA/PB-EO(90)57, rev.6, Principles of the Provision of ERS data to users 1994.
GMES Issues and Need for Action Report from the 1st GMES forum Brussels, 15–17 July 2002.
International Charter on Space and Major Disasters, Executive Secretariat, 6th Annual Report January–December 2006.
Iridium 33-Cosmos 2251 Collision, Fact Sheet, Secure World Foundation, http://www.secureworldfoundation.org/siteadmin/images/files/file_273.pdf
Joint FAO/WHO Food Standards Programme Codex Alimentations Commission Annex VI to Report of the Twentieth Session of the Codex Committee on General Principles, 2004, ALINORM 04/27/33A.
NPA Group, *Final Report BNSC Sectors Studies Programme Applications of Earth Observation to Legal Sector*, BNSC, London, 2001.
RESTEC, World of Remote Sensing, Kokinshoin Press, Tokyo, 2005.
Report on the legal aspects of the privatisation of space activities Remote Sensing and national space legislation, Berlin Conference 2004, International Law Bar Association.
Office of the United Nations Disaster Relief Coordinator (UNDRO), *Space Application and Dissemination of Disaster-Related Data*, Expert Meeting Geneva, 14-17 June 1983.
Tampere Convention: saving lives through emergency telecommunications, *UN Press Release*, IHA/983 L/I4385/SAG/309, 7 January, 2005.
World Disaster Report 2005, International Federation of Red Cross and Red Crescent Societies.
World Disaster Report 2007, International Federation of Red Cross and Red Crescent Societies.

VI. Websites of Organisations

http://www.asc-csa.gc.ca
http://www.cnes.fr
http://www.cnsa.gov.cn

http://www.conae.gov.ar
http://www.digitalglobe.com
http://www.disasterscharter.org
http://www.dlr.de
http://www.dmcii.com
http://www.earthobservations.org
http://www.esa.int
http://www.earthobservations.org
http://europa.eu
http://www.geoeye.com
http://www.gmes.info
http://www.icj-cij.org/
http://www.isro.org
http://www.jaxa.jp/
http://www.landsat.org
http://www.mdacorporation.com
http://www.nasa.gov
http://www.noaa.gov
http://ochaonline.un.org
http://sertit.u-strasbg.fr/
http://www.spotimage.fr
http://www.sstl.co.uk
http://www.unep.org
http://www.unesco.org
http://www.unhcr.org
http://www.unoosa.org
http://www.usgs.gov/
http://www.wipo.int
http://www.wmo.int

Index

STUDIES IN SPACE LAW

ISSN 1871-7659

1. Dunk, F.G. von der and M.M.T.A. Brus (eds). *The International Space Station*. Commercial Utilisation from a European Legal Perspective. 2006. ISBN 978 90 04 15256 4

2. Goh, Gérardine Meishan. *Dispute Settlement in International Space Law*. A Multi-Door Courthouse for Outer Space. 2007. ISBN 978 90 04 15545 9

3. Viikari, Lotta. *The Environmental Element in Space Law: Assessing the Present and Charting the Future*. Assessing the Present and Charting the Future. 2008. ISBN 978 90 04 16744 5

4. Tronchetti, Fabio. *The Exploitation of Natural Resources of the Moon and Other Celestial Bodies*. A Proposal for a Legal Regime. 2009. ISBN 978 90 04 17535 8

5. Ito, Atsuyo. *Legal Aspects of Satellite Remote Sensing*. 2011. ISBN 978 90 04 19032 0

www.ingramcontent.com/pod-product-compliance
Lightning Source LLC
La Vergne TN
LVHW081514300926
849161LV00006B/1172

* 9 7 8 9 0 0 4 1 9 0 3 2 0 *